Weak solutions of the Boussinesq equations in domains with rough boundaries

Vom Fachbereich Mathematik
der Technischen Universität Darmstadt
zur Erlangung des Grades eines
Doktors der Naturwissenschaften
(Dr. rer. nat.)
genehmigte

Dissertation

von

Dipl.-Math. Christian Komo

aus Offenbach am Main

Referent:	Prof. Dr. R. Farwig
Korreferenten:	PD. Dr. Matthias Geißert
	Prof. Dr. Werner Varnhorn
Tag der Einreichung:	6. Juni 2013
Tag der mündlichen Prüfung:	11. Juli 2013

Darmstadt 2013

D 17

Bibliografische Information der Deutschen Nationalbibliothek

Die Deutsche Nationalbibliothek verzeichnet diese Publikation in der
Deutschen Nationalbibliografie; detaillierte bibliografische Daten sind
im Internet über http://dnb.d-nb.de abrufbar.

ISBN 978-3-8325-3504-9

Logos Verlag Berlin GmbH
Comeniushof, Gubener Str. 47,
10243 Berlin
Tel.: +49 (0)30 42 85 10 90
Fax: +49 (0)30 42 85 10 92
INTERNET: http://www.logos-verlag.de

Acknowledgements

I would like to thank all of those who have contributed to this work and who have supported me in various ways during its progress.

Foremost, I want to express my deep gratitude to my supervisor Prof. Dr. Reinhard Farwig for his kind and valuable support during the last years. I am very grateful for his patience and confidence in me during this long period of time. His guidance and suggestions helped me invaluably to complete this thesis.

Furthermore, I want to thank PD. Dr. Matthias Geißert and Prof. Dr. Werner Varnhorn for being the co-referees for this thesis.

I am grateful to Prof. Dr. Šárka Nečasová and Prof. Dr. Eduard Feireisl from Prague for giving me the opportunity to visit Nečas Center for Mathematical Modeling. They supported my work with many valuable ideas and hints.

I want to thank the whole Analysis PDE group at TU Darmstadt for the good and helpful working environment. I really enjoyed working there. Especially I want to thank Dr. Tobias Hansel, Dr. Nataliya Kraynyukova, Dr. Sergiy Nesenenko, Dr. Felix Riechwald, Dr. Veronika Rosteck and Dr. Raphael Schulz for many productive discussions and their friendship. Moreover I am grateful to PD. Dr. Mads Kyed for his hints and suggestions for my thesis. I am thankful to Prof. Dr. Karl Heinrich Hofmann for helping me to improve the English of this text.

Last but not least I thank my family for their support and encouragement during my studies and this thesis.

Zusammenfassung

In der vorliegenden Arbeit betrachten wir die *Boussinesq-Gleichungen*

$$u_t - \nu\Delta u + (u \cdot \nabla)u + \frac{1}{\rho_0}\nabla p = \beta\theta g + f_1 \quad \text{in }]0,T[\times\Omega \,,$$
$$\text{div } u = 0 \quad \text{in }]0,T[\times\Omega \,,$$
$$\theta_t - \kappa\Delta\theta + (u \cdot \nabla)\theta = f_2 \quad \text{in }]0,T[\times\Omega \,, \qquad (0.1)$$
$$u(0) = u_0 \quad \text{in } \Omega \,,$$
$$\theta(0) = \theta_0 \quad \text{in } \Omega \,,$$
$$\text{geeignete Randbedingungen für } u,\theta \text{ auf }]0,T[\times\partial\Omega$$

für ein viskoses, inkompressibles Fluid mit Geschwindigkeitsfeld u, Druck p und Temperatur θ auf einem Gebiet $\Omega \subseteq \mathbb{R}^n$, $n \in \{2,3\}$, und einem Zeitintervall $[0,T[$. Wir beweisen Existenz von schwachen Lösungen der Boussinesq-Gleichungen, wobei u die no-slip Haftbedingung und θ eine Robin-Randbedingung auf dem Rand $\partial\Omega$ erfüllt. Dabei wird angenommen, dass Ω einer gleichmäßigen Lipschitz-Bedingung genügt.

Weiterhin beschäftigen wir uns mit Existenz und Eindeutigkeit von starken Lösungen der Boussinesq-Gleichungen (0.1) mit Dirchlet-Randbedingungen für u,θ, wobei das betrachtete Gebiet $\Omega \subseteq \mathbb{R}^3$ völlig beliebig ist. Wir zeigen, dass diese Lösungen glatt sind, falls die Daten es sind. Diese Resultate werden verwendet, um Regularitätskriterien für schwache Lösungen der Boussinesq-Gleichungen zu formulieren.

Im nächsten Teil der Arbeit wird der Einfluss der Oberflächenrauigkeit auf schwache Lösungen der Boussinesq-Gleichungen untersucht. In unserem Model wird das "ideale Gebiet" Ω durch eine Folge $(\Omega_k)_{k\in\mathbb{N}}$ von Gebieten mit "rauen Rändern" ersetzt, die gegen Ω konvergiert. Die Motivation für diesen Ansatz ist die Annahme, dass der Rand $\partial\Omega$ von jedem, selbst glatten Gebiet Ω mikroskopische Unebenheiten besitzt. Seien (u_k,θ_k), $k \in \mathbb{N}$, schwachen Lösungen der Boussinesq-Gleichungen in $[0,T[\times\Omega_k$, die den Energieungleichungen genügen. Wir nehmen an, dass u_k eine Navier-slip Randbedingung und dass θ_k eine Robin-Randbedingung auf $]0,T[\times\partial\Omega_k$ erfüllt. Sei (u,θ) ein schwacher Grenzwert von (u_k,θ_k) in $[0,T[\times\Omega$. Wir werden beobachten, dass θ eine Robin-Randbedingung auf $]0,T[\times\partial\Omega$ mit einer Gewichtsfunktion Λ erfüllt, die einen zusätzlichen Wäremübergangskoeffizienten beschreibt, der aufgrund der Rauigkeit der Ränder entsteht. Falls "genug Randrauigkeit" vorhanden ist, wird gezeigt, dass u die no-slip Haftbedingung auf $]0,T[\times\partial\Omega$ erfüllt. Unser Vorgehen beruht auf der Konstruktion von Youngschen Maßen, welche die Oszillationen von $\partial\Omega_k$ beschreiben.

Die Optimierung der Wärmeübertragung durch eine Oberfläche ist eine wichtige Anwendung der Fluidmechanik. Diese Wärmeübertragung wird durch die Oberflächenrauigkeit beeinflusst. Wir beschäftigen uns mit der Optimierung der zusätzlichen Gewichtsfunktion Λ, so dass die durch $\partial\Omega$ übertragene Wärmeenergie maximal bzw. minimal wird.

Abstract

In this thesis we consider the *Boussinesq equations*

$$u_t - \nu \Delta u + (u \cdot \nabla)u + \frac{1}{\rho_0}\nabla p = \beta \theta g + f_1 \quad \text{in }]0, T[\times \Omega,$$
$$\text{div } u = 0 \quad \text{in }]0, T[\times \Omega,$$
$$\theta_t - \kappa \Delta \theta + (u \cdot \nabla)\theta = f_2 \quad \text{in }]0, T[\times \Omega, \qquad (0.2)$$
$$u(0) = u_0 \quad \text{in } \Omega,$$
$$\theta(0) = \theta_0 \quad \text{in } \Omega,$$

suitable boundary conditions for u, θ on $]0, T[\times \partial\Omega$

for a viscous, incompressible fluid with velocity u, pressure p and temperature θ on a domain $\Omega \subseteq \mathbb{R}^n$, $n \in \{2, 3\}$, and a time interval $[0, T[$. We will prove existence of weak solutions of the Boussinesq equations with no slip boundary condition for u and Robin boundary condition for θ in domains satisfying a uniform Lipschitz condition.

Further, we deal with existence and uniqueness of strong solutions of (0.2) with Dirichlet boundary conditions for u, θ in arbitrary domains $\Omega \subseteq \mathbb{R}^3$. It will be shown that these solutions are smooth if the data are smooth. These results will be used to formulate regularity criteria for weak solutions of the Boussinesq equations.

The next part investigates the influence of surface roughness to weak solutions of the Boussinesq equations. In our model the 'ideal' domain Ω is replaced by a sequence $(\Omega_k)_{k\in\mathbb{N}}$ of domains with 'rough boundaries' converging to Ω. The motivation for this approach is the idea that every, even smooth domain Ω is covered by microscopic asperities. Let $(u_k, \theta_k)_{k\in\mathbb{N}}$ be a sequence of weak solutions of the Boussinesq equations in $[0, T[\times \Omega_k$ satisfying the energy inequalities. We assume that u_k satisfies a Navier slip boundary condition and that θ_k fulfils a Robin boundary condition on $]0, T[\times \partial\Omega_k$. We will observe that θ fulfils a Robin boundary condition on $]0, T[\times \partial\Omega$ with a weight function Λ describing an additional heat transfer coefficient to the exterior which is due to the rugosity of the boundaries. Furthermore, if there is 'enough surface roughness' we will show that u satisfies the no slip boundary condition. The main tool of our approach is the construction of Young measures describing the character of oscillations of $\partial\Omega_k$.

Optimizing the heat transfer through a surface is an important application of fluid mechanics. This heat transfer is influence by surface roughness. We address the problem of optimizing the additional weight function Λ such that the heat energy transferred through $\partial\Omega$ becomes maximal/minimal.

Contents

1 Introduction

Let $\Omega \subseteq \mathbb{R}^n$, $n \in \{2,3\}$, be a domain and let $[0,T[$ be a time interval. In this thesis we consider the *Boussinesq equations* which are the following system of coupled partial differential equations:

$$u_t - \nu \Delta u + (u \cdot \nabla)u + \frac{1}{\rho_0}\nabla p = \beta \theta g + f_1 \quad \text{in }]0,T[\times\Omega\,,$$
$$\text{div } u = 0 \qquad \text{in }]0,T[\times\Omega\,, \qquad (1.1)$$
$$\theta_t - \kappa \Delta \theta + (u \cdot \nabla)\theta = f_2 \qquad \text{in }]0,T[\times\Omega\,.$$

In (1.1) we denote by $u : [0,T[\times\Omega \to \mathbb{R}^n$ the velocity, by $\theta : [0,T[\times\Omega \to \mathbb{R}$ the temperature and by $p : [0,T[\times\Omega \to \mathbb{R}$ the pressure. To be more precise θ is the difference of the absolute temperature to a constant reference temperature. Let us denote by $f_1 : [0,T[\times\Omega \to \mathbb{R}^n$ the external force per unit mass, by $g : [0,T[\times\Omega \to \mathbb{R}^n$ the gravitational force and by $f_2 : [0,T[\times\Omega \to \mathbb{R}$ the external thermal radiation per heat capacity. Further, we consider the following positive, real constants: ρ_0 is the density, ν the kinematic viscosity, β the coefficient of thermal expansion and κ the thermal diffusivity. For a derivation of the Boussinesq equations we refer to Chapter 3.

The Boussinesq equations constitute a model of motion of a viscous, incompressible buoyancy-driven fluid flow coupled with heat convection.

These equations have to be completed by initial and boundary conditions. Throughout this thesis we consider the following initial condition:

$$u(0) = u_0 \quad \text{in } \Omega\,,$$
$$\theta(0) = \theta_0 \quad \text{in } \Omega\,, \qquad (1.2)$$

where $u_0 : \Omega \to \mathbb{R}^n$ and $\theta_0 : \Omega \to \mathbb{R}$ are the given initial values.

The Boussinesq equations have been investigated by several researchers. Existence and uniqueness criteria of weak solutions of the Boussinesq equations with mixed Dirichlet/Neumann boundary conditions in smooth bounded or in smooth exterior domains can be found in the papers in [CKM92, Mo92, Ka93]. For the construction of regular solutions of the Boussinesq equations with homogeneous or non homogeneous Dirichlet boundary conditions in smooth domains or in the whole space we refer to [BrSc12, CaDiB80, Hi91, HiYa92, SaTa04].

The intention of this introduction is to motivate the topics studied in the thesis and to explain the main results. We want to give an overview about the thesis and consequently, we try to avoid presenting the mathematical details. For a rigorous mathematical formulation we refer to the following chapters. Especially we assume that $f_1, f_2, g, h_0, u_0, \theta_0$ are suitable data with integrability properties depending on the problem to be considered.

1.1 Weak solutions

Throughout this section we supplement the Boussinesq system (1.1), (1.2) with the *no slip boundary condition*

$$u = 0 \quad \text{on }]0, T[\times\partial\Omega \tag{1.3}$$

and the *Robin boundary condition*

$$\frac{\partial\theta}{\partial N} + \Lambda(\theta - h_0) = 0 \quad \text{on }]0, T[\times\partial\Omega \tag{1.4}$$

where h_0 denotes the exterior temperature on $\partial\Omega$ and where $\Lambda : \partial\Omega \to]0, \infty[$ is a positive, scalar function describing the ratio of heat transfer to the temperature difference $\theta - h_0$. In (1.4) we denote by N the exterior unit normal vector on $\partial\Omega$ and by $\frac{\partial\theta}{\partial N} := \nabla\theta \cdot N$ the normal derivative. In the following we motivate our definition of a *weak solution*. Let (u, θ) be a smooth solution of (1.1), (1.2) with boundary conditions (1.3), (1.4). Multiplying the first equation in (1.1) by u and integrating the resulting equation in space and time leads to

$$\frac{1}{2}\|u(t)\|_2^2 + \nu \int_0^t \|\nabla u\|_2^2 \, d\tau = \frac{1}{2}\|u_0\|_2^2 + \beta \int_0^t \langle\theta g, u\rangle_\Omega \, d\tau + \int_0^t \langle f_1, u\rangle_\Omega \, d\tau, \tag{1.5}$$

for all $t \in [0, T[$. Analogously, we multiply the third equation in (1.1) with θ and integrate by parts to obtain

$$\begin{aligned}
\frac{1}{2}\|\theta(t)\|_2^2 + \kappa \int_0^t \|\nabla\theta\|_2^2 \, d\tau + \kappa \int_0^t \|\sqrt{\Lambda}\,\theta\|_{2,\partial\Omega}^2 \, d\tau \\
= \frac{1}{2}\|\theta_0\|_2^2 + \kappa \int_0^t \langle\Lambda h_0, \theta\rangle_{\partial\Omega} \, d\tau + \int_0^t \langle f_2, \theta\rangle_\Omega \, d\tau
\end{aligned} \tag{1.6}$$

for all $t \in [0, T[$. The equations (1.5), (1.6) are called the *energy equalities*. By Young's inequality and Gronwall's inequality

$$\begin{aligned}
u \in L^\infty(0, T; L_\sigma^2(\Omega)), \quad \nabla u \in L^2(0, T; L^2(\Omega)^{3\times3}), \\
\theta \in L^\infty(0, T; L^2(\Omega)), \quad \nabla\theta \in L^2(0, T; L^2(\Omega)^3).
\end{aligned} \tag{1.7}$$

The relation above will be the starting point for our definition of a weak solution of the Boussinesq system. We call (u, θ) a weak solution of this system if the following properties are satisfied.

1. (u, θ) satisfies (1.7).

2. (u, θ) fulfils (1.1) in the sense of distributions with an associated pressure p.

3. The initial conditions (1.2) and boundary conditions (1.3), (1.4) are satisfied.

We remark that the condition (1.7) we require for a weak solution of the Boussinesq equations is strongly influenced by the concept of a weak solution of the instationary Navier-Stokes equations in the sense of *Leray-Hopf*. For a detailed definition of weak solutions we refer to Definition 4.7.

A detailed mathematical analysis of weak solutions with boundary conditions (1.3) and (1.4) in domains satisfying the uniform Lipschitz condition is the content of Chapter 4. For the definition of the class of domains satisfying the *uniform Lipschitz condition* and further properties of these domains we refer to Section 4.1. Especially, the trace operator $T : H^1(\Omega) \to L^2(\partial\Omega)$ exists as a continuous, linear operator. In the following we sketch the main results about weak solutions.

In Theorem 4.18 we will prove existence of a weak solution (u, θ) satisfying the *energy inequalities*, i.e., (u, θ) fulfils (1.5), (1.6) as an inequality for almost all $t \in [0, T[$. The proof is based on the *Galerkin method* which is an often used tool to prove existence of solutions of partial differential equations. Furthermore, for the case $n = 2$, we are able to prove uniqueness of weak solutions (see Theorem 4.23). Since it is not known whether weak solutions of the Navier-Stokes equations in the three-dimensional case are unique, we cannot expect uniqueness for weak solutions of the Boussinesq system for $n = 3$ (see the discussion in the following section).

1.2 Strong solutions

In this section we complete the Boussinesq system (1.1), (1.2) with the Dirichlet boundary conditions

$$\begin{aligned}
u(t, x) &= 0 \quad \text{on }]0, T[\times \partial\Omega, \\
\theta(t, x) &= 0 \quad \text{on }]0, T[\times \partial\Omega.
\end{aligned} \tag{1.8}$$

As a motivation for the definition of strong solutions of the Boussinesq equations let us first consider the instationary Navier-Stokes equations with Dirichlet boundary condition

$$\begin{aligned}
u_t - \Delta u + u \cdot \nabla u + \nabla p &= f \quad \text{in }]0, T[\times \Omega, \\
\operatorname{div} u &= 0 \quad \text{in }]0, T[\times \Omega, \\
u &= 0 \quad \text{on }]0, T[\times \partial\Omega, \\
u &= u_0 \quad \text{at } t = 0,
\end{aligned} \tag{1.9}$$

in the space time cylinder $[0, T[\times \Omega$ with external force f and initial value u_0.

Since the pioneering work of J. Leray [Le34] the biggest open question in mathematical fluid dynamics is the existence of global smooth solutions to (1.9). Due to the importance of this question in mathematical fluid dynamics and in the theory of partial differential equations in general it became one of the seven Millennium Problems of Clay Mathematics Institute. The researcher who solves this *important classical question that has resisted solutions over years* gets one million dollar.

The following problems for weak solutions u of (1.9) are still unsolved:

1. Is u smooth in the sense $u \in C^{\infty}(]0, T[\times \Omega)^3$ if Ω, f, u_0 are sufficiently smooth?

2. Is u unique in the Leray-Hopf class $L^{\infty}(0, T; L^2_{\sigma}(\Omega)) \cap L^2(0, T; W^{1,2}_{0,\sigma}(\Omega))$?

At the moment it is not known whether a weak solution in the sense of Leray-Hopf satisfies $u \in C^{\infty}(]0, T[\times \Omega)^3$ if the data are smooth. Let us recall the concept of a strong solution: A weak solution u of (1.9) is a strong solution if the *Serrin condition* $u \in L^s(0, T; L^q(\Omega)^3)$ where $2 < s < \infty, 3 < q < \infty$ with $\frac{2}{s} + \frac{3}{q} = 1$, is satisfied. It is well known (see [So01, Chapter V, Theorem 1.8.1]) that a strong solution is smooth if the data are sufficiently smooth. Up to now the existence of a strong solution u of (1.9) could only be proven on a sufficiently small interval $[0, T'[, 0 < T' \leq T$, and under additional assumptions on Ω, f, and u_0.

The construction of a strong solution in the Serrin class $L^s(0, T; L^q(\Omega)^3), \frac{2}{s} + \frac{3}{q} = 1$, is based on the L^q-theory of the Stokes operator A_q. However, in general domains only the L^2-approach to the Stokes operator is available. Therefore, in general domains, which may have several exits to infinity or may have edges and corners, only strong solutions $u \in L^8(0, T; L^4(\Omega)^3)$ can be constructed. For the construction of strong solutions of the instationary Navier-Stokes system (1.9) in general domains we refer to [FSV09, FS10] and to [So01, Chapter V, Section 4.2].

Further, our definition of a strong solution is based on the observation that the Navier-Stokes equations can be reduced to the Boussinesq equations. In more detail, consider an external force f and an initial value u_0 for the instationary Navier-Stokes equations (1.9) in $\Omega \times [0, T[$. Define $f_1 := f, f_2 := 0$ and $\theta_0 := 0$. Then (u, p) is a solution of the Boussinesq system (1.1), (1.2), (1.8), if and only if, (u, p, θ) with $\theta := 0$ is a solution of the instationary Navier-Stokes equations (1.9) with data f, u_0.

Due to the discussion about strong solutions of the Navier-Stokes equations (1.9) we define a strong solution of the Boussinesq system (1.1), (1.2), (1.8) as a weak solution (u, θ) satisfying the additional condition $u \in L^8(0, T; L^4(\Omega)^3)$. No additional condition is required for θ. For a detailed definition we refer to Definition 5.1.

Our strategy to construct a strong solution of the Boussinesq system is to rewrite this system as a nonlinear fixed point equation in an infinite dimensional Banach space and to use a fixed point argument. For a sufficient criterion for the existence of a strong solution (u, θ) we refer to Theorem 5.8. Furthermore (see Theorem 5.7) strong solutions of (1.1) are unique. In Section 5.3 we will prove that these strong solutions are smooth if the data are smooth. The main difficulty in these proofs is the fact that we have required no additional integrability condition for θ in the definition of a strong solution. In Section 5.4 these results will be applied to prove regularity of a weak solution (u, θ) of (1.1) satisfying the strong energy inequalities (see (5.144), (5.145)). Finally, Leray's structure theorem will be formulated for the Boussinesq system in general domains $\Omega \subseteq \mathbb{R}^3$.

1.3 The influence of surface roughness to the boundary behaviour to weak solutions

The effect of surface roughness to weak solutions of the Navier-Strokes equations was observed by several researchers. The common idea is that the 'physical' domain $\Omega \subseteq \mathbb{R}^3$ is covered by microscopic asperities and is therefore replaced by a sequence $(\Omega_k)_{k \in \mathbb{N}}$ of domains with rough boundaries where $\Omega_k \to \Omega$ holds in some sense. Let Γ_k be a 'rough' part of the boundary of Ω_k and let Γ be the part of the boundary of Ω which is approximated by Γ_k. Let $(u_k)_{k \in \mathbb{N}}$ be a sequence of weak solutions of the Navier-Stokes equations on Ω_k fulfilling the impermeability condition $u_k \cdot N = 0$ on $]0, T[\times \Gamma_k$.

In the pioneering work [CFS03] the authors proved for the case of the stationary Navier-Stokes equations and for periodically, smooth oscillating Γ_k that a weak limit u of u_k in Ω fulfils the no slip condition $u = 0$ on Γ. In [BFNW08] Bucur et al. removed the periodicity assumption on Γ_k and proved the corresponding result for the instationary Navier-Stokes equations. For further results and approaches we refer to [Br10, BFN08, BFN10, CLS10, FeNe08]. This 'rugosity effect' (see [BoBu12]) can be seen as the mathematical justification of the no slip boundary condition.

Motivated by these results we want to determine the effect on boundary rugosity to weak solutions (u, θ) of the Boussinesq equations (1.1) in an impermeable domain with θ satisfying a Robin boundary condition. A detailed mathematical analysis of this topic is the content of Chapter 7.

Let $n \in \{2, 3\}$, consider the domains

$$\Omega := \left\{ (x', x_n) \in \mathbb{R}^n; \ x_n > 0; \ x' \in \mathbb{R}^{n-1} \right\}, \tag{1.10}$$

$$\Omega_k := \left\{ (x', x_n) \in \mathbb{R}^n; \ x_n > -h_k(x'); \ x' \in \mathbb{R}^{n-1} \right\}, \quad k \in \mathbb{N}, \tag{1.11}$$

where $(h_k)_{k \in \mathbb{N}}$ are non-negative, *equi-Lipschitz* continuous functions with $h_k \to 0$ uniformly on $\mathbb{R}^{n-1}$. The equi-Lipschitz continuity means that functions $(h_k)_{k \in \mathbb{N}}$ share the same Lipschitz constant. Consider a sequence $(u_k, \theta_k)_{k \in \mathbb{N}}$ of weak solutions of the Boussinesq equations (1.1) in $[0, T[\times \Omega_k$ satisfying the energy inequalities where u_k satisfies the impermeability condition

$$u_k \cdot N = 0 \quad \text{on }]0, T[\times \partial \Omega_k$$

and θ_k satisfies the Robin boundary condition

$$\frac{\partial \theta_k}{\partial N} + (\theta_k - h_0) = 0 \quad \text{on }]0, T[\times \partial \Omega_k. \tag{1.12}$$

Following the approach in [BFNW08] we will introduce a *Young measure* ν which describes the character of oscillations of $(\nabla h_k)_{k \in \mathbb{N}}$. The use of Young measures will be one of the main tools of this thesis. For the needed information about this topic we refer to Chapter 6. To investigate the influence of surface roughness let (u, θ) be a weak limit of (u_k, θ_k) in $]0, T[\times \Omega$. We will show that (u, θ) satisfy the Boussinesq equations (1.1) in $[0, T[\times \Omega$. Moreover, it is proved that θ satisfies the Robin boundary condition

$$\frac{\partial \theta}{\partial N} + \Lambda(\theta - h_0) = 0 \quad \text{on }]0, T[\times \partial \Omega \tag{1.13}$$

with a weight function $\Lambda : \partial \Omega \to [1, \infty[$ describing an additional heat transfer coefficient to the exterior. This function is due to the rugosity of the oscillations of the boundaries and can be explicitly computed using the Young measure ν, see (7.124). Moreover, if there is 'enough boundary rugosity' (in the sense that the non-degeneracy condition holds) we will show that u satisfies the no slip boundary condition

$$u = 0 \quad \text{on }]0, T[\times \partial \Omega.$$

This means that the microscopic asperities prevent the fluid from slipping, i.e. u adheres completely to the boundary.

1.4 Construction of some special domains with rough boundaries

To explain the main ideas of Chapter 8 we consider the case $n = 2$ throughout this section. Consequently, weak solutions of (1.1)-(1.4) are uniquely determined. Motivated from the transition of (1.12) to (1.13) due to surface roughness we consider the following situation: Assume that $\Lambda : \partial \Omega \to [1, \infty[$ is a bounded function and let Ω be the two dimensional half space (1.10). Consider a weak solution (u, θ) of the Boussinesq equations in $[0, T[\times \Omega$ satisfying the no slip boundary condition $u = 0$ on $]0, T[\times \partial \Omega$ and the Robin boundary condition

$$\frac{\partial \theta}{\partial N} + \Lambda(\theta - h_0) = 0 \quad \text{on }]0, T[\times \partial \Omega. \tag{1.14}$$

By Theorem 8.4 we can construct a sequence of domains $(\Omega_k)_{k \in \mathbb{N}}$ as in (1.11) and a sequence $(u_k, \theta_k)_{k \in \mathbb{N}}$ of weak solutions of (1.1), (1.2) satisfying the no slip boundary condition $u_k = 0$ on $]0, T[\times \partial \Omega_k$ and the Robin boundary condition

$$\frac{\partial \theta_k}{\partial N} + (\theta_k - h_0) = 0 \quad \text{on }]0, T[\times \partial \Omega_k$$

such that $u_k \to u$ and $\theta_k \to \theta$ converge in a weak sense in $]0, T[\times \Omega$. This result makes essentially use of the Young measures constructed in Theorem 8.1.

1.5 Optimizing the heat energy transferred to the exterior

Maximizing or minimizing the transferred heat energy through a surface is an important problem in the application of fluid mechanics. Our fluid could act as a coolant and therefore we try to minimize the heat transfer to the exterior. Conversely, the heating of a fluid via the contact with the surface of a heating device results in the maximization of this quantity. Our aim is to optimize the surface roughness such that the heat energy transferred to the exterior becomes maximal. In the following we develop the model which will be the basis of the considered optimization problem. Let $\Lambda : \partial\Omega \to \mathbb{R}$ with $\Lambda(x) \geq 1$, for a.a. $x \in \partial\Omega$, be a scalar function describing the heat transfer coefficient to the exterior. Consider a solution (u, θ) of (1.1), (1.2) such that u satisfies the no slip boundary condition $u = 0$ on $]0, T[\times\partial\Omega$ and θ satisfies the Robin boundary condition

$$\chi\frac{\partial\theta}{\partial N} + \Lambda(\theta - h_0) = 0 \quad \text{on }]0, T[\times\partial\Omega \tag{1.15}$$

where $\chi > 0$ denotes the thermal conductivity. Let $n \in \{2, 3\}$, let $\Omega \subseteq \mathbb{R}^n$ be a domain satisfying the uniform Lipschitz condition, let $\Gamma \subseteq \partial\Omega$ be a suitable bounded part of $\partial\Omega$. Let $\dot{Q}$ be the heat flux and let q be the heat flux density. Using the Fourier law of thermal conduction there holds $q = -\chi\nabla\theta$. Let $E = E(u, \theta, \Lambda)$ denote the total heat energy transferred through Γ to the exterior. Using (1.15) there holds

$$E = \int_0^T \dot{Q}\, dt = \int_0^T \int_\Gamma q \cdot N\, dS\, dt = \int_0^T \int_\Gamma \Lambda(\theta - h_0)\, dS\, dt. \tag{1.16}$$

Consider fixed data $f_1, f_2, g, h_0, u_0, \theta_0$ and a fixed constant $L > 0$. The goal of Chapter 9 is to determine (u, θ, Λ) where $1 \leq \Lambda(x) \leq L$ holds for a.a. $x \in \partial\Omega$ and (u, θ) is a weak solution of the Boussinesq equations with no slip condition $u = 0$ on $]0, T[\times\partial\Omega$ and (1.15) such that the heat energy E in (1.16) becomes maximal or minimal.

2 Preliminaries

2.1 Notation, function spaces and basic inequalities

Basic notation

Let $\mathbb{R}^n$, $n \in \mathbb{N}$, denote the n-dimensional euclidean space. The vectors $x \in \mathbb{R}^n$ will be written in the form $x = (x_1, \ldots, x_n)$ where $x_i \in \mathbb{R}$, $i = 1, \ldots, n$. A domain $\Omega \subseteq \mathbb{R}^n$, $n \in \mathbb{N}$, is an open, connected and nonempty subset of $\mathbb{R}^n$. The notation " := " always means "equal by definition". We denote by $\mathbb{N} := \{1, 2, 3, \ldots\}$ the set of positive integers and we set $\mathbb{N}_0 := \mathbb{N} \cup \{0\}$. The sets of integers, real and complex numbers are denoted by $\mathbb{Z}$, $\mathbb{R}$ and $\mathbb{C}$, respectively. We use the notation $\mathbb{R}_0^+ := \{x \in \mathbb{R}; x \geq 0\}$ for the set of non-negative, real numbers. Further, the expression "x is a positive real number" means $x \in \mathbb{R}$, $x > 0$. Unless explicitly stated the function spaces and Banach spaces considered in this work are real-valued.

Throughout this thesis we use generic constants $c > 0$ which may change from line to line. Their dependence on parameters is expressed only if necessary. In these cases we write $c = c(\alpha, \beta, \ldots)$ if the constant depends on the parameters $\alpha, \beta, \ldots$.

For $p \in]1, \infty[$ we define the *Hölder conjugate exponent* $p' \in]1, \infty[$ of p by the relation $\frac{1}{p} + \frac{1}{p'} = 1$. If $p = 1$ define $p' := \infty$ and if $p = \infty$ we set $p' := 1$.

For vectors $x, y \in \mathbb{R}^n$ we write $x \cdot y := \sum_{k=1}^n x_k y_k$ for the usual euclidean scalar product where $x = (x_1, \ldots, x_n)$, $y = (y_1, \ldots, y_n)$. Let $|x| := \sqrt{x \cdot x}$ denote the euclidean norm of $x \in \mathbb{R}^n$. Correspondingly we define $A : B := \sum_{j,k=1}^n a_{j,k} b_{j,k}$ for two real matrices $A = (a_{j,k})_{j,k=1}^n$, $B = (b_{j,k})_{j,k=1}^n$. The notation $B_r(x) := \{y \in \mathbb{R}^n; |y - x| < r\}$ is reserved for an open ball with radius $r > 0$ centred at x. If $\Omega \subseteq \mathbb{R}^n$ is a domain with sufficiently smooth boundary we denote by $N(x)$ the outer unit normal vector in $x \in \partial\Omega$. If X is a Banch space we denote by X' the dual space and by $[x', x]_{X',X}$ the application of the functional $x' \in X'$ on X.

Consider $a, b \in \mathbb{R}$ with $a < b$. We use the notation $[a, b]$ for the closed, real interval with endpoints a, b and $]a, b[$ denotes the open, real interval with these endpoints. Consequently, the half open intervals are denoted by $]a, b]$, $[a, b[$.

The support of a function $u : \Omega \to \mathbb{R}$ is defined by $\operatorname{supp}(u) := \overline{\{x \in \Omega; u(x) \neq 0\}}$ where the latter closure is taken with respect to the euclidean norm of $\mathbb{R}^n$. Let $(X, \|\cdot\|)$ be a normed vector space. We write $\overline{M}^{\|\cdot\|}$ for the closure of a set $M \subseteq X$. When there is no possibility of confusion we will just write $\overline{M}$.

For two domains Ω_1, Ω_2 we write $\Omega_1 \subseteq\subseteq \Omega_2$ if $\overline{\Omega_1}$ is compact and $\overline{\Omega_1} \subseteq \Omega_2$. Let $A, B \subseteq \mathbb{R}^n$ be sets. The expression "let $A \subseteq B$ be a compact set" means that $K \subseteq \mathbb{R}^n$ is a compact set and $A \subseteq B$. Further if $(X_n)_{n \in \mathbb{N}}$ is a sequence of Banach spaces we will write that a sequence $(v_n)_{n \in \mathbb{N}}$ with $v_n \in X_n, n \in \mathbb{N}$, is bounded in $(X_n)_{n \in \mathbb{N}}$ if there is a constant $M > 0$ such that $\|v_n\|_{X_n} \leq M$ for all $n \in \mathbb{N}$.

Lebesgue and Sobolev spaces

Let $n \in \mathbb{N}$, let $\Omega \subseteq \mathbb{R}^n$ be an open set, let $1 \leq p \leq \infty$, and $k \in \mathbb{N}$. We denote by $L^p(\Omega), W^{k,p}(\Omega), W_0^{k,p}(\Omega)$ the usual Lebesgue and Sobolev spaces with norm $\|\cdot\|_{L^p(\Omega)} = \|\cdot\|_p$ and $\|\cdot\|_{W^{k,p}(\Omega)} = \|\cdot\|_{W^{k,p}}$, respectively. We identify in $L^p(\Omega), W^{k,p}(\Omega), W_0^{k,p}(\Omega)$ functions that are equal almost everywhere in Ω; the elements of these spaces are thus equivalence classes of measurable functions. By an abuse of notation, we identify a measurable function $u : \Omega \to \mathbb{R}$ with its equivalence class $[u]$. We set $H^k(\Omega) := W^{k,2}(\Omega)$ and $H_0^k(\Omega) := W_0^{k,2}(\Omega)$. Furthermore define $H^0(\Omega) := L^2(\Omega)$ and $H^{-1}(\Omega) := W_0^{1,2}(\Omega)'$. For the corresponding spaces of vector-valued functions we will use the symbol $L^p(\Omega)^n$ etc. For two measurable functions f, g with the property $f \cdot g \in L^1(\Omega)$ where $f \cdot g$ means the usual scalar product of scalar, vector or matrix fields, we set

$$\langle f, g \rangle_\Omega := \int_\Omega f(x) \cdot g(x) \, dx.$$

Further

$$L_{\text{loc}}^p(\Omega) := \{\, u : \Omega \to \mathbb{R} \text{ is Lebsgue measurable}$$
$$\text{and } u \in L^p(K) \text{ for all compact sets } K \subseteq \Omega \,\}$$

and

$$L_{\text{loc}}^p(\overline{\Omega}) := \{\, u : \Omega \to \mathbb{R} \text{ is Lebsgue measurable}$$
$$\text{and } u \in L^p(\Omega \cap B) \text{ for all open balls } B \subseteq \Omega \text{ with } B \cap \Omega \neq \varnothing \,\}.$$

Definition 2.1. Let $\Omega \subseteq \mathbb{R}^n, n \in \mathbb{N}$, be an open set. Let $0 < s < 1$. Then we define the *Slobodeckij space*

$$W^{s,2}(\Omega) := \left\{ u \in L^2(\Omega); \int_\Omega \int_\Omega \frac{|u(x) - u(y)|^2}{|x - y|^{n+2s}} \, dx \, dy < \infty \right\} \tag{2.1}$$

and equip it with the norm

$$\|u\|_{W^{s,2}(\Omega)}^2 := \|u\|_{2,\Omega}^2 + \int_\Omega \int_\Omega \frac{|u(x) - u(y)|^2}{|x - y|^{n+2s}} \, dx \, dy \tag{2.2}$$

for $u \in W^{s,2}(\Omega)$.

The space $W^{s,2}(\Omega)$ is a separable Banach space (see [Wl82, Satz 3.1]).

Definition 2.2. Let $n \in \mathbb{N}, n \geq 2$, let $\Omega \subseteq \mathbb{R}^n$ be a bounded Lipschitz domain and let dS denote the surface measure on $\partial\Omega$. Let us define for $0 < s < 1$ the space

$$W^{s,2}(\partial\Omega) := \left\{ u \in L^2(\partial\Omega); \int_{\partial\Omega} \int_{\partial\Omega} \frac{|u(x) - u(y)|^2}{|x - y|^{n-1+2s}} \, dS(x) \, dS(y) < \infty \right\} \tag{2.3}$$

and equip it with the norm

$$\|u\|^2_{W^{s,2}(\partial\Omega)} := \|u\|^2_{2,\partial\Omega} + \int_{\partial\Omega} \int_{\partial\Omega} \frac{|u(x) - u(y)|^2}{|x - y|^{n-1+2s}} \, dS(x) \, dS(y)$$

for $u \in W^{s,2}(\partial\Omega)$.

Spaces of continuous functions

Let $C^k(\Omega), k = 0, 1, \ldots, \infty$, denote the usual space of functions for which all partial derivatives of finite order $|\alpha| \leq k$ exist and are continuous. As usual, $C_0^k(\Omega)$ is the set of all functions from $C^k(\Omega)$ with compact support in Ω and let $C_0^\infty(]0, T[\times\Omega)$ denote the space of smooth function with compact support in $]0, T[\times\Omega$. Define $C_0^k(\overline{\Omega}) := \{u|_{\overline{\Omega}}; u \in C_0^k(\mathbb{R}^n)\}$ if $k \in \mathbb{N}_0 \cup \{\infty\}$.

Further we introduce the following spaces of solenoidal vector fields:

$$C_{0,\sigma}^\infty(\Omega) := \{ \phi \in C_0^\infty(\Omega)^n; \, \mathrm{div}\phi = 0 \},$$
$$L_\sigma^2(\Omega) := \overline{C_{0,\sigma}^\infty(\Omega)}^{\|\cdot\|_2},$$
$$W_{0,\sigma}^{1,2}(\Omega) := \overline{C_{0,\sigma}^\infty(\Omega)}^{\|\cdot\|_{H^1(\Omega)}}.$$

Moreover, $C_0^\infty(]0, T[; C_{0,\sigma}^\infty(\Omega))$ is the space of smooth, solenoidal vector fields with compact support in $]0, T[\times\Omega$ and

$$C_0^\infty([0, T[; C_{0,\sigma}^\infty(\Omega)) := \{ v|_{[0,T[\times\Omega} \, ; \, v \in C_0^\infty(] - 1, T[; C_{0,\sigma}^\infty(\Omega)) \},$$
$$C_0^\infty([0, T[\times\overline{\Omega}) := \{ w|_{[0,T[\times\overline{\Omega}} \, ; \, w \in C_0^\infty(] - 1, T[\times\mathbb{R}^n) \}.$$

Let $\Omega \subseteq \mathbb{R}^n, n \in \mathbb{N}$, be an open set, and let $j \in \mathbb{N}_0$. Define

$$C^j(\overline{\Omega}) := \{u \in C^j(\Omega); u \text{ has bounded, uniformly continuous derivatives}$$
$$\text{up to order } j \text{ on } \Omega \},$$
$$C_{\mathrm{loc}}^j(\overline{\Omega}) := \{u \in C^j(\Omega); u \text{ has bounded, uniformly continuous derivatives}$$
$$\text{up to order } j \text{ on all bounded, open sets } U \subseteq \Omega \}.$$

Furthermore define

$$C^\infty(\overline{\Omega}) := \{u \in C^\infty(\Omega); u \in C^j(\overline{\Omega}) \text{ for all } j \in \mathbb{N}\}, \tag{2.4}$$
$$C_{\mathrm{loc}}^\infty(\overline{\Omega}) := \{u \in C^\infty(\Omega); u \in C_{\mathrm{loc}}^j(\overline{\Omega}) \text{ for all } j \in \mathbb{N}\}. \tag{2.5}$$

Definition 2.3. Let $n, m \in \mathbb{N}$, let $A \subseteq \mathbb{R}^n$ be an arbitrary set, let $f : \mathbb{R}^n \to \mathbb{R}^m$ be a function.

(i) We say that f is *Lipschitz continuous* if there is a constant $L > 0$ such that

$$|f(x) - f(y)| \leq L|y - x| \tag{2.6}$$

for all $x, y \in A$. Define

$$\mathrm{Lip}(f) := \sup \left\{ \frac{|f(x) - f(y)|}{|x - y|}; \, x, y \in A; x \neq y \right\}. \tag{2.7}$$

(ii) We say that f is *locally Lipschitz continuous* if for each compact set $K \subseteq A$ there is a constant $C_K > 0$ such that

$$|f(x) - f(y)| \leq C_K|x - y| \tag{2.8}$$

for all $x, y \in K$.

Remark. In this context we mention *Kirszbraun's Theorem* (see [Fe69, Theorem 2.10.43]): Consider a Lipschitz continuous function $f : A \to \mathbb{R}^m$ where $A \subseteq \mathbb{R}^n$ is an arbitrary set. Then there exists a Lipschitz continuous function $\bar{f} : \mathbb{R}^n \to \mathbb{R}^m$ with $\bar{f} = f$ on A and $\mathrm{Lip}(\bar{f}) = \mathrm{Lip}(f)$.

Partial derivatives and differential operators

Let $\Omega \subseteq \mathbb{R}^n, n \in \mathbb{N}$, let $\phi : \Omega \to \mathbb{R}$ be a real-valued function. We assume that the partial derivates of the following defining equations exist in the classical or the weak sense. We use the notation

$$\partial_i \phi := \frac{\partial \phi}{\partial x_i}, \quad \partial_i \partial_j \phi := \frac{\partial^2 \phi}{\partial x_i \partial x_j}, \quad D^\alpha \phi := \frac{\partial^{|\alpha|} \phi}{\partial x_1^{\alpha_1} \dots \partial x_n^{\alpha_n}} \tag{2.9}$$

where $i, j \in \{1, \dots, n\}$ and where $\alpha = (\alpha_1, \dots, \alpha_n) \in \mathbb{N}_0^n$ is a multi-index and $|\alpha| := \sum_{k=1}^n |\alpha_k|$. Let $\nabla \phi := (\partial_1 \phi, \dots, \partial_n \phi)$ denote the *gradient* and let $\Delta \phi := \sum_{k=1}^n \partial_k \partial_k \phi$ denote the *Laplacian*. Further, let $u, w : \Omega \to \mathbb{R}^n$ be vector fields. Then

$$\mathrm{div} u := \sum_{k=1}^n \partial_k u_k, \quad \nabla u := (\partial_i u_j)_{i,j=1}^n, \quad \Delta u := (\Delta u_1, \dots, \Delta u_n).$$

Define $u \cdot \nabla \phi := \sum_{k=1}^n u_k \partial_k \phi$ and $u \cdot \nabla w := (u \cdot \nabla w_1, \dots, u \cdot \nabla w_n)$. We set

$$u \otimes w = \begin{pmatrix} u_1 w_1 & \dots & u_1 w_n \\ \dots & \dots & \dots \\ u_n w_1 & \dots & u_n w_n \end{pmatrix}.$$

Imbeddings

Theorem 2.4. *Let $n \in \mathbb{N}, n \geq 2$, let $\Omega \subseteq \mathbb{R}^n$ be a domain satisfying the uniform Lipschitz condition, and let $\Omega_0 \subseteq \Omega$ be an arbitrary bounded domain. Then the imbedding*

$$H^1(\Omega) \hookrightarrow L^2(\Omega_0) \tag{2.10}$$

is compact.

Proof. A proof for this classic compactness result can be found in [AdFo03, Theorem 6.3]. □

Theorem 2.5. *Let $n \in \mathbb{N}, n \geq 2$, let $\Omega \subseteq \mathbb{R}^n$ be a bounded Lipschitz domain, and let $0 \leq s_2 < s_1 \leq 1$. Then the imbedding*

$$W^{s_1,2}(\Omega) \hookrightarrow W^{s_2,2}(\Omega) \tag{2.11}$$

is compact.

Proof. See [Wl82, Satz II.7.9]. □

Theorem 2.6. *Let $n \in \mathbb{N}, n \geq 2$, let $\Omega \subseteq \mathbb{R}^n$ be a bounded Lipschitz domain. Then the imbedding*

$$H^1(\Omega) \hookrightarrow L^2(\partial\Omega) \tag{2.12}$$

is compact.

Proof. Choose any $\frac{1}{2} < s < 1$. Then Theorem 2.5 and Theorem 2.8 below applied to the chain of imbeddings

$$H^1(\Omega) \underset{\text{compact}}{\hookrightarrow} W^{s,2}(\Omega) \underset{\text{continuous}}{\hookrightarrow} W^{s-\frac{1}{2},2}(\partial\Omega) \underset{\text{continuous}}{\hookrightarrow} L^2(\partial\Omega)$$

yield the result. □

For the approximation of functions in $W^{m,p}(\Omega)$ by smooth functions the following lemma will useful.

Lemma 2.7. *Let $n \in \mathbb{N}, n \geq 2$, let $\Omega \subseteq \mathbb{R}^n$, be a Lipschitz domain, let $m \in \mathbb{N}_0$ and $1 \leq p < \infty$. Then $C_0^\infty(\overline{\Omega})$ is dense in $W^{m,p}(\Omega)$.*

Proof. See [Ga98, Theorem III.2.1]. It is easily seen that Ω fulfils the segment property. □

The next theorem proves the existence of the trace operator acting on functions in $W^{s,2}(\Omega)$.

Theorem 2.8. *Let $n \in \mathbb{N}, n \geq 2$, let $\Omega \subseteq \mathbb{R}^n$ be a bounded Lipschitz domain. Then for $\frac{1}{2} < s \leq 1$ there exists a continuous, linear trace operator*

$$T : W^{s,2}(\Omega) \to W^{s-\frac{1}{2},2}(\partial\Omega) \tag{2.13}$$

with the property

$$T\phi = \phi|_{\partial\Omega} \quad \text{for } \phi \in C^1(\overline{\Omega}). \tag{2.14}$$

Proof. See [Wl82, Satz II.8.7]. $\qquad\square$

Remark. Looking at (2.14), (4.16) we see that for $\phi \in H^1(\Omega)$ the trace in Theorem 2.8 and in Theorem 4.6 coincides.

Lemma 2.9. *Let $n \in \mathbb{N}, n \geq 2$, let $h : \mathbb{R}^{n-1} \to \mathbb{R}$ be a Lipschitz continuous function, assume that Ω is given by*

$$\Omega := \{(x', x_n); x_n > h(x'); x' \in \mathbb{R}^{n-1}\}.$$

Then we have

$$W^{1,2}_{0,\sigma}(\Omega) = \{v \in H^1(\Omega)^n; \operatorname{div} v = 0, v|_{\partial\Omega} = 0\}, \tag{2.15}$$

$$H^1_0(\Omega) = \{\phi \in H^1(\Omega); \phi|_{\partial\Omega} = 0\}. \tag{2.16}$$

Proof. For a proof of (2.15) we refer to [Ga98, Chapter III, Section 4.3]. $\qquad\square$

2.2 The Bochner integral and spaces of Bochner integrable functions

Let $\mathcal{L}(\mathbb{R}^n), n \in \mathbb{N}$, denote the σ-algebra of *Lebesgue measurable* subsets of $\mathbb{R}^n$. Let λ_n denote the *Lebesgue measure*. We call a set $E \subseteq \mathbb{R}^n$ *measurable* if $E \in \mathcal{L}(\mathbb{R}^n)$. Moreover we will write $|E| := \lambda_n(E)$ for the Lebesgue measure of E. The sets $N \in \mathcal{L}(\mathbb{R}^n)$ with $\lambda_n(N) = 0$ are called *null sets*.

Let $\Omega \subseteq \mathbb{R}^n$ be a measurable set. Define $\mathcal{L}(\Omega) := \{A \subseteq \Omega; A \in \mathcal{L}(\mathbb{R}^n)\}$. The restriction of λ_n to $\mathcal{L}(\Omega)$ will also be denoted by λ_n. The measure space $(\Omega, \mathcal{L}(\Omega), \lambda_n)$ is *complete* which means that every subset of a null set is still Lebesgue measurable. Moreover λ_n is σ-*finite* on $\mathcal{L}(\Omega)$ which means that Ω is the countable union of sets with finite Lebesgue measure.

Definition 2.10. Let $\Omega \subseteq \mathbb{R}^n, n \in \mathbb{N}$, be a measurable set, let $(X, \|\cdot\|)$ be a Banach space. Let $f : \Omega \to X$ be a function.

  (i) f is called *step function* if the image $f(\Omega)$ is finite and if for each $x \in f(\Omega)$ there holds $f^{-1}(\{x\}) := \{y \in \Omega; f(y) = x\} \in \mathcal{L}(\Omega)$.

(ii) f is called *almost separably-valued* if there exists a null set $N \subseteq \Omega$ such that $f(\Omega \setminus N) \subseteq X$ is separable.

(iii) f is called *strongly measurable* if there exist a sequence $(s_k)_{k \in \mathbb{N}} : \Omega \to X$ of step functions such that $s_k(y) \to f(y)$ as $k \to \infty$ for almost all $y \in \Omega$.

(iv) f is called *weakly measurable* if for all $x' \in X'$ the function $x' \circ f : \Omega \to \mathbb{R}$ is measurable in the usual sense of real-valued functions.

Remark. Let $(Z, \mathcal{S}, \mu)$ be a measure space with a measure μ which is not necessarily σ-finite. In this case the strongly measurable functions on Z are defined as the pointwise almost everywhere limit of countably-valued step functions, i.e. of functions $s : Z \to X$ with a countable image $s(Z)$ such that $s^{-1}(x) \in \mathcal{S}, x \in X$. Since the measure space $(\Omega, \mathcal{L}(\Omega), \lambda_n)$ is σ-finite, it is equivalent to define (as in Definition 2.10) the strongly measurable functions as the pointwise almost everywhere limit of step functions having a finite image.

Theorem 2.11 (Pettis). *Let $\Omega \subseteq \mathbb{R}^n, n \in \mathbb{N}$, be a measurable set, let X be a Banach space, and let $f : \Omega \to X$. Then f is strongly measurable if and only if f is weakly measurable and almost separably-valued.*

Proof. See [HiPh57, Theorem 3.5.3] or [Yos80, § V.4]. $\qquad\qquad\qquad\qquad \square$

Definition 2.12. Let $\Omega \subseteq \mathbb{R}^n, n \in \mathbb{N}$, be a measurable set, let $(X, \|\cdot\|)$ be a Banach space.

(i) A step function $f : \Omega \to X$ is called (Bochner) integrable if $\int_\Omega \|f(y)\| \, dy < \infty$. In this case, the *(Bochner) integral* $\int_\Omega f(y) \, dy \in X$ is defined in the following way. Let $A_i \in \mathcal{L}(\Omega), 1 \le i \le k$ with $A_i \cap A_j$ for $i \ne j$, and let $x_i \in X, 1 \le i \le k$, such that

$$f(y) = \sum_{i=1}^k 1_{A_i}(y) \, x_i, \qquad \text{for almost all } y \in \Omega. \tag{2.17}$$

Define

$$\int_\Omega f(y) \, dy := \sum_{i=1}^k |E_i| \, x_i. \tag{2.18}$$

It can be shown that (2.18) is independent of the representation in (2.17)

(ii) Let $f : \Omega \to X$ be a general function. Then f is called *(Bochner) integrable* if there exists a sequence $(s_k)_{k \in \mathbb{N}}$ of integrable step functions $s_k : \Omega \to X$ such that $s_k(y)$ converges to $f(y)$ for almost all $y \in \Omega$ and such that $\lim_{k \to \infty} \int_\Omega \|f - s_k\| \, dy = 0$ is satisfied. We define the *(Bochner) integral* by

$$\int_\Omega f(y) \, dy := \lim_{k \to \infty} \int_\Omega s_k(y) \, dy.$$

Theorem 2.13 (Bochner). *Let $\Omega \subseteq \mathbb{R}, n \in \mathbb{N}$, be a measurable set, let $(X, \|\cdot\|)$ be a Banach space, and let $f : \Omega \to X$ be strongly measurable function. Then f is Bochner integrable if and only if $\int_{\Omega} \|f(y)\|\, dy < \infty$.*

Proof. See [HiPh57, Theorem 3.7.4] or [Yos80, §V, Theorem 5.1]. $\qquad\square$

Given a Banach space X, an interval $[0, T[, 0 < T \leq \infty$, and $1 \leq p \leq \infty$, we denote by $L^p(0, T; X)$ the Banach space of equivalence classes of strongly measurable functions $f :]0, T[\to X$ such that

$$\|f\|_{L^p(0,T;X)} := \left(\int_0^T \|f(t)\|_X^p \, dt \right)^{\frac{1}{p}} < \infty$$

if $p < \infty$ and

$$\|f\|_{L^\infty(0,T;X)} := \operatorname{ess\,sup}_{[0,T[} \|f(\cdot)\|_X ,$$

if $p = \infty$. As in the scalar-valued case we identify functions with their equivalence classes. In the case that $X = L^q(\Omega), 1 \leq q \leq \infty$ the norm in the space $L^p(0, T; L^q(\Omega))$ will be denoted by $\|\cdot\|_{q,p;\Omega;T}$. Furthermore, if $1 \leq p \leq \infty$ and $0 < T \leq \infty$ we define

$$L^p_{\text{loc}}([0, T[; X) := \{u : [0, T[\to X \text{ strongly measurable};$$
$$\text{for all } 0 < T' < T \text{ there holds } u \in L^p(0, T'; X) \}.$$

If X is a reflexive Banach space and if $1 \leq p < \infty$ there holds

$$(L^p(0, T; X))' \cong L^{p'}(0, T; X') \tag{2.19}$$

where p' is the Hölder conjugate of p. To simplify the notation we use the abbreviation $L^p(0, T; X)'$ for the expression $(L^p(0, T; X))'$. For a proof of (2.19) we refer to ([FoLe07, Theorem 2.112]. Especially if $X = L^q(\Omega), 1 < q < \infty$ there holds

$$L^p(0, T; L^q(\Omega))' \cong L^{p'}(0, T; L^{q'}(\Omega)).$$

Consider $n \in \mathbb{N}$ and an open set $\Omega \subseteq \mathbb{R}^n$. It is well known that

$$L^p(0, T; L^p(\Omega)) \cong L^p(]0, T[\times \Omega) \quad \text{if } 1 \leq p < \infty.$$

In the case $p = \infty$ there holds $L^\infty(0, T; L^\infty(\Omega)) \subseteq L^\infty(]0, T[\times \Omega)$ with a strict inclusion if $\Omega \neq \emptyset$. Since the strong measurability which has to be fulfilled by the elements of the space $L^\infty(0, T; L^\infty(\Omega))$ is very restrictive, we will assume $g \in L^\infty(]0, T[\times \Omega)^n$ for the gravitational force. Therefore we formulate the following lemma. In the progress of this thesis we will often make use of this result without referring to it at every single occurrence.

Lemma 2.14. *Let $0 < T < \infty$, let $\Omega \subseteq \mathbb{R}^n$, $n \in \mathbb{N}$, be an open set, let $s, q \in [1, \infty[$, let $\theta \in L^s(0, T; L^q(\Omega))$, and let $g \in L^\infty(]0, T[\times\Omega)^n$. Then*

$$\theta g \in L^s(0, T; L^q(\Omega)^n). \tag{2.20}$$

Furthermore $g(t) \in L^\infty(\Omega)^n$ and $\|g(t)\|_\infty \leq \|g\|_{L^\infty(]0,T[\times\Omega)^n}$ for almost all $t \in]0, T[$. Consequently

$$\|\theta g\|_{q,s;\Omega;T} \leq \|g\|_{L^\infty(]0,T[\times\Omega)^n} \|\theta\|_{q,s;\Omega;T} \tag{2.21}$$

Proof. It is well known that $\theta :]0, T[\times\Omega \to \mathbb{R}$ is Lebesgue measurable. Consequently $\theta g :]0, T[\times\Omega \to \mathbb{R}^n$ is Lebesgue measurable.

By definition there is a null set $N \in \mathcal{L}(]0, T[\times\Omega)$ such that $|g(t, x)| \leq \|g\|_{L^\infty(]0,T[\times\Omega)^n}$ for all $(t, x) \in]0, T[\times\Omega \setminus N$. Using [El05, Chapter V, Aufgabe 1.4] it follows that for almost all $t \in]0, T[$ the set $N_t := \{x \in \Omega; (t, x) \in N\}$ is a null set with respect to λ_n. Furthermore, for almost all $t \in]0, T[$ the function $g(t, \cdot) : \Omega \to \mathbb{R}^n$ is Lebesgue measurable (with respect to λ_n). Thus $g(t, \cdot) \in L^\infty(\Omega)^n$ and $\|g(t, \cdot)\|_\infty \leq \|g\|_{L^\infty(]0,T[\times\Omega)^n}$ for almost all $t \in]0, T[$. Especially $(\theta g)(t) \in L^q(\Omega)^n$ and

$$\|(\theta g)(t)\|_q \leq \|\theta(t)\|_q \|g\|_{L^\infty(]0,T[\times\Omega)^n} \tag{2.22}$$

for almost all $t \in]0, T[$.

Combining Bochner's Theorem, Pettis' Theorem (see Theorems 2.13 and 2.11) and the separability of $L^q(\Omega)$ it follows that (2.20) is satisfied if the following two conditions are fulfilled:

(i) $\theta g :]0, T[\to L^q(\Omega)^n$ is weakly measurable.

(ii) $\displaystyle\int_0^T \|(\theta g)(t)\|_q^s \, dt < \infty.$

Fix $w \in L^{q'}(\Omega)^n \cong (L^q(\Omega)^n)'$. From (2.22), Hölder's inequality and $0 < T < \infty$ we obtain that the function $(t, x) \mapsto (\theta g)(t, x) \cdot w(x)$ lies in $L^1(]0, T[\times\Omega)$. It follows from Fubini's theorem that for almost all $t \in]0, T[$ the function $x \mapsto (\theta g)(t, x) \cdot w(x)$, $x \in \Omega$, is Lebesgue integrable with respect to λ_n. Moreover we get that the real-valued function $t \mapsto \langle (\theta g)(t), w \rangle_\Omega$, which is well defined for almost all $t \in]0, T[$, is Lebesgue measurable. Therefore (i) holds. From (2.22) we get

$$\int_0^T \|(\theta g)(t)\|_q^s \, dt \leq \|g\|_{L^\infty(]0,T[\times\Omega)^n}^s \int_0^T \|\theta(t)\|_q^s \, dt < \infty.$$

Thus (ii) holds as well. $\qquad\square$

Lemma 2.15. *Let $\Omega \subseteq \mathbb{R}^n$, $n \in \mathbb{N}$, be a measurable set, let X, Y be Banach spaces, and $B : \mathcal{D}(B) \subseteq X \to Y$ be a closed operator. Let $f : \Omega \to X$ be a Bochner integrable function with $f(t) \in \mathcal{D}(B)$ for almost all $t \in [0, T[$ and such that $Bf : \Omega \to Y$ is Bochner integrable. Then*

$$\int_\Omega f(s) \, ds \in \mathcal{D}(B), \quad \text{and} \quad B \int_\Omega f(s) \, ds = \int_\Omega (Bf)(s) \, ds. \tag{2.23}$$

Proof. For a proof we refer to [HiPh57, Theorem 3.7.12]. $\qquad\square$

Weak derivatives and absolute continuity

We begin with the following definition of the weak derivative of a Banach space-valued function.

Let $(X, \| \cdot \|)$ be a Banach space, let $0 < T \leq \infty$, and let $u \in L^1_{\text{loc}}(0, T; X)$. Then a function $g \in L^1_{\text{loc}}(0, T; X)$ is called *weak derivative* of u, if

$$\int_0^T \eta'(t) u(t) \, dt = -\int_0^T \eta(t) g(t) \, dt \tag{2.24}$$

for all $\eta \in C_0^\infty(]0, T[)$. It can be shown that g is uniquely determined (for almost all $t \in]0, T[$) by (2.24). Therefore we can write $u' = g$.

Let $1 \leq s \leq \infty$. Let us define the Sobolev space

$$W^{1,s}(0, T; X) := \{ u \in L^s(0, T; X); u' \in L^s(0, T; X) \}.$$

Lemma 2.16. *Let X be a Banach space, let $0 < T \leq \infty, 1 \leq s < \infty$, and let $u, g \in L^s(0, T; X)$. The following statements are equivalent:*

(i) $u \in W^{1,s}(0, T; X)$ and $u' = g$.

(ii) There exist $u_0 \in X$ such that

$$u(t) = u_0 + \int_0^t g(\tau) \, d\tau \tag{2.25}$$

for almost all $t \in [0, T[$.

(iii) There is a dense subspace $D \subseteq X'$ such that

$$\frac{d}{dt}[f, u(t)]_{X';X} = [f, g(t)]_{X';X} \tag{2.26}$$

holds for all $f \in D$ in the usual weak sense of real-valued functions in $]0, T[$.

Proof. See [So01, Chapter IV, Lemma 1.3.1]. □

Remark. (i) The equation (2.26) means explicitly

$$\int_0^T \eta'(t)[f, u(t)]_{X';X} \, dt = -\int_0^T \eta(t)[f, g(t)]_{X';X} \, dt$$

for all $\eta \in C_0^\infty(]0, T[)$.

(ii) Consider $u \in W^{1,s}(0, T; X)$. There holds that u_0 in (2.25) is well defined. Consequently, we can define the *initial value* of u by $u(0) := u_0$.

Next we define the notation of a real-valued absolutely continuous function. Let $a, b \in \mathbb{R}$ with $a < b$. A function $u : [a, b] \to \mathbb{R}$ is called *absolutely continuous*, if for every $\epsilon > 0$ there exists a $\delta > 0$ such that

$$\sum_{i=1}^{k} |u(\beta_i) - u(\alpha_i)| \leq \epsilon$$

is satisfied for every $k \in \mathbb{N}$ and pairwise disjoint open intervals $]\alpha_i, \beta_i[\,, i = 1, \dots, k$, with $]\alpha_i, \beta_i[\,\subseteq I \ (i = 1, \dots, k)$ such that

$$\sum_{i=1}^{k} |\beta_i - \alpha_i| \leq \delta.$$

If $I \subseteq \mathbb{R}$ is a (bounded or unbounded) interval, we call a function $u : I \to \mathbb{R}$ *locally absolutely continuous* on I if $u : [a, b] \to \mathbb{R}$ is absolutely continuous on every interval $[a, b] \subseteq I$.

Theorem 2.17. *Let $I = [a, b]$ with $a, b \in \mathbb{R}, a < b$, and let $u : [a, b] \to \mathbb{R}$ be an absolutely continuous function. Then the classical derivative $u'(t) \in \mathbb{R}$ exists for almost all $t \in [a, b]$, there holds $u' \in L^1([a, b])$ and*

$$u(t) = u(t_0) + \int_{t_0}^{t} u'(s)\, ds \tag{2.27}$$

for all $t_0, t \in [a, b]$.

Proof. See [El05, Kapitel 6 , Satz 4.14] □

Lemma 2.18. *Let $I = [a, b]$ with $a, b \in \mathbb{R}, a < b$, and let $f, g : [a, b] \to \mathbb{R}$ be absolutely continuous functions. Then*

$$\int_a^b f'(x)g(x)\, dx = f(x)g(x)\Big|_a^b - \int_a^b f(x)g'(x)\, dx. \tag{2.28}$$

Proof. See [El05, Satz VII, 4.16]. □

Theorem 2.19. *Let $n \in \mathbb{N}, n \geq 2$, let $U \subseteq \mathbb{R}^n$, be an open set and let $1 \leq p < \infty$. Then $u \in W^{1,p}(U)$ if and only if u is, after a redefinition on a null set of U, locally absolutely continuous on almost all line segments parallel to the coordinate axes and whose (classical) partial derivatives belong to $L^p(U)$.*

Proof. This is [Zi89, Theorem 2.1.4]. □

Lemma 2.20. *Let X, Y be Banach spaces, let $0 < T < \infty, 1 \leq s < \infty$, and let $B : \mathcal{D}(B) \subseteq X \to Y$ be a closed operator. Further let $u \in W^{1,s}(0, T; X)$ such that $Bu, B(u_t) \in L^s(0, T; Y)$. Then $(Bu)_t \in L^s(0, T; Y)$ with $(Bu)_t = B(u_t)$.*

Proof. By Lemma 2.15 there holds

$$\int_0^t u'(s)\, ds \in \mathcal{D}(B)\,, \quad \text{and } B \int_0^T u'(s)\, ds = \int_0^T B(u'(s)))\, ds \qquad (2.29)$$

for all $t \in [0, T[$. Moreover (see Lemma 2.16), after a redefinition on a null set of $[0, T[$, there holds that $u : [0, T[\to X$ is continuous and there is an initial value $u_0 \in X$ such that

$$u(t) = u_0 + \int_0^t u'(s)\, ds \qquad (2.30)$$

for all $t \in [0, T[$. Especially $u(t) \in X$ is well defined for all $t \in [0, T[$. Since $u(t) \in \mathcal{D}(B)$ for a.a. $t \in [0, T[$, we obtain from (2.29), (2.30) that $u_0 \in \mathcal{D}(B)$ holds and

$$Bu(t) = Bu_0 + \int_0^t B(u'(s))\, ds$$

is fulfilled for a.a $t \in [0, T[$. □

We need the following version of the *Lebesgue differentiation theorem*:

Theorem 2.21. *Let $\Omega \subseteq \mathbb{R}^n$, $n \in \mathbb{R}^n$, be an open set, let $f \in L^1_{loc}(\Omega)$. Then*

$$\lim_{r \to 0} \frac{1}{|B(x,r)|} \int_{B(x,r)} |f(y) - f(x)|\, dy = 0 \qquad (2.31)$$

for almost all $x \in \Omega$.

Proof. This is [St70, Chapter 1, identity (11)]. □

Lipschitz continuous functions and $W^{1,\infty}(\Omega)$

The next remarkable theorem is known as *Rademacher's Theorem*.

Theorem 2.22. *Let $n, m \in \mathbb{N}$, let $U \subseteq \mathbb{R}^n$ be open, let $f : U \to \mathbb{R}^m$ be a locally Lipschitz continuous function. Then f is differentiable for a.a. $x \in U$.*

Proof. See [Zi89, Theorem 2.2.1] combined with Kirszbraun's Theorem. □

Lemma 2.23. *Let $\Omega \subseteq \mathbb{R}^n$, $n \in \mathbb{N}$, be an open set, and let $f : \Omega \to \mathbb{R}$ be a Lipschitz continuous function. Then the classical partial derivatives $[\frac{\partial f}{\partial x_i}]$, $i = 1, \ldots, n$, exist for a.a. $x \in \Omega$. There holds that all weak derivatives $\frac{\partial f}{\partial x_i} \in L^\infty(\Omega)$, $i = 1, \ldots, n$ exist and we have*

$$\left[\frac{\partial f}{\partial x_i}\right](x) = \frac{\partial f}{\partial x_i}(x)$$

for almost all $x \in \Omega$ and all $i = 1, \ldots, n$.

Proof. See [Wl82, Satz 1.8]. □

Lemma 2.24. *Let $n \in \mathbb{N}$, let $f \in W^{1,\infty}(\mathbb{R}^n)$ be given. Then, after a redefinition on a null set of $\mathbb{R}^n$, there holds that $f : \mathbb{R}^n \to \mathbb{R}$ is Lipschitz continuous and*

$$\mathrm{Lip}(f) \leq c\|\nabla f\|_\infty$$

with an absolute constant $c > 0$.

Proof. This can be proven analogously to [Ev98, Section 5.8 Theorem 4]. □

Gelfand triples and Aubin-Lions compactness theorems

Let X, Y be normed spaces. We call an injective function $f : X \to Y$ an *imbedding*. An imbedding $f : X \to Y$ is called *dense* if its range $f(X)$ is dense in Y.

Theorem 2.25 (Aubin-Lions). *Let X_0, X_1 be reflexive Banach spaces and let X be a Banach space. Assume there is a compact imbedding $X_0 \hookrightarrow X$ and a continuous imbedding $X \hookrightarrow X_1$. Let $0 < T < \infty$ and $\alpha_0, \alpha_1 \in \mathbb{R}$ with $1 < \alpha_i < \infty$ for $i = 0, 1$. Assume that $(u_k)_{k\in\mathbb{N}}$ is a bounded sequence in $L^{\alpha_0}(0, T; X_0)$ such that its weak derivative $(u'_k)_{k\in\mathbb{N}}$ is bounded in $L^{\alpha_1}(0, T; X_1)$. Then there exist $u \in L^{\alpha_0}(0, T; X)$ and a subsequence $(m_k)_{k\in\mathbb{N}}$ with*

$$u_{m_k} \underset{k\to\infty}{\to} u \quad \text{strongly in } L^{\alpha_0}(0, T; X).$$

Proof. See [Te77, Chapter 3, Theorem 2.1]. □

Let X be a complex Hilbert space and let $u : \mathbb{R} \to X$ be an integrable function. Then we define the Fourier transform $\hat{u}$ of u by

$$\hat{u}(\tau) := \int_{\mathbb{R}} e^{-2\pi i t \tau} u(t) \, d\tau, \quad \tau \in \mathbb{R}.$$

Exactly as in the case of a scalar-valued function we can prove that the *Parseval identity* holds: For every function $f \in L^2(\mathbb{R}; X) \cap L^1(\mathbb{R}; X)$ we have

$$\|\hat{f}\|_{L^2(\mathbb{R};X)} = \|f\|_{L^2(\mathbb{R};X)}. \tag{2.32}$$

Theorem 2.26 (Fourier-valued Aubin-Lions). *Let X_0, X, X_1 be complex Hilbert spaces with a compact imbedding $X_0 \hookrightarrow X$ and a continuous imbedding $X \hookrightarrow X_1$. Let $(u_k)_{k\in\mathbb{N}}$ be a bounded sequence in $L^2(\mathbb{R}; X_0)$ such that the following two properties are satisfied:*

1. *There exists a compact set $K \subseteq \mathbb{R}$ with $\mathrm{supp}(u_k) \subseteq K$ for all $k \in \mathbb{N}$.*

2. *There exists $0 < \gamma < \infty$ and $0 < M < \infty$ such that*

$$\int_{\mathbb{R}} |\tau|^{2\gamma} \|\widehat{u}_k(\tau)\|_{X_1}^2 \, d\tau \leq M \tag{2.33}$$

for all $k \in \mathbb{N}$.

Then there exist $u \in L^2(\mathbb{R}; X)$ and a subsequence $(u_{m_k})_{k \in \mathbb{N}}$ with

$$u_{m_k} \underset{k \to \infty}{\to} u \quad \text{strongly in } L^2(\mathbb{R}; X).$$

Proof. See [Te77, Chapter 3, Theorem 2.2]. $\qquad \square$

Definition 2.27. Let X be a reflexive Banach space, Y be a Hilbert space with scalar product $\langle \cdot, \cdot \rangle_Y$. Assume there is a continuous, dense imbedding $i : X \hookrightarrow Y$. Let $i' : Y' \to X'$ denote the dual (or adjoint) operator. Then it can be proved that i' is also a continuous imbedding with dense range $i'(Y')$ in X'. Using the theorem of Riesz we identify Y with its dual Y'. Then we have the continuous and dense imbedding scheme

$$X \overset{i}{\hookrightarrow} Y \cong Y' \overset{i'}{\hookrightarrow} X' \tag{2.34}$$

which is called *Gelfand triple.*

Note that the imbedding $X \hookrightarrow X'$ is given by

$$x \in X \mapsto F_x(v) := \langle ix, iv \rangle_Y, v \in X. \tag{2.35}$$

Moreover, for $1 \leq \alpha < \infty$ we introduce the space

$$W^\alpha(0, T; X, Y) := \{ y \in L^2(0, T; X); \ y' \in L^\alpha(0, T; X') \} \tag{2.36}$$

equipped with the norm

$$\|y\|_{W^\alpha(0,T;X,Y)} := \left(\|y\|_{L^2(X)}^2 + \|y'\|_{L^\alpha(X')}^2 \right)^{1/2}.$$

In the next Lemma we formulate a criterion for $y' \in L^1(0, T; X')$. Here the reflexivity of X is used.

Lemma 2.28. *Let X, Y be a Gelfand triple as in (2.34). Let $u \in L^1(0, T; X)$ and $g \in L^1(0, T; X')$. We identify u with $u \in L^1(0, T; X')$ by (2.34). Then $u' = g$ in $L^1(0, T; X')$ if and only if*

$$\frac{d}{dt} \langle iu, ix \rangle_Y = [g, x]_{X';X} \tag{2.37}$$

for all $x \in X$ in the weak sense of real-valued functions.

Proof. By definition, we have to show

$$\frac{d}{dt}[x'', iu]_{X'';X'} = [x'', g]_{X'';X'} \quad \forall x'' \in X''. \tag{2.38}$$

Let $J : X \to X''$, $x \mapsto J(x)$ with $(Jx)(x') := [x', x]_{X';X}$, $x' \in X'$, denote the canonical imbedding into X''. By definition of reflexivity, J is surjective. Let $x'' \in X''$. Then there exists a unique $x \in X$ such that $Jx = x''$. This means

$$[x', x]_{X';X} = [x'', x']_{X'';X'} \tag{2.39}$$

for all $x' \in X'$. Using (2.39) we see that (2.37) is equivalent to

$$\frac{d}{dt}[u, x]_{X';X} = [g, x]_{X';X} \quad \forall x \in X. \tag{2.40}$$

Finally, with (2.35) we see that (2.37) is satisfied if and only if

$$\frac{d}{dt}\langle iu, ix \rangle_Y = [g, x]_{X';X} \quad \forall x \in X.$$

$\square$

Theorem 2.29. *Let X, Y be a Gelfand triple as in (2.34). Then $W^2(0, T; X, Y)$ is a Banach space and we have the continuous imbedding*

$$W^2(0, T; X, Y) \hookrightarrow C([0, T]; Y). \tag{2.41}$$

Moreover for all $y, v \in W^2(0, T; X, Y)$, the integration by parts formula (after a redefinition on a null set)

$$\langle y(t), v(t) \rangle_Y - \langle y(s), v(s) \rangle_Y = \int_s^t \left(\langle y'(\tau), v(\tau) \rangle_{X';X} + \langle v'(\tau), y(\tau) \rangle_{X';X} \right) d\tau \tag{2.42}$$

holds for all $t, s \in [0, T]$.

Proof. See [HPUU09, Theorem 1.32]. $\square$

Lemma 2.30. *Consider a Gelfand triple X, Y as in (2.34) and $u \in W^2(0, T; X, Y)$. Then u coincides almost everywhere with a function in $C([0, T]; Y)$. Moreover the identity*

$$\frac{1}{2}\frac{d}{dt}\|u\|_Y^2 = [u', u]_{X';X} \tag{2.43}$$

holds in the weak sense of real valued functions on $]0, T[$.

Proof. Define $v := u$ and apply Theorem 2.29 combined with Lemma 2.16. $\square$

Definition 2.31. Let X be a Banach space, let $0 < T \leq \infty$, let $u : [0, T[\to X$. Then u is *weakly continuous*, if for all $f \in X'$ there holds that the function

$$t \mapsto [f, u(t)]_{X';X}, \quad t \in [0, T[,$$

is continuous.

Lemma 2.32. *Let X be a reflexive Banach space, let $0 < T < \infty$, let Y be a Banach space and assume there is a continuous imbedding $X \hookrightarrow Y$. Let $u \in L^\infty(0, T; X)$ and assume that $u : [0, T[\to Y$ is weakly continuous. Then we have that $u : [0, T[\to X$ is weakly continuous.*

Proof. See the proof of [Te77, Chapter 3, Lemma 1.4]. □

2.2.1 Positive definite, self-adjoint operators on Hilbert spaces

Let H be a (real) Hilbert space with norm $\| \cdot \|$, and let $B : \mathcal{D}(B) \subseteq H \to H$ be a densely defined, linear operator with dual operator $B' : \mathcal{D}(B') \subseteq H \to H$.

 (i) B is called *self-adjoint*, if $\mathcal{D}(B) = \mathcal{D}(B')$ and $Bv = B'v$ for all $v \in \mathcal{D}(B)$.

 (ii) B is called *positive definite*, if

$$\langle Bv, v \rangle > 0 \quad \text{for all } v \in H, v \neq 0.$$

Throughout this subsection let H be a (real) Hilbert space with norm $\| \cdot \|$ and let $B : \mathcal{D}(B) \subseteq H \to H$ be a self-adjoint, positive definite operator.

Remark. (i) Since B is positive definite, it follows that B is *accretive*. Moreover, due to the self-adjointness of B, we get that B is in fact *m-accretive*.

 (ii) Since B is self-adjoint and densely defined it follows that B is closed. The positive definiteness of B implies that B is injective.

Fractional powers

Definition 2.33. A *resolution of identity on $[0, \infty[$* is a family of continuous, self-adjoint projections $E_\lambda : H \to H, \lambda \geq 0$, satisfying the following properties:

 (i) $E_\lambda E_\mu = E_\mu E_\lambda = E_\lambda$ for all $\lambda \leq \mu$,

 (ii) $\lim_{t \searrow 0} E_{\lambda + t} v = E_\lambda v$ for all $\lambda \geq 0, v \in H$,

 (iii) $E_0 v = 0$ and $\lim_{\lambda \to \infty} E_\lambda v = v$ for all $v \in H$.

Since B is a positive definite self-adjoint operator, there exists a uniquely determined resolution of identity $\{E_\lambda; \lambda \geq 0\}$ on $[0, \infty[$ such that

$$\mathcal{D}(B) = \left\{ v \in H; \int_0^\infty \lambda^2 \, d\|E_\lambda v\|^2 < \infty \right\}, \quad Bv = \int_0^\infty \lambda \, dE_\lambda v \text{ for all } v \in H. \tag{2.44}$$

The integrals in (2.44) have to be understood in the sense of improper Riemann-Stieltjes integrals in $\mathbb{R}$ respectively in H.

Let $\alpha \geq 0$. Define the fractional power B^α by

$$\mathcal{D}(B^\alpha) = \left\{ v \in H; \int_0^\infty \lambda^{2\alpha} \, d\|E_\lambda v\|^2 < \infty \right\}, \tag{2.45}$$

$$B^\alpha v := \int_0^\infty \lambda^\alpha \, dE_\lambda v, \quad v \in \mathcal{D}(B^\alpha). \tag{2.46}$$

Let $\alpha > 0$. Define an operator $B^{-\alpha}$ by

$$\mathcal{D}(B^{-\alpha}) := \left\{ v \in H; \int_0^\infty \lambda^{-2\alpha} \, d\|E_\lambda v\|^2 < \infty \right\},$$

$$B^{-\alpha} v := \int_0^\infty \lambda^{-\alpha} \, dE_\lambda v, \quad v \in \mathcal{D}(B^{-\alpha}). \tag{2.47}$$

Since the function $\lambda \mapsto \lambda^{-\alpha}$ is not continuous on the whole interval $[0, \infty[$ by definition

$$\int_0^\infty \lambda^{-\alpha} \, dE_\lambda v := \lim_{a \to 0} \int_a^\infty \lambda^{-\alpha} \, dE_\lambda v, \quad v \in \mathcal{D}(B^{-\alpha}),$$

It can be proved that B^{-1} as defined in (2.47) coincides with the inverse of B so the notation is consistent. Moreover B^α is injective and

$$\mathcal{D}(B^{-\alpha}) = \mathcal{R}(B^\alpha) \text{ and } B^{-\alpha} = (B^{-1})^\alpha = (B^\alpha)^{-1}.$$

The content of the following lemma is a useful interpolation inequality of fractional powers.

Lemma 2.34. *Let $0 \leq \alpha \leq 1$. Then*

$$\|B^\alpha v\| \leq \|Bv\|^\alpha \|v\|^{1-\alpha} \leq \alpha \|Bv\| + (1-\alpha)\|v\| \tag{2.48}$$

for all $v \in \mathcal{D}(B)$.

Proof. See [So01, Chapter II, Lemma 3.2.2]. □

Maximal Regularity

Since B is m-accretive, the Lumer-Phillips Theorem implies that $-B$ is the generator of a strongly continuous contraction semigroup $e^{-tB}, t > 0$. Using the resolution of identity we get the following representation of this semigroup: For $t > 0$ we have

$$\mathcal{D}(e^{-tB}) = H, \qquad e^{-tB}v = \int_0^\infty e^{-t\lambda} dE_\lambda v. \tag{2.49}$$

As in section 1.1 these integrals are meant as improper Riemann Stieltjes integrals in $\mathbb{R}$ respectively in H.

Since B is a positive definite, self-adjoint operator on a Hilbert space, we can apply [EnNa99, Chapter II, Corollary 4.8] to obtain that $e^{-tB}, t > 0$, is a uniformly bounded analytic semigroup. For the definition and properties of analytic semigroups we refer to [EnNa99, Chapter II, Section 4].

The next theorem states that the operator B has *maximal regularity*.

Theorem 2.35. *Let H be a Hilbert space, let $B : \mathcal{D}(B) \subseteq H \to H$ be a positive definite, self-adjoint operator. Further, let $1 < s < \infty, 0 < T \le \infty$, let $f \in L^s(0,T;H)$, and let $u_0 \in H$ such that $\int_0^T \|Be^{-tB}u_0\|_H^s \, dt < \infty$. Then the evolution equation*

$$\begin{aligned} u_t + Bu &= f \quad in \ ]0, T[, \\ u(0) &= u_0 , \end{aligned} \tag{2.50}$$

has a unique (strong) solution $u \in L^s_{loc}(0,T;\mathcal{D}(B))$ with $u_t \in L^s(0,T;H)$ and $u \in C([0,T[;H)$. Moreover, u satisfies the maximal regularity estimate

$$\|u_t\|_{L^s(0,T;H)} + \|Bu\|_{L^s(0,T;H)} \le c \left[\left(\int_0^T \|Be^{-tB}u_0\|_H^s \, dt \right)^{1/s} + \|f\|_{L^s(0,T;H)} \right] \tag{2.51}$$

with a constant $c > 0$ independent of f, T and has the representation

$$u(t) = e^{-tB}u_0 + \int_0^t e^{-(t-\tau)B} f(\tau) \, d\tau \tag{2.52}$$

for all $t \in [0,T[$. In the case $T < \infty$ there even holds $u \in L^s(0,T;\mathcal{D}(B))$.

Proof. We already know that $-B$ is the generator of a uniformly bounded, analytic semigroup $e^{-tB}, t > 0$, on a Hilbert space. Therefore, B has maximal regularity in the following sense: For every $f \in L^s(0,T;H)$ the function

$$v(t) := \int_0^t e^{-(t-\tau)B} f(\tau) \, d\tau , \quad t \in [0,T[,$$

satisfies $v', Bv \in L^s(0,T;H)$ and v is a solution of (2.50) with $u_0 = 0$. Further v satisfies the estimate (2.51) with $u_0 = 0$. The first proof of this very important fact can be found in [deS64, Theorem 4.4].

Consider now $u_0 \in H$ and define $w(t) := e^{-tB}u_0$, $t > 0$. Due to the analyticity of the semigroup we have $w'(t) = -Bw(t)$ for all $t > 0$. The assumption on u_0 yields $Bw \in L^s(0, T; H)$ and consequently, the equation $w' + Bw = 0$ holds in $L^s(0, T; H)$. Altogether $u(t) := v(t) + w(t)$ satisfies (2.50), (2.51). $\qquad\square$

2.2.2 The Stokes operator and the Laplace operator: Definition and Properties

Throughout this subsection let $\Omega \subseteq \mathbb{R}^3$ be a general domain.

Definition 2.36. The operator $(-\Delta)$ is defined by

$$\mathcal{D}(-\Delta) := \left\{ \phi \in H_0^1(\Omega); \exists f \in L^2(\Omega) \text{ with } \langle \nabla \phi, \nabla \psi \rangle = \langle f, \psi \rangle_\Omega \quad \forall \psi \in H_0^1(\Omega) \right\},$$
$$\langle (-\Delta)\phi, \psi \rangle_\Omega = \langle \nabla \phi, \nabla \psi \rangle_\Omega \quad \forall \phi \in \mathcal{D}(-\Delta) \, \forall \psi \in H_0^1(\Omega).$$

The Stokes operator A is defined by

$$\mathcal{D}(A) := \left\{ u \in W_{0,\sigma}^{1,2}(\Omega); \exists f \in L_\sigma^2(\Omega) \text{ with } \langle \nabla u, \nabla w \rangle_\Omega = \langle f, w \rangle_\Omega \quad \forall w \in W_{0,\sigma}^{1,2}(\Omega) \right\},$$
$$\langle Au, w \rangle_\Omega = \langle \nabla u, \nabla w \rangle_\Omega \quad \forall u \in \mathcal{D}(A) \, \forall w \in W_{0,\sigma}^{1,2}(\Omega).$$

It is well known (see [So01, Chapter II, Lemma 3.2.1]) that $(-\Delta) : \mathcal{D}(-\Delta) \subseteq L^2(\Omega) \to L^2(\Omega)$ and $A : \mathcal{D}(A) \subseteq L_\sigma^2(\Omega) \to L_\sigma^2(\Omega)$ are positive definite, self-adjoint operators. Consequently all results from subsection 2.2.1 can be applied to $(-\Delta)$ and A.

Fix $\alpha \in [-1, 1]$. Define the fractional powers A^α, $(-\Delta)^\alpha$ as in (2.45), (2.46), (2.47). There holds that $A^\alpha : \mathcal{D}(A^\alpha) \to L_\sigma^2(\Omega)$ with dense domain $D(A^\alpha) \subseteq L_\sigma^2(\Omega)$ and dense range $\mathcal{R}(A^\alpha) \subseteq L_\sigma^2(\Omega)$ is a well defined, injective, closed, positive definite, self-adjoint operator such that

$$(A^\alpha)^{-1} = A^{-\alpha}, \quad \mathcal{R}(A^\alpha) = \mathcal{D}(A^{-\alpha}).$$

The same properties hold for $(-\Delta)^\alpha$, $\alpha \in [-1, 1]$. Especially, for $\alpha = \frac{1}{2}$ we have (see [So01, Chapter II, Lemma 2.2.1])

$$\mathcal{D}((-\Delta)^{1/2}) = H_0^1(\Omega) \quad \text{and} \quad \|(-\Delta)^{1/2}\phi\|_2 = \|\nabla\phi\|_2, \tag{2.53}$$
$$\mathcal{D}(A^{1/2}) = W_{0,\sigma}^{1,2}(\Omega) \quad \text{and} \quad \|A^{1/2}u\| = \|\nabla u\|_2, u \in \mathcal{D}(A^{1/2}). \tag{2.54}$$

In general $\mathcal{D}((-\Delta)^\alpha)$ will be equipped with the graph norm $\|\phi\|_2 + \|(-\Delta)^\alpha \phi\|_2$ which makes $\mathcal{D}((-\Delta)^\alpha)$ to a Banach space since $(-\Delta)^\alpha$ is closed. Analogously $\mathcal{D}(A^\alpha)$ becomes a Banach space when equipped with the graph norm $\|u\|_2 + \|A^\alpha u\|_2$.

There holds that Δ generates a uniformly bounded, analytic semigroup $\{e^{t\Delta}; t \geq 0\}$ on $L^2(\Omega)$ and that $-A$ generates a uniformly bounded, analytic semigroup $\{e^{-tA}; t \geq 0\}$ on $L_\sigma^2(\Omega)$. The decay estimates

$$\|(-\Delta)^\alpha e^{t\Delta}\phi\|_2 \leq t^{-\alpha}\|\phi\|_2, \quad t > 0, \phi \in L^2(\Omega), \tag{2.55}$$
$$\|A^\alpha e^{-tA}u\|_2 \leq t^{-\alpha}\|u\|_2, \quad t > 0, u \in L_\sigma^2(\Omega), \tag{2.56}$$

are satisfied for all $\alpha \in [0,1]$. For a proof of (2.56) we refer to [So01, Chapter IV, equation (1.5.15)]. This proof uses only the properties of the resolution of identity of a positive definite, self-adjoint operator on a Hilbert space. Thus (2.55) holds true with the same proof.

We introduce the Helmholtz projection which will play an essential role in our functional analytic approach of the Boussinesq equations.

Lemma 2.37. *Let $\Omega \subseteq \mathbb{R}^n$, $n \in \mathbb{N}$, be a general domain. There exists a unique linear, bounded, orthogonal projection $P_\Omega : L^2(\Omega)^n \to L^2_\sigma(\Omega)$ with $P_\Omega f = f$, $f \in L^2_\sigma(\Omega)$, and null space $\mathcal{N}(P_\Omega) = \{\nabla p \in L^2(\Omega); p \in L^2_{loc}(\overline{\Omega})\}$. There holds $\|P_\Omega\| = 1$ and P_Ω is self-adjoint.*

Proof. See [So01, Lemma II.2.5.2]. □

When there is no possibility of confusion we simply write P instead of P_Ω. We need that the imbeddings

$$\|\phi\|_4 \leq K\|(-\Delta)^{3/8}\phi\|_2 \quad \forall \phi \in \mathcal{D}((-\Delta)^{3/8}),\tag{2.57}$$

$$\|u\|_4 \leq K\|A^{3/8}u\|_2 \quad \forall u \in \mathcal{D}(A^{3/8}),\tag{2.58}$$

are satisfied with an absolute constant $K > 0$ (independent of Ω, u, ϕ). For a proof of (2.58) we refer to [So01, Lemma 2.4.2]. The proof of (2.57) is analogous to the proof of (2.58).

In this section we will make extensive use of the *Hardy-Littlewood inequality*.

Theorem 2.38. *Let $0 < \alpha < 1$, $1 < r < q < \infty$ with $\alpha + \frac{1}{q} = \frac{1}{r}$ and let $f \in L^r(\mathbb{R})$. Then the integral*

$$v(t) := \int_{\mathbb{R}} |t - \tau|^{\alpha-1} f(\tau)\, d\tau\tag{2.59}$$

converges absolutely for almost all $t \in \mathbb{R}$ and

$$\|v\|_{L^q(\mathbb{R})} \leq c\|f\|_{L^r(\mathbb{R})}\tag{2.60}$$

with a constant $c = c(\alpha, r) > 0$.

Proof. See [St70, Theorem Chapter V, Theorem 1] or [Tri78, 1.18.9, Theorem 3]. □

We proceed with a definition of $A^{-1/2}P\mathrm{div}$ and of $(-\Delta)^{-1/2}\mathrm{div}$.

Lemma 2.39. *Let $\Omega \subseteq \mathbb{R}^n$, $n \in \mathbb{N}$, be a general domain. The following statements are fulfilled:*

(i) Let $F \in L^2(\Omega)^{n \times n}$. Then there exists a unique element in $L^2_\sigma(\Omega)$ to be denoted by $A^{-1/2}P\mathrm{div}F$ such that

$$\langle A^{-1/2}P\mathrm{div}F, A^{1/2}w\rangle_\Omega = -\langle F, \nabla w\rangle_\Omega \quad \forall w \in W^{1,2}_{0,\sigma}(\Omega). \tag{2.61}$$

The estimate

$$\|A^{-1/2}P\mathrm{div}F\|_2 \leq \|F\|_2 \tag{2.62}$$

holds for all $F \in L^2(\Omega)^{n \times n}$.

(ii) Let $f \in L^2(\Omega)^n$. Then there exists a unique element in $L^2(\Omega)$ to be denoted by $(-\Delta)^{-1/2}\mathrm{div}f$ such that

$$\langle(-\Delta)^{-1/2}\mathrm{div}f, (-\Delta)^{1/2}\phi\rangle_\Omega = -\langle f, \nabla\phi\rangle_\Omega \quad \forall \phi \in H^1_0(\Omega). \tag{2.63}$$

The estimate

$$\|(-\Delta)^{-1/2}\mathrm{div}f\|_2 \leq \|f\|_2 \tag{2.64}$$

is satisfied for all $f \in L^2(\Omega)^n$.

Proof. The proof of statement (i) can be found in [So01, Chapter III, Lemma 2.6.1]. The proof of statement (ii) is analogous. $\qquad\square$

The following technical lemma is needed for the proof of Theorem 2.41 and for Section 5.3. Especially we need that the choice $s_1 = 1$ is allowed in this lemma.

Lemma 2.40. *Let $\Omega \subseteq \mathbb{R}^3$ be a general domain, let $1 \leq s_1 < \infty, 1 < s_2 < \infty$, let $0 < T < \infty$, and let $f \in L^{s_1}(0, T; L^2(\Omega))$, $F \in L^{s_2}(0, T; L^2(\Omega)^3)$. Assume that for a.a. $t \in [0, T[$ there holds $f(t) = \mathrm{div}(F(t))$ in the sense of distributions in Ω, i.e. $\langle f(t), w\rangle_\Omega = -\langle F(t), \nabla w\rangle$ for all $w \in C^\infty_0(\Omega)$. Then*

$$\int_0^t e^{(t-\tau)\Delta}f(\tau)\,d\tau = (-\Delta)^{1/2}\int_0^t e^{(t-\tau)\Delta}(-\Delta)^{-1/2}\mathrm{div}F(\tau)\,d\tau \tag{2.65}$$

for almost all $t \in [0, T[$.

Proof. Looking carefully at the proof of [So01, Chapter IV, Lemma 1.6.1 a)] we see that in this proof only properties of a self-adjoint, positive definite operator on a Hilbert space are used. Consequently, the left hand side of (2.65) defines a continuous function $[0, T[\to L^2(\Omega)$; especially this expression is well defined in $L^2(\Omega)$ for all $t \in [0, T[$.

For $t \in [0, T[$ define the function $F^t(\tau) := e^{(t-\tau)\Delta}(-\Delta)^{-1/2}\mathrm{div}F(\tau), \tau \in [0, t[$. It follows from (2.64) and the uniform boundedness of the semigroup generated by Δ, that $F^t \in L^{s_2}(0, t; L^2(\Omega))$ for all $t \in]0, T[$. From (2.55), (2.64) we obtain

$$\int_0^t \|(-\Delta)^{1/2}F^t(\tau)\|_2\,d\tau \leq \int_0^T |t - \tau|^{-1/2}\|F(\tau)\|_2\,d\tau$$

for almost all $t \in [0, T[$. Consequently, by Theorem 2.38 and $0 < T < \infty$ we conclude

$$\int_0^t \| F^t(\tau) \|_2 \, d\tau < \infty \tag{2.66}$$

for almost all $t \in [0, T[$ and consequently (see Bochner's Theorem) $(-\Delta)^{1/2} F^t \in L^1(0, t; L^2(\Omega))$. Thus, by Lemma 2.15 we obtain

$$(-\Delta)^{1/2} \int_0^t e^{(t-\tau)\Delta} (-\Delta)^{-1/2} \mathrm{div} F(\tau) \, d\tau = \int_0^t (-\Delta)^{1/2} e^{(t-\tau)\Delta} (-\Delta)^{-1/2} \mathrm{div} F(\tau) \, d\tau \tag{2.67}$$

for almost all $t \in [0, T[$.

Fix $w \in L^2(\Omega)$. Due to the analyticity of the semigroup $e^{t\Delta}$, $t > 0$, there holds $e^{t\Delta} w \in \mathcal{D}(-\Delta)$ for all $t > 0$; especially $e^{t\Delta} w \in H_0^1(\Omega)$. We use (2.63), (2.67), the self-adjointness of $(-\Delta)^{1/2}$ and the relation $f(t) = \mathrm{div} F(t)$ a.a. $t \in [0, T[$, and obtain

$$\left\langle (-\Delta)^{1/2} \int_0^t e^{(t-\tau)\Delta} (-\Delta)^{-1/2} \mathrm{div} F(\tau) \, d\tau, w \right\rangle_\Omega$$
$$= \int_0^t \left\langle (-\Delta)^{1/2} e^{(t-\tau)\Delta} (-\Delta)^{-1/2} \mathrm{div} F(\tau), w \right\rangle_\Omega \, d\tau$$
$$= \int_0^t \left\langle (-\Delta)^{-1/2} \mathrm{div} F(\tau), (-\Delta)^{1/2} e^{(t-\tau)\Delta} w \right\rangle_\Omega \, d\tau$$
$$= -\int_0^t \left\langle F(\tau), \nabla e^{(t-\tau)\Delta} w \right\rangle_\Omega \, d\tau \tag{2.68}$$
$$= \int_0^t \left\langle f(\tau), e^{(t-\tau)\Delta} w \right\rangle_\Omega \, d\tau$$
$$= \left\langle \int_0^t e^{(t-\tau)\Delta} \, d\tau, w \right\rangle_\Omega$$

for a.a. $t \in [0, T[$. Therefore, (2.65) holds for almost all $t \in [0, T[$. $\qquad\square$

The lemma above will be a central tool in the proof of Theorem 5.16. For the notion of the initial value of a function in the space $W^{1,s}(0, T; X)$ we refer to the remark following Lemma 2.16.

Lemma 2.41. *Let $\Omega \subseteq \mathbb{R}^3$ be a general domain, let $1 < s < \infty$, $0 < T < \infty$. Then the following statements hold:*

(i) Let $f \in W^{1,s}(0, T; L^2(\Omega))$ with initial value $f(0) = 0$. Define

$$u(t) := \int_0^t e^{(t-\tau)\Delta} f(\tau) \, d\tau, \quad t \in [0, T[. \tag{2.69}$$

Then $u \in W^{2,s}(0, T; L^2(\Omega)) \cap W^{1,s}(0, T; D(-\Delta))$ and

$$u'(t) = \int_0^t e^{(t-\tau)\Delta} f'(\tau) \, d\tau, \quad \text{for a.a. } t \in [0, T[. \tag{2.70}$$

(ii) *Let $F \in W^{1,s}(0,T;L^2(\Omega)^3)$ with initial value $F(0) = 0$. Define*

$$v(t) := (-\Delta)^{1/2} \int_0^t e^{(t-\tau)\Delta}(-\Delta)^{-1/2}\mathrm{div}F(\tau)\,d\tau\,, \quad t \in [0,T[. \tag{2.71}$$

Then $v \in W^{1,s}(0,T;L^2(\Omega))$ and we have

$$v'(t) = (-\Delta)^{1/2} \int_0^t e^{(t-\tau)\Delta}(-\Delta)^{-1/2}\mathrm{div}\big(F'(\tau)\big)\,d\tau \quad \text{for a.a. } t \in [0,T[. \tag{2.72}$$

Moreover $\big((-\Delta)^{-1/2}(v_t)\big)_t, (-\Delta)^{1/2}(v_t) \in L^s(0,T;L^2(\Omega))$.

Proof. Proof of (i). Define

$$g(t) := \int_0^t e^{(t-\tau)\Delta}f'(\tau)\,d\tau \quad t \in [0,T[. \tag{2.73}$$

The maximal regularity property of $(-\Delta)$ implies $u, g \in W^{1,s}(0,T;L^2(\Omega))$. Using Lemma 2.16 and the density of $\mathcal{D}(-\Delta)$ in $L^2(\Omega)$ we see that (2.70) is satisfied if the following assertion holds:

Assertion. *We have*

$$\langle u(\rho), w \rangle = \int_0^\rho \langle g(t), w \rangle\,dt$$

for all $w \in \mathcal{D}(-\Delta)$ and all $0 \leq \rho < T$.

Proof of the assertion. Fix $w \in \mathcal{D}(-\Delta)$ and $0 \leq \rho < T$. We get with integration by parts (see ([So01, Chapter IV, Lemma 1.3.2]), $f(0) = 0$ and Fubini's theorem

$$\int_0^\rho \langle g(t), w \rangle\,dt = \int_0^\rho \int_0^t \langle f'(\tau), e^{(t-\tau)\Delta}w \rangle\,d\tau\,dt$$

$$= \int_0^\rho \langle f(t), w \rangle\,dt - \int_0^\rho \int_0^t \langle f(\tau), \frac{\partial}{\partial\tau}e^{(t-\tau)\Delta}w \rangle\,d\tau\,dt$$

$$= \int_0^\rho \langle f(t), w \rangle\,dt + \int_0^\rho \int_0^t \langle f(\tau), \frac{\partial}{\partial t}e^{(t-\tau)\Delta}w \rangle\,d\tau\,dt$$

$$= \int_0^\rho \langle f(t), w \rangle\,dt + \int_0^\rho \int_\tau^\rho \langle f(\tau), \frac{\partial}{\partial t}e^{(t-\tau)\Delta}w \rangle\,dt\,d\tau$$

$$= \int_0^\rho \langle f(t), w \rangle\,dt + \int_0^\rho \Big(\langle f(\tau), e^{(\rho-\tau)\Delta}w \rangle - \langle f(\tau), w \rangle\Big)\,d\tau$$

$$= \int_0^\rho \langle e^{(\rho-\tau)\Delta}f(\tau), w \rangle\,d\tau$$

$$= \langle u(\rho), w \rangle.$$

$\square$

Thus, the maximal regularity property of $(-\Delta)$ applied to (2.70) yields

$$u \in W^{2,s}(0,T;L^2(\Omega)) \cap W^{1,s}(0,T,D(-\Delta)).$$

Proof of (ii). Define

$$\phi(t) := \int_0^t e^{(t-\tau)\Delta}(-\Delta)^{-1/2}\mathrm{div}\big(F(\tau)\big)\, d\tau\,, \quad t \in [0, T[. \tag{2.74}$$

With the help of Lemma 2.20 it follows $\big((-\Delta)^{-1/2}\mathrm{div}F\big)_t = (-\Delta)^{-1/2}\mathrm{div}\big(F_t\big)$ as identity in $L^s(0, T; L^2(\Omega))$. We make use of statement (i) of this theorem to get $\phi \in W^{2,s}(0, T; L^2(\Omega))$ and

$$\phi'(t) = \int_0^t e^{(t-\tau)\Delta}(-\Delta)^{-1/2}\mathrm{div}\big(F'(\tau)\big)\, d\tau \tag{2.75}$$

for almost all $t \in [0, T[$. By the maximal regularity of $(-\Delta)$ it follows $\phi_t, (-\Delta)(\phi_t) \in L^s(0, T; L^2(\Omega))$, and consequently, by interpolation $(-\Delta)^\alpha(\phi_t) \in L^s(0, T; L^2(\Omega))$ for all $0 \le \alpha \le 1$. Further, by Lemma 2.20, there holds $\big((-\Delta)^{1/2}\phi\big)_t = (-\Delta)^{1/2}(\phi_t)$. Since $v = (-\Delta)^{1/2}\phi$ we get

$$v'(t) = (-\Delta)^{1/2}\int_0^t e^{(t-\tau)\Delta}(-\Delta)^{-1/2}\mathrm{div}\big(F'(\tau)\big)\, d\tau \tag{2.76}$$

for almost all $t \in [0, T[$. □

Remark. The proof of Lemma 2.40 Lemma 2.41 is based on the properties of a positive, self-adjoint operator on a Hilbert space and the relations (2.63), (2.64). Therefore, a careful inspection of the proofs of these results shows that an analogous version of these lemmata hold if $(-\Delta)$ is replaced by A and $L^2(\Omega)$ is replaced by $L^2_\sigma(\Omega)$.

3 Modelisation

The goal of this chapter is to develop the Boussinesq approximation for the motion of a viscous, incompressible, heat conducting fluid in a domain $\Omega \subseteq \mathbb{R}^3$ and a time interval $[0, T[$. We denote by $u = u(t, x) = (u_1(t, x), u_2(t, x), u_3(t, x))$ the velocity of the fluid at $(t, x) \in [0, T[\times \Omega$, by $\vartheta = \vartheta(t, x)$ the absolute temperature of the fluid at (t, x) and by $p = p(t, x)$ the pressure at (t, x).

We can interpret u in the following sense: Let $X \in \Omega$ be a particle. Then the evolution in time of X is given by $x(t) = x(t; t_0; x_0)$ where x is the unique solution of the differential equation system

$$\dot{x}(t) = u(t, x(t)), \quad x(t_0) = X. \tag{3.1}$$

Especially we assume that $x(t)$ exists on $[0, \infty[$ for all $X \in \Omega$ and $t_0 \in [0, \infty[$.

We call a bounded C^1-domain $V \subseteq \mathbb{R}^n$ with $V \subseteq \Omega$ a control volume. Let $t_0 \in [0, \infty[$, let V_0 be a fixed control volume and let $V(t)$ be the material volume defined by

$$V(t) := \{ x(t; t_0; x_0); x_0 \in V_0 \}, \quad t \in [0, \infty[. \tag{3.2}$$

Especially $V(t_0) = V_0$.

Let $m(t, V)$ be the mass of $V \subseteq \Omega$ at time t. Let us assume that the density function $\rho \in C^1([0, \infty[\times \Omega; \mathbb{R})$ is well defined by the relation

$$\rho(t, x) = \lim_{R \searrow 0} \frac{m(t, B_R(x))}{|B_R(x)|} \tag{3.3}$$

for $x \in \Omega$ and $t \geq 0$.

Assume that the following data are given:

- $f_1 : [0, T[\times \Omega \to \mathbb{R}^n$ is the external force per unit mass. We assume that $f_1 = \tilde{f}_1 + g$ where g is the gravitational force per unit mass and $\tilde{f}_1$ denotes the part of the external force which is different from g.

- $\tilde{f}_2 : [0, T[\times \Omega \to \mathbb{R}$ is the external thermal radiation per unit mass and $\sigma \in]0, \infty[$ is the specific heat capacity. We define $f_2 := \frac{1}{\sigma} \tilde{f}_2$. Consequently, f_2 is the external thermal radiation per heat capacity.

- $\rho_0 \in]0, \infty[$ is the fixed reference density and $\vartheta_0 \in]0, \infty[$ is a fixed reference temperature

- Further we introduce the following real, positive constants: ν is the kinematic viscosity, β is the coefficient of thermal expansion and κ is the thermal diffusivity

Define

$$\theta := \vartheta - \vartheta_0. \tag{3.4}$$

In the following section we use the conservation of mass, volume, energy and the balance of momentum to obtain the system of equations (3.28)-(3.31). For more detailed information and also for more general models we refer to [Li96, Pa11]. In Section 3.2 we describe how the Boussinesq approximation can be obtained from these equations. We remark that this section is purely heuristic. For further information about the Boussinesq approximation we refer to [Ze03].

3.1 Conservation equations

In the following we describe the assumptions we need to obtain the Boussinesq approximation. For this reason we need the Reynolds' transport theorem.

Theorem 3.1. *Let $I \subseteq \mathbb{R}$ be an open interval, let $G \subseteq \mathbb{R}^n$, $n \in \mathbb{N}$, be an open set, and let $u : I \times G \to \mathbb{R}^n$ be a continuously differentiable and bounded function. For $t_0 \in I$ and $x_0 \in G$ let $x(t) = x(t; t_0; x_0)$ denote the solution of the ODE-system*

$$\dot{x}(t) = u(t, x(t)), \quad x(t_0) = x_0. \tag{3.5}$$

Let $V_0 \subseteq G$ be a bounded domain with C^1-boundary ∂V_0. Define

$$V(t) = \{ x(t; t_0; x_0); x_0 \in V_0 \}. \tag{3.6}$$

Then the following statements are satisfied.

(i) There exists a $\delta > 0$ such that $x(t; t_0; x_0)$ exists on $]t_0 - \delta, t_0 + \delta[$ for all $x_0 \in V_0$ and consequently $V(t)$ is well defined on $]t_0 - \delta, t_0 + \delta[$.

(ii) Let $f : I \times G \to \mathbb{R}$ be a continuously differentiable function. Then

$$\frac{d}{dt} \int_{V(t)} f(t, x) \, dV(x) \bigg|_{t=s} = \int_{V(s)} \left(\frac{\partial f}{\partial t} + \operatorname{div}(fu) \right)(s, x) \, dx \tag{3.7}$$

for all $s \in]t_0 - \delta, t_0 + \delta[$.

Proof. This is a classical result; a proof can be found in [BeMa02, Proposition 1.3]. $\square$

Assumption 1 (Conservation of mass). Let $t_0 \in]0, \infty[$, let V_0 be a control volume and let $V(t)$ be the material volume defined by (3.2). Then

$$\frac{d}{dt} \int_{V(t)} \rho(t, x) \, dx \bigg|_{t=t_0} = 0. \tag{3.8}$$

Using Reynolds' transport theorem and $V(t_0) = V_0$ it follows

$$\int_{V_0} \Big(\rho_t + \operatorname{div}(\rho u) \Big)(t_0, x)\, dx = 0. \tag{3.9}$$

For $x_0 \in \Omega$ we get from Lebesgue's differentiation theorem (see Theorem 2.21) that

$$0 = \lim_{r \searrow 0} \frac{1}{|B_r(x_0)|} \int_{B_r(x_0)} \Big(\rho_t + \operatorname{div}(\rho u) \Big)(t_0, x)\, dx = \rho_t(t_0, x_0) + (\operatorname{div}(\rho u))(t_0, x_0).$$

Altogether we obtain the continuity equation

$$\rho_t + \operatorname{div}(\rho u) = 0 \quad \text{in }]0, T[\times \Omega. \tag{3.10}$$

From Newton's second law we obtain the following

Assumption 2 (Balance of momentum). Let $t_0 \in]0, \infty[$, let V_0 be a control volume and let $V(t)$ be the material volume defined by (3.2). Then

$$\frac{d}{dt} \int_{V(t)} (\rho u)(t, x)\, dx \Big|_{t=t_0} = \int_{V_0} (\rho f_1)(t_0, x)\, dx + \int_{\partial V_0} T(t_0, x) N\, dS(x). \tag{3.11}$$

The meaning of T is that

$$\int_{\partial V_0} T(t_0, x) N\, dS(x) \tag{3.12}$$

equals the total surface force exerted by the fluid in $\Omega \setminus \overline{V_0}$ onto the fluid in V_0.

We get

$$\int_{V_0} \Big((\rho u)_t + \operatorname{div}(\rho u \otimes u) \Big)(t_0, x)\, dV(x) = \int_{V_0} (\rho f_1)(t_0, x)\, dV(x) + \int_{V_0} \operatorname{div} T(t_0, x)\, dV(x). \tag{3.13}$$

With Lebesgue's differentiation theorem it follows from (3.11) that

$$\Big((\rho u)_t + \operatorname{div}(\rho u \otimes u) \Big)(t_0, x) = (\rho f_1)(t_0, x) + \operatorname{div} T(t_0, x) \tag{3.14}$$

for all $x \in \Omega$.

Furthermore we need the following assumption.

Assumption 3 (Assumptions on the stress tensor). We assume that T is given by

$$T := (-p + \lambda \operatorname{div} u)I + 2\mu D(u) \tag{3.15}$$

where $D(u) = \frac{1}{2}(\partial_i u_j + \partial_j u_i)_{i,j=1}^3$ denotes the symmetric part of the gradient and with real constants λ, μ such that $\mu \geq 0$ and $3\lambda + 2\mu \geq 0$. The constant μ is called the dynamic viscosity.

Due to the incompressibility of the fluid we obtain the following

Assumption 4. Let $t_0 \in]0, \infty[$, let V_0 be a fixed control volume and let $V(t)$ be the material volume as defined in (3.2). The principle of *conservation of volume* states

$$\frac{d}{dt} \int_{V(t)} 1 \, dx \Big|_{t=t_0} = 0. \tag{3.16}$$

By Reynolds' transport theorem and Lebesgue's differentiation theorem we obtain from (3.16)

$$\mathrm{div}\, u = 0 \quad \text{in }]0, T[\times \Omega. \tag{3.17}$$

Consequently

$$\mathrm{div}\, T = -\nabla p + \mu \Delta u. \tag{3.18}$$

A short computation yields

$$(\rho u)_t + \mathrm{div}(\rho u \otimes u) = \rho(u_t + u \cdot \nabla u). \tag{3.19}$$

Making use of (3.18), (3.19) we obtain from (3.14) the equation

$$\rho\big(u_t + u \cdot \nabla u\big) - \mu \Delta u + \nabla p = \rho f_1 \quad \text{in }]0, T[\times \Omega. \tag{3.20}$$

The total energy of a material volume is given by the sum $E_{\text{tot}} = E_{\text{kin}} + U$ with the kinetic energy E_{kin} and the internal energy U. The conservation of energy applied to our system can be formulated as follows. *In every material volume the change in time of the kinetic energy plus the internal energy equals the sum of mechanical power exerted on the system plus the rate of heat transfer supplied to the system.* In a formula

$$\frac{d}{dt}(E_{\text{kin}} + U) = L + \dot{Q}. \tag{3.21}$$

In (3.21) we denote by L the mechanical power done at the system and $\dot{Q}$ denotes the rate of heat transfer supplied to the system.

Let e denote the internal energy per unit mass.

Assumption 5 (Conservation of energy). Let $t_0 \in]0, \infty[$, let V_0 be a control volume and let $V(t)$ be the material volume defined by (3.2). Then

$$\frac{d}{dt} \int_{V(t)} \big(\tfrac{1}{2}\rho|u|^2 + \rho e\big)(t, x) \, dx \Big|_{t=t_0}$$
$$= \int_{V_0} \big(\rho f_1 \cdot u + \rho \widetilde{f_2}\big)(t_0, x) \, dx + \int_{\partial V_0} \big((Tu) \cdot N - q \cdot N\big)(t_0, x) \, dS(x). \tag{3.22}$$

In (3.22) q denotes the heat flux per unit area. The integral

$$- \int_{\partial V_0} (q \cdot N)(t_0, x) \, dS(x)$$

describes the rate of heat transfer supplied to the system through the boundary ∂V_0. Using Reynolds' transport theorem and the arbitrary choice of t_0, V_0 we obtain the equation

$$\partial_t(E_{tot}) + \operatorname{div}(E_{tot} u) = \rho f_1 \cdot u + \rho \widetilde{f_2} - \operatorname{div} q + \operatorname{div}(Tu) \quad \text{in }]0, T[\times \Omega. \qquad (3.23)$$

From (3.27) below we get the equation

$$\rho(e_t + u \cdot \nabla e) = \rho \widetilde{f_2} - \operatorname{div} q + 2\mu D(u) : D(u) \quad \text{in }]0, T[\times \Omega. \qquad (3.24)$$

In the following we will show that (3.24) is indeed fulfilled. There holds

$$\partial_t(E_{tot}) = \rho_t(\frac{1}{2}|u|^2 + e) + \rho u \cdot u_t + \rho e_t$$

and

$$\operatorname{div}(E_{tot} u) = (\frac{1}{2}|u|^2 + e)\operatorname{div}(\rho u) + \rho u \cdot \nabla(\frac{1}{2}|u|^2 + e)$$

$$= (\frac{1}{2}|u|^2 + e)\operatorname{div}(\rho u) + \rho u \cdot (u \cdot \nabla u + \nabla e).$$

Especially we get with (3.10)

$$\partial_t(E_{tot}) + \operatorname{div}(E_{tot} u)$$

$$= (\rho_t + \operatorname{div}(\rho u))(\frac{1}{2}|u|^2 + e) + u \cdot (\rho u \cdot \nabla u + \rho u_t) + \rho e_t + \rho u \cdot \nabla e \qquad (3.25)$$

$$= u \cdot (\rho u_t + \rho u \cdot \nabla u) + \rho(e_t + u \cdot \nabla e).$$

We insert (3.25) in (3.23) and obtain the equation

$$u \cdot (\rho u_t + \rho u \cdot \nabla u) + \rho(e_t + u \cdot \nabla e) = \rho f_1 \cdot u + \rho \widetilde{f_2} - \operatorname{div} q + \operatorname{div}(Tu).$$

Moreover from (3.17), (3.18) it follows

$$\operatorname{div}(Tu) = T : \nabla u + (\operatorname{div} T) \cdot u$$

$$= \big(-pI + \lambda \operatorname{div} uI + 2\mu D(u) \big) : \nabla u + (\operatorname{div} T) \cdot u \qquad (3.26)$$

$$= 2\mu D(u) : D(u) + (\operatorname{div} T) \cdot u.$$

Then we obtain with (3.26)

$$u \cdot (\rho u_t + \rho u \cdot \nabla u - \rho f_1 - \operatorname{div} T) + \rho(e_t + u \cdot \nabla e) = 2\mu D(u) : D(u) - \operatorname{div} q + \rho \widetilde{f_2}.$$

Finally, with (3.18), (3.20) we get

$$\rho(e_t + u \cdot \nabla e) = 2\mu D(u) : D(u) - \operatorname{div} q + \rho \widetilde{f_2} \quad \text{in }]0, T[\times \Omega. \qquad (3.27)$$

3.2 The Boussinesq approximation

From (3.10), (3.17), (3.20), (3.24) we obtain the following system of equations:

$$\rho_t + \operatorname{div}(\rho u) = 0, \tag{3.28}$$

$$\operatorname{div} u = 0, \tag{3.29}$$

$$\rho\big(u_t + u \cdot \nabla u\big) - \mu \Delta u + \nabla p = \rho(\widetilde{f_1} + g), \tag{3.30}$$

$$\rho(e_t + u \cdot \nabla e) = \rho \widetilde{f_2} - \operatorname{div} q + 2\mu D(u) : D(u). \tag{3.31}$$

In the Boussinesq approximation we neglect the term $2\mu D(u) : D(u)$. The Boussinesq approximation is based on the assumption that variations in density can be neglected unless they get multiplied with the gravitational force. In more detail we assume $\rho \approx \rho_0$ with the exception

$$\rho g = \rho_0(1 - \beta(\vartheta - \vartheta_0))g. \tag{3.32}$$

Define

$$\theta := \vartheta_0 - \vartheta. \tag{3.33}$$

In the following we will apply these assumptions to the system of equations (3.28)-(3.31). By the Fourier law of thermal conduction there holds

$$q = -\chi \nabla \vartheta \tag{3.34}$$

with the thermal conductivity χ. Furthermore, we need the relation

$$e = \sigma \vartheta \tag{3.35}$$

where the constant σ denotes the specific heat capacity. We make use of $\rho \approx \rho_0$ and (3.34), (3.35) to obtain from (3.31) the equation

$$\sigma \rho_0 \vartheta_t - \chi \Delta \vartheta + \sigma \rho_0 \operatorname{div}(\vartheta u) = \rho_0 \widetilde{f_2}. \tag{3.36}$$

Define the thermal diffusivity by $\kappa := \frac{\chi}{\sigma \rho_0}$. We replace ϑ by θ and use $\operatorname{div} u = 0$ in the equation above to obtain

$$\theta_t - \kappa \Delta \theta + u \cdot \nabla \theta = \frac{1}{\sigma} \widetilde{f_2}. \tag{3.37}$$

We obtain from (3.30)

$$\rho_0 u_t - \mu \Delta u + \rho_0 u \cdot \nabla u + \nabla p = \rho_0 \widetilde{f_1} + \rho_0(1 - \beta(\vartheta - \vartheta_0))g. \tag{3.38}$$

Then

$$u_t - \frac{\mu}{\rho_0} \Delta u + u \cdot \nabla u + \frac{1}{\rho_0} \nabla p = (\widetilde{f_1} + g) + \beta(\vartheta_0 - \vartheta)g.$$

We define the kinematic viscosity by $\nu := \frac{\mu}{\rho_0}$. We obtain the equation

$$u_t - \nu \Delta u + u \cdot \nabla u + \frac{1}{\rho_0} \nabla p = f_1 + \beta \theta g. \tag{3.39}$$

Finally, from (3.29), (3.37), (3.39) we obtain the following system of equations:

$$
\begin{aligned}
u_t - \nu \Delta u + (u \cdot \nabla) u + \frac{1}{\rho_0} \nabla p &= \beta \theta g + f_1 && \text{in }]0, T[\times \Omega\,, \\
\operatorname{div} u &= 0 && \text{in }]0, T[\times \Omega\,, \\
\theta_t - \kappa \Delta \theta + (u \cdot \nabla) \theta &= f_2 && \text{in }]0, T[\times \Omega\,.
\end{aligned}
\tag{3.40}
$$

4 Weak solutions of the Boussinesq equations

We consider the Boussinesq equations

$$u_t - \nu \Delta u + (u \cdot \nabla)u + \frac{1}{\rho_0}\nabla p = \beta \theta g + f_1 \quad \text{in }]0,T[\times\Omega,$$
$$\text{div } u = 0 \quad \text{in }]0,T[\times\Omega,$$
$$\theta_t - \kappa \Delta \theta + (u \cdot \nabla)\theta = f_2 \quad \text{in }]0,T[\times\Omega, \qquad (4.1)$$
$$u(0) = u_0 \quad \text{in } \Omega,$$
$$\theta(0) = \theta_0 \quad \text{in } \Omega$$

for a viscous, incompressible fluid with velocity u, pressure p, and temperature θ on a domain $\Omega \subseteq \mathbb{R}^n$, $n \in \{2,3\}$, and a finite time interval $[0,T[$. We assume that the following boundary conditions hold: u satisfies the no slip boundary condition

$$u = 0 \quad \text{on }]0,T[\times\partial\Omega \qquad (4.2)$$

and θ satisfies the Robin boundary condition

$$\frac{\partial\theta}{\partial N} + \Lambda(\theta - h_0) = 0 \quad \text{on }]0,T[\times\partial\Omega \qquad (4.3)$$

with a positive, scalar function $\Lambda : \partial\Omega \to]0,\infty[$ describing the heat transfer to the exterior.

4.1 Domains satisfying the uniform Lipschitz condition and the surface measure

Let $n \in \mathbb{N}$, and let $0 \leq s < \infty$. We denote by $\mathcal{H}^s$ the s-dimensional Hausdorff measure on $\mathbb{R}^n$ as defined in [Ev98, Section 2.1]. Especially (see [EvGa92, Section 2.1, Theorem 1]) it follows that $\mathcal{H}^s$ is a positive measure (in the sense of Definition 6.2 (ii)) on the Borel σ-algebra $\mathcal{B}(\mathbb{R}^n)$. We denote by $\mathcal{L}(\mathbb{R}^n)$ the σ-algebra of Lebesgue measurable subsets of $\mathbb{R}^n$ and by λ_n the Lebesgue measure (see Section 2.2). By [EvGa92, Section 2.2, Theorem 2] there holds $\lambda_n(A) = \mathcal{H}^n(A)$ for all $A \in \mathcal{L}(\mathbb{R}^n)$.

Fix $n \in \mathbb{N}, n \geq 2$. Let $\Gamma \in \mathcal{B}(\mathbb{R}^n)$ be given. It is easily seen that $\mathcal{B}(\Gamma) = \{A \subseteq \Gamma; A \in \mathcal{B}(\mathbb{R}^n)\}$ holds. Define

$$\mathcal{L}^{n-1}(\Gamma) := \{A \cup B; \ A \in \mathcal{B}(\Gamma), B \subseteq \mathbb{R}^n \text{ with } \mathcal{H}^{n-1}(B) = 0\}. \tag{4.4}$$

Consequently we have that the measure space $(\Gamma, \mathcal{L}^{n-1}(\Gamma), \mathcal{H}^{n-1})$ is the completion of the measure space $(\Gamma, \mathcal{B}(\Gamma), \mathcal{H}^{n-1})$.

Lemma 4.1. *Let $n \in \mathbb{N}, n \geq 2$, let $U' \subseteq \mathbb{R}^{n-1}$ be an open set, assume $\phi : \mathbb{R}^n \to \mathbb{R}^n$ is a Cartesian coordinate system with origin at $x_0 \in \mathbb{R}^n$, assume $h : \mathbb{R}^{n-1} \to \mathbb{R}$ is a Lipschitz continuous function. Define*

$$\Gamma := \{y \in \mathbb{R}^n; y = \phi(x', h(x')) \text{ with } x' \in U'\}.$$

(i) Then

$$\mathcal{H}^{n-1}(\Gamma) = \int_{U'} \sqrt{1 + |\nabla h(x')|^2} \, d\lambda_{n-1}(x'). \tag{4.5}$$

(ii) Fix a measurable function $u : (\Gamma, \mathcal{L}^{n-1}(\Gamma)) \to \mathbb{R}$. Then $u : \Gamma \to \mathbb{R}$ is integrable with respect to the Hausdorff measure $\mathcal{H}^{n-1}$ if and only if the function $x' \mapsto u(\phi(x', h(x'))\sqrt{1 + |\nabla h(x')|^2}$, which is well defined for a.a. $x' \in U'$, is integrable with respect to the Lebesgue measure λ_{n-1}. In this case there holds

$$\int_{U'} u(\phi(x', h(x'))\sqrt{1 + |\nabla h(x')|^2} \, d\lambda_{n-1}(x') = \int_{\Gamma} u(y) \, d\mathcal{H}^{n-1}(y). \tag{4.6}$$

Proof. Due to the invariance of the Hausdorff measure with respect to affine isometries (see [EvGa92, Section 2.1, Theorem 2]) there holds $\mathcal{H}^{n-1}(\phi(A)) = \mathcal{H}^{n-1}(A)$ for all sets $A \subseteq \mathbb{R}^n$. Consequently we can assume without loss of generality that $\phi = \text{id}$ holds. Define the injective Lipschitz continuous function by

$$f : \mathbb{R}^{n-1} \to \mathbb{R}^n, \quad f(x') := (x', h(x')). \tag{4.7}$$

Then $f(U') = \Gamma$.

Proof of (i). Combine the *Area Formula* (see [EvGa92, Section 3.3, Theorem 1]) and [EvGa92, Section 3.3.4 , Application B] to obtain

$$\mathcal{H}^{n-1}(\Gamma) = \int_{U'} \sqrt{1 + |\nabla h(x')|^2} \, d\lambda_{n-1}(x').$$

Proof of (ii). It follows immediately that $\mathcal{H}^{n-1}(f^{-1}(A)) = 0$ is satisfied for all sets $A \subseteq \mathbb{R}^n$ with $\mathcal{H}^{n-1}(A) = 0$. Consequently, using the definition of $\mathcal{L}(\Gamma)$, we prove that $u \circ f : U' \to \mathbb{R}$ is measurable with respect to the Lebesgue σ-algebra $\mathcal{L}(U')$.

First, let us consider the case that u is additionally a non-negative function. Let $M \subseteq \mathbb{R}^{n-1}$ be the set of all $x' \in \mathbb{R}^{n-1}$ such that $\nabla h(x') \in \mathbb{R}^{n-1}$ is well defined. Define

$$g(x') := \begin{cases} u(f(x'))\sqrt{1 + |\nabla h(x')|^2} & \text{if } x' \in M \cap U', \\ 0, & \text{otherwise.} \end{cases}$$

It follows that $g : (\mathbb{R}^{n-1}, \mathcal{L}(\mathbb{R}^{n-1}))$ is a non-negative measurable function. We obtain from [EvGa92, Section 3.3]

$$\int_{\mathbb{R}^{n-1}} g(x') \sqrt{1 + |\nabla h(x')|^2}\, d\lambda_{n-1}(x') = \int_{\mathbb{R}^n} g(f^{-1}(y))\, d\mathcal{H}^{n-1}(y). \tag{4.8}$$

Thus

$$\int_{U'} (u \circ f)(x') \sqrt{1 + |\nabla h(x')|^2}\, d\lambda_{n-1}(x') = \int_{\Gamma} u(y)\, d\mathcal{H}^{n-1}(y). \tag{4.9}$$

Next, consider the general case that u is not necessarily non-negative. We introduce $u^+ := \max\{u, 0\}$ and $u^- := \min\{u, 0\}$. Applying (4.9) to the non-negative, measurable functions u^+, u^- and making use of $u = u^+ + u^-$ it follows that the conclusions stated in Lemma 4.1 (ii) are fulfilled. □

Let $A \subseteq \mathbb{R}^n, n \in \mathbb{N}$. Then a collection $(U_j)_{j \in I}$ (I set of indices) of open sets $U_j \subseteq \mathbb{R}^n, j \in I$, is called a *locally finite open cover* of A if $A \subseteq \bigcup_{j \in I} U_j$ and every compact set $K \subseteq \mathbb{R}^n$ can intersect at most a finite number of the sets U_j.

Let $n \in \mathbb{N}, n \geq 2$, let $x_0 \in \mathbb{R}^n$, let $\alpha, \beta \in]0, \infty[$, assume that $\phi : \mathbb{R}^n \to \mathbb{R}^n$ is a Cartesian coordinate system with origin at x_0, and that $h : \mathbb{R}^{n-1} \to \mathbb{R}$ is a Lipschitz continuous function. We define the open neighbourhood $U_{x_0,\alpha,\beta,h,\phi}$ of x_0 by

$$U_{x_0,\alpha,\beta,h,\phi} := \{\, x \in \mathbb{R}^n; x = \phi(y', y_n) \text{ where } h(y') - \beta < y_n < h(y') + \beta; |y'| < \alpha \,\}. \tag{4.10}$$

Now we can formulate the following

Definition 4.2. Let $n \in \mathbb{N}, n \geq 2$, let $\Omega \subseteq \mathbb{R}^n$ be an arbitrary domain. We say that Ω fulfils the *uniform Lipschitz condition* if there is $N \in \mathbb{N}$, constants $\alpha, \beta, L \in]0, \infty[$, a set $M \subseteq \mathbb{N}$ and a locally finite open cover $(U_j)_{j \in M}$ of $\partial\Omega$ such that the following two properties are satisfied:

(i) Let $j \in M$. Then $U_j = U_{x_j,\alpha,\beta,h_j,\phi_j}$ where $x_j \in \partial\Omega$, where $\phi_j : \mathbb{R}^n \to \mathbb{R}^n$ is a Cartesian coordinate system with origin at x_j and where $h_j : \mathbb{R}^{n-1} \to \mathbb{R}$ is a Lipschitz continuous function with $\mathrm{Lip}(h_j) \leq L$ such that the conditions

$$\begin{aligned}
U_j \cap \Omega &= \{\, x \in \mathbb{R}^n; x = \phi_j(y', y_n) \text{ where } h(y') < y_n < h(y') + \beta, |y'| < \alpha \,\}, \\
U_j \cap \partial\Omega &= \{\, x \in \mathbb{R}^n; x = \phi_j(y', y_n) \text{ where } y_n = h(y'), |y'| < \alpha \,\}, \\
U_j \cap \Omega^c &= \{\, x \in \mathbb{R}^n; x = \phi_j(y', y_n) \text{ where } h(y') - \beta < y_n < h(y'), |y'| < \alpha \,\}
\end{aligned} \tag{4.11}$$

are satisfied.

(ii) Every collection of $N + 1$ different sets of the U_j has an empty intersection.

For a bounded domain $\Omega \subseteq \mathbb{R}^n$ satisfying the uniform Lipschitz condition, we will say that Ω is a bounded Lipschitz domain.

Definition of the measure space $(\partial\Omega, \mathcal{L}(\partial\Omega), dS)$. Let $n \in \mathbb{N}, n \geq 2$, let $\Omega \subseteq \mathbb{R}^n$ be a domain satisfying the uniform Lipschitz condition. Then $\partial\Omega \in \mathcal{B}(\mathbb{R}^n)$. Define the σ-algebra $\mathcal{L}^{n-1}(\partial\Omega)$ as in (4.4) where Γ replaced by $\partial\Omega$. Motivated by (4.5), we define the *surface measure* $dS : \mathcal{L}^{n-1}(\partial\Omega) \to [0, \infty]$ by $dS(A) = \mathcal{H}^{n-1}(A)$ for all $A \in \mathcal{L}^{n-1}(\partial\Omega)$. By construction the surface measure is a complete measure on the σ-algebra $\mathcal{L}(\partial\Omega) := \mathcal{L}^{n-1}(\partial\Omega)$.

We define $L^2(\partial\Omega)$ as the L^2-space with respect to the measure space $(\partial\Omega, \mathcal{L}(\partial\Omega), dS)$. Endow $L^2(\partial\Omega)$ with the usual scalar product $\langle \cdot, \cdot \rangle_{\partial\Omega}$. To reduce the integration of a function $u : \partial\Omega \to \mathbb{R}$ to a local situation we introduce a smooth partition of unity.

Definition 4.3. Let $\Omega \subseteq \mathbb{R}^n, n \in \mathbb{N}$, be a domain, let $(U_j)_{j \in M}$ be a locally finite open cover of $\partial\Omega$. Then we call functions $\xi_j \in C_0^\infty(U_j)$, $j \in M$, a *smooth partition of unity subordinate to* $(U_j)_{j \in \mathbb{N}}$ if the following conditions are satisfied:

(i) There holds $0 \leq \xi_j(x) \leq 1$ for all $x \in U_j$ and $j \in M$,

(ii) There holds $\sum_{j \in M} \xi_j(x) = 1$ for all $x \in \partial\Omega$.

The next lemma tells us that in a domain satisfying the uniform Lipschitz condition we can always find such a smooth partition of unity.

Lemma 4.4. *Let $n \in \mathbb{N}, n \geq 2$, let $\Omega \subseteq \mathbb{R}^n$ be a domain satisfying the uniform Lipschitz condition, let $(U_j)_{j \in M}$ be a locally finite open cover as in Definition 4.2. Then there exists a smooth partition of unity subordinate to $(U_j)_{j \in M}$.*

Proof. This follows from [Zi89, Lemma 2.3.1] applied to a locally finite open cover. $\qquad\square$

The next lemma shows how to compute the integral $\int_{\partial\Omega} u(x)\, dS(x)$ of functions $u : \partial\Omega \to \mathbb{R}$ with respect to the surface measure dS in domains satisfying the uniform Lipschitz condition.

Lemma 4.5. *Let $n \in \mathbb{N}, n \geq 2$, let $\Omega \subseteq \mathbb{R}^n$ be a domain satisfying the uniform Lipschitz condition, let $U_j = U_{x_j, \alpha, \beta, h_j, \phi_j}$, $j \in M$, be a locally finite open cover of $\partial\Omega$ as in Definition 4.2, and let $\xi_j \in C_0^\infty(U_j)$, $j \in M$, be a smooth partition of unity subordinate to $(U_j)_{j \in M}$. Let $u : \partial\Omega \to \mathbb{R}$ be an integrable function with respect to the surface measure dS. Then*

$$
\begin{aligned}
\int_{\partial\Omega} u(y)\, dS(y) &= \sum_{j \in M} \int_{\partial\Omega \cap U_j} (\xi_j\, u)(y)\, dS(y) \\
&= \sum_{j \in M} \int_{\{x' \in \mathbb{R}^{n-1}; |x'| < \alpha\}} (\xi_j\, u)(\phi_j(x', h_j(x')))\, \sqrt{1 + |\nabla h_j(x')|^2}\, dx'.
\end{aligned}
\tag{4.12}
$$

Proof. Fix $j \in M$. Define

$$
\Gamma_j := U_j \cap \partial\Omega = \{y \in \mathbb{R}^n; y = \phi_j(x', h_j(x'))\text{ where } x' \in \mathbb{R}^{n-1}; |x'| < \alpha\}.
$$

By (4.6) there holds

$$\int_{\{x' \in \mathbb{R}^{n-1}; |x'| < \alpha\}} (\xi_j u)(\phi_j(x', h_j(x'))) \sqrt{1 + |\nabla h_j(x')|^2} \, dx' = \int_{U_j \cap \partial\Omega} (\xi_j u)(y) \, d\mathcal{H}^{n-1}(y).$$

$$(4.13)$$

Further, by the definition of a partition of unity there holds

$$\int_{\partial\Omega} u(y) \, dS(y) = \sum_{j \in M} \int_{\partial\Omega} (\xi_j u)(y) \, dS(y) = \sum_{j \in M} \int_{\partial\Omega \cap U_j} (\xi_j u)(y) \, dS(y). \qquad (4.14)$$

Combining (4.13) and (4.13) yields (4.12). $\qquad \square$

The following theorem shows existence and continuity of the trace operator. Furthermore, the estimate (4.17) will play a crucial role in this thesis.

Theorem 4.6. *Let $n \in \mathbb{N}, n \geq 2$, let $\Omega \subseteq \mathbb{R}^n$ be a domain satisfying the uniform Lipschitz condition with parameters (α, β, N, L). Then there exists a continuous, linear trace operator*

$$T : H^1(\Omega) \to L^2(\partial\Omega) \qquad (4.15)$$

with the property

$$Tu = u|_{\partial\Omega} \quad \text{for all } u \in C_0^1(\overline{\Omega}). \qquad (4.16)$$

There is a constant $c = c(\beta, L, N) > 0$ such that

$$\|Tu\|_{L^2(\partial\Omega)} \leq c \|u\|_{H^1(\Omega)} \qquad (4.17)$$

for all $u \in H^1(\Omega)$.

Remark. In the special case that

$$\Omega = \{ (x', x_n) \in \mathbb{R}^n; x_n > h(x'), x' \in \mathbb{R}^{n-1} \} \qquad (4.18)$$

where $h : \mathbb{R}^n \to \mathbb{R}$ is a Lipschitz continuous function, it follows that there is a constant $c = c(\mathrm{Lip}(h)) > 0$ such that

$$\|Tu\|_{L^2(\partial\Omega)} \leq c \|u\|_{H^1(\Omega)} \qquad (4.19)$$

for all $u \in H^1(\Omega)$.

Proof. We refer to the appendix for a proof. $\qquad \square$

We will use the notation $u|_{\partial\Omega}$ for the trace of a function $u \in H^1(\Omega)$. When there is no possibility of confusion we will identify $u \in H^1(\Omega)$ with its trace and just write u instead of $u|_{\partial\Omega}$. Consequently, if $\phi : \partial\Omega \to \mathbb{R}$ is a function then we will write $\langle u, \phi \rangle_{\partial\Omega}$ instead of $\langle u|_{\partial\Omega}, \phi \rangle_{\partial\Omega}$ if the latter integral exists.

4.2 Definition of weak solutions

Multiplying the Boussinesq system (4.1) with test functions w, ϕ, using integration by parts and (4.2), (4.3) we obtain the following definition of a weak solution:

Definition 4.7. Let $\Omega \subseteq \mathbb{R}^n$, $n \in \{2, 3\}$, be a domain satisfying the uniform Lipschitz condition, let $0 < T < \infty$, $g \in L^\infty(]0, T[\times\Omega)^n$, let $h_0 \in L^2(0, T; L^2(\partial\Omega))$, $\Lambda \in L^\infty(\partial\Omega)$ with $\Lambda(x) \geq 0$ for almost all $x \in \partial\Omega$. Further assume $f_1 \in L^2(0, T; L^2(\Omega)^n)$, assume $f_2 \in L^2(0, T; L^2(\Omega))$, assume $u_0 \in L^2_\sigma(\Omega)$, $\theta_0 \in L^2(\Omega)$, and assume that ν, β, κ are positive, real constants.

A pair

$$
\begin{aligned}
&u \in L^\infty(0, T; L^2_\sigma(\Omega)) \cap L^2(0, T; W^{1,2}_{0,\sigma}(\Omega)), \\
&\theta \in L^\infty(0, T; L^2(\Omega)) \cap L^2(0, T; H^1(\Omega))
\end{aligned} \tag{4.20}
$$

is called a *weak solution* of the Boussinesq equations (4.1) with no slip boundary condition (4.2) and Robin boundary condition (4.3) if there holds

$$
\begin{aligned}
&-\int_0^T \langle u, w_t \rangle_\Omega \, dt + \nu \int_0^T \langle \nabla u, \nabla w \rangle_\Omega \, dt + \int_0^T \langle u \cdot \nabla u, w \rangle_\Omega \, dt \\
&= \beta \int_0^T \langle \theta g, w \rangle_\Omega \, dt + \int_0^T \langle f_1, w \rangle_\Omega \, dt + \langle u_0, w(0) \rangle_\Omega
\end{aligned} \tag{4.21}
$$

for all $w \in C_0^\infty([0, T[; C^\infty_{0,\sigma}(\Omega))$ and

$$
\begin{aligned}
&-\int_0^T \langle \theta, \phi_t \rangle_\Omega \, dt + \kappa \int_0^T \langle \nabla\theta, \nabla\phi \rangle_\Omega \, dt + \int_0^T \langle u \cdot \nabla\theta, \phi \rangle_\Omega \, dt \\
&= -\kappa \int_0^T \langle \Lambda(\theta - h_0), \phi \rangle_{\partial\Omega} \, dt + \int_0^T \langle f_2, \phi \rangle_\Omega \, dt + \langle \theta_0, \phi(0) \rangle_\Omega
\end{aligned} \tag{4.22}
$$

for all $\phi \in C_0^\infty([0, T[\times\overline{\Omega})$.

Definition 4.8. Consider data as in Definition 4.7, let (u, θ) be a weak solution of (4.1) with (4.2), (4.3). Then we say that (u, θ) satisfies the *energy inequalities*, if the inequalities

$$
\begin{aligned}
&\frac{1}{2}\|u(t)\|_2^2 + \nu \int_0^t \|\nabla u\|_2^2 \, d\tau \\
&\leq \frac{1}{2}\|u_0\|_2^2 + \beta \int_0^t \langle \theta g, u \rangle_\Omega \, d\tau + \int_0^t \langle f_1, u \rangle_\Omega \, d\tau,
\end{aligned} \tag{4.23}
$$

$$
\begin{aligned}
&\frac{1}{2}\|\theta(t)\|_2^2 + \kappa \int_0^t \|\nabla\theta\|_2^2 \, d\tau + \kappa \int_0^t \|\sqrt{\Lambda}\,\theta\|_{2,\partial\Omega}^2 \, d\tau \\
&\leq \frac{1}{2}\|\theta_0\|_2^2 + \kappa \int_0^t \langle \Lambda h_0, \theta \rangle_{\partial\Omega} \, d\tau + \int_0^t \langle f_2, \theta \rangle_\Omega \, d\tau
\end{aligned} \tag{4.24}
$$

are satisfied for almost all $t \in [0, T[$.

The construction of an associated pressure p is given by the following lemma.

Lemma 4.9. *Consider data as in Definition 4.7, let (u, θ) be a a weak solution of (4.1) with (4.2), (4.3), and let $s = \frac{4}{n}$. Then there exists a function $\hat{p} \in L^s(0, T; L^2_{loc}(\Omega))$ such that the time derivative $p = \frac{\partial}{\partial t}\hat{p}$ is an associated pressure of u This means that p satisfies the equation*

$$u_t - \nu \Delta u + u \cdot \nabla u + \nabla p = \beta \theta g + f_1$$

in the sense of distributions in $[0, T[\times \Omega$. In more detail, the identity above means

$$-\int_0^T \langle u, w_t \rangle_\Omega \, dt + \nu \int_0^T \langle \nabla u, \nabla w \rangle_\Omega \, dt + \int_0^T \langle u \cdot \nabla u, w \rangle_\Omega \, dt + \int_0^T \langle \hat{p}, \partial_t \text{div} w \rangle_\Omega \, dt$$

$$= \beta \int_0^T \langle \theta g, w \rangle_\Omega \, dt + \int_0^T \langle f_1, w \rangle_\Omega \, dt + \langle u_0, w(0) \rangle_\Omega$$

for all $w \in C_0^\infty([0, T[\times \Omega)$.

Proof. This follows from [So01, Chapter V, Theorem 1.7.1] with $\tilde{f} := f_1 + \beta \theta g$. $\square$

4.3 Formulation of the Boussinesq system as a nonlinear operator equation

The main goal of this section is to write the Boussinesq equations (4.1) with boundary conditions (4.2), (4.3) as a nonlinear operator equation in the form

$$u_t - \nu \Delta u + B(u, u) = f_1 + \beta \theta g \quad \text{in } L^{4/3}(0, T; W_{0,\sigma}^{1,2}(\Omega)'),$$
$$\theta_t - \kappa \Delta \theta + D(u, \theta) + \kappa S(\Lambda(\theta - h_0)) = f_2 \quad \text{in } L^{4/3}(0, T; H^1(\Omega)'),$$
$$u(0) = u_0 \quad \text{in } W_{0,\sigma}^{1,2}(\Omega)', \qquad (4.25)$$
$$\theta(0) = \theta_0 \quad \text{in } H^1(\Omega)'.$$

For this reason we have to give a precise meaning to every term in (4.25).

In the definition of weak solutions of the Boussinesq system (4.1) we have to test the nonlinear terms $(u \cdot \nabla)u$ and $(u \cdot \nabla)\theta$ with test functions w and ϕ, respectively. We begin with a definition for the stationary case.

Definition 4.10. Let $\Omega \subseteq \mathbb{R}^n, n \in \{2, 3\}$, be a Lipschitz domain, let $0 < T < \infty$. Then we define

$$b(u, v, w) := \langle u \cdot \nabla v, w \rangle_\Omega = \sum_{i,j=1}^n \int_\Omega u_i \, \partial_i v_j \, w_j \, dx \qquad (4.26)$$

for $u, v, w \in W^{1,2}_{0,\sigma}(\Omega)$. Moreover, define

$$d(u, \theta, \phi) := \langle u \cdot \nabla \theta, \phi \rangle_{\Omega} = \sum_{i=1}^{n} \int_{\Omega} u_i \, \partial_i \theta \, \phi \, dx \qquad (4.27)$$

for $u \in W^{1,2}_{0,\sigma}(\Omega)$ and $\theta, \phi \in H^1(\Omega)$.

The definition implies that b is a trilinear form on $W^{1,2}_{0,\sigma}(\Omega)$ and that $d : W^{1,2}_{0,\sigma}(\Omega) \times H^1(\Omega) \times H^1(\Omega) \to \mathbb{R}$ is linear in each of its arguments.

The next lemma studies elementary properties of b and d.

Lemma 4.11. *Let $\Omega \subseteq \mathbb{R}^n, n \in \{2, 3\}$, be a Lipschitz domain. We have*

$$b(u, v, w) = -b(u, w, v) \quad \forall u, v, w \in W^{1,2}_{0,\sigma}(\Omega), \qquad (4.28)$$
$$d(u, \theta, \phi)) = -d(u, \phi, \theta) \quad \forall u \in W^{1,2}_{0,\sigma}(\Omega), \quad \forall \theta, \phi \in H^1(\Omega). \qquad (4.29)$$

Proof. (i) Fix $u, v, w \in C^\infty_{0,\sigma}(\Omega)$. Then

$$\begin{aligned}
b(u, v, w) &= \sum_{i,j=1}^{n} \int_{\Omega} u_i \, \partial_i v_j \, w_j \, dx \\
&= -\sum_{i,j=1}^{n} \int_{\Omega} \partial_i(u_i \, w_j) \, v_j \, dx \\
&= -\sum_{i,j=1}^{n} \int_{\Omega} u_i \, \partial_i w_j \, v_j \, dx \\
&= -b(u, w, v).
\end{aligned} \qquad (4.30)$$

By density, we see that (4.30) is still valid for all $u, v, w \in W^{1,2}_{0,\sigma}(\Omega)$. Thus (4.28) holds.

(ii) Fix $u \in C^\infty_{0,\sigma}(\Omega)$ and $\theta, \phi \in C^\infty_0(\overline{\Omega})$. Then, analogously to (4.30) we show

$$d(u, \theta, \phi) = -d(u, \phi, \theta).$$

By density, we get that (4.29) is satisfied.

$\square$

Lemma 4.12. *(i) Let $\Omega \subseteq \mathbb{R}^n, n \in \{2, 3\}$, be an arbitrary domain. Then the following estimates are satisfied with a constant $c > 0$ independent of the domain Ω:*

$$n=2: \quad \|u\|_4 \leq c\|u\|_2^{1/2}\|\nabla u\|_2^{1/2}, \quad \forall u \in H^1_0(\Omega)^2. \qquad (4.31)$$
$$n=3: \quad \|u\|_4 \leq c\|u\|_2^{1/4}\|\nabla u\|_2^{3/4}, \quad \forall u \in H^1_0(\Omega)^3. \qquad (4.32)$$

(ii) *Let $\Omega \subseteq \mathbb{R}^n, n \in \{2, 3\}$, be a domain satisfying the uniform Lipschitz condition. Then the following estimates are satisfied:*

$$n=2: \quad \|\phi\|_4 \leq c(\Omega)\|\phi\|_2^{1/2}\|\phi\|_{H^1(\Omega)}^{1/2}, \quad \forall \phi \in H^1(\Omega). \tag{4.33}$$

$$n=3: \quad \|\phi\|_4 \leq c(\Omega)\|\phi\|_2^{1/4}\|\phi\|_{H^1(\Omega)}^{3/4}, \quad \forall \phi \in H^1(\Omega). \tag{4.34}$$

Consider the special case that

$$\Omega = \{\, x', x_n) \in \mathbb{R}^n; \, x_n > h(x')\,\} \tag{4.35}$$

where $h : \mathbb{R}^{n-1} \to \mathbb{R}$ is a Lipschitz continuous function. Then we have that the estimates (4.33), (4.34) can be satisfied with a constant $c = c(n, \mathrm{Lip}(h)) > 0$.

Proof. Equation (4.31) is [Te77, Chapter III, Lemma 3.3]. If $n = 3$ the interpolation inequality

$$\|u\|_4 \leq \|u\|_2^{1/4}\|u\|_6^{3/4} \tag{4.36}$$

is satisfied for all $u \in H^1(\Omega)^n$. To obtain (4.32) use in (4.36) the Sobolev inequality [AdFo03, Theorem 4.31].

To prove statement (ii) we need a result which will be described in the following. Let $G \subseteq \mathbb{R}^n$ be a domain satisfying the *cone condition* (defined as in [AdFo03, Section 4.6]). By [AdFo03, Theorem 5.8] the estimates

$$\begin{aligned} n=2: \quad & \|\phi\|_4 \leq c\|\phi\|_2^{1/2}\|\phi\|_{H^1(G)}^{1/2}, \quad \forall \phi \in H^1(G), \\ n=3: \quad & \|\phi\|_4 \leq c\|\phi\|_2^{1/4}\|\phi\|_{H^1(U)}^{3/4}, \quad \forall \phi \in H^1(G), \end{aligned} \tag{4.37}$$

are fulfilled with constants $c > 0$ depending only on n and the dimensions of the cone providing the cone condition for G. A careful inspection of [Wl82, Satz 2.1] shows that every domain Ω satisfying the uniform Lipschitz condition fulfils the cone condition and the dimensions of the cone providing the cone condition for Ω can be chosen to depend only on the parameters (α, β, L) appearing in Definition 4.2. Therefore, we get from (4.37) that the statements of Lemma 4.12 (ii) hold. $\qquad\square$

Lemma 4.13. (i) *Let $\Omega \subseteq \mathbb{R}^2$ be an arbitrary domain. Then*

$$|b(u, v, w)| \leq c\|u\|_2^{1/2}\|\nabla u\|_2^{1/2}\|\nabla v\|_2\|w\|_2^{1/2}\|\nabla w\|_2^{1/2} \tag{4.38}$$

for all $u, v, w \in W_{0,\sigma}^{1,2}(\Omega)$ with a constant $c > 0$ independent of the domain Ω.

(ii) *Let $\Omega \subseteq \mathbb{R}^3$ be an arbitrary domain. Then*

$$|b(u, v, w)| \leq c\|u\|_2^{1/4}\|\nabla u\|_2^{3/4}\|\nabla v\|_2\|w\|_2^{1/4}\|\nabla w\|_2^{3/4} \tag{4.39}$$

for all $u, v, w \in W_{0,\sigma}^{1,2}(\Omega)$ with a constant $c > 0$ independent of the domain Ω.

(iii) Let $\Omega \subseteq \mathbb{R}^2$ be a domain satisfying the uniform Lipschitz condition. Then

$$|d(u,\theta,\phi)| \le c\|u\|_2^{1/2}\|\nabla u\|_2^{1/2}\|\nabla\theta\|_2\|\phi\|_2^{1/2}\|\phi\|_{H^1(\Omega)}^{1/2} \qquad (4.40)$$

for all $u \in W_{0,\sigma}^{1,2}(\Omega)$ and all $\theta,\phi \in H^1(\Omega)$ with a constant $c = c(\Omega) > 0$.

(iv) Let $\Omega \subseteq \mathbb{R}^3$ be a domain satisfying the uniform Lipschitz condition. Then

$$|d(u,\theta,\phi)| \le c\|u\|_2^{1/4}\|\nabla u\|_2^{3/4}\|\nabla\theta\|_2\|\phi\|_2^{1/4}\|\phi\|_{H^1(\Omega)}^{3/4} \qquad (4.41)$$

for all $u \in W_{0,\sigma}^{1,2}(\Omega)$ and all $\theta,\phi \in H^1(\Omega)$ with a constant $c = c(\Omega) > 0$.

Proof. Use the estimate

$$|b(u,v,w)| \le \|u\|_4\|\nabla v\|_2\|w\|_4$$

and Lemma 4.12. Analogously we proceed with $|d(u,\theta,\phi)|$. $\qquad\square$

Now a precise definition of $B(u,u)$ and $D(u,\theta)$ in (4.25) can be formulated.

Lemma 4.14. Let $\Omega \subseteq \mathbb{R}^n, n \in \{2,3\}$, be a domain satisfying the uniform Lipschitz condition, and let $0 < T < \infty$.

(i) Let

$$u,v \in L^\infty(0,T;L_\sigma^2(\Omega)) \cap L^2(0,T;W_{0,\sigma}^{1,2}(\Omega)).$$

Then for a.a. $t \in]0,T[$ the term $B(u,v)(t) \in W_{0,\sigma}^{1,2}(\Omega)'$ is well defined by

$$\langle B(u,v)(t),w\rangle_{W_{0,\sigma}^{1,2}(\Omega)';W_{0,\sigma}^{1,2}(\Omega)} := b(u(t),v(t),w)\,, \quad w \in W_{0,\sigma}^{1,2}(\Omega)\,, \qquad (4.42)$$

there holds $t \mapsto B(u,v)(t) \in L^{4/3}(0,T;W_{0,\sigma}^{1,2}(\Omega)')$ and the estimate

$$\|B(u,v)\|_{L^{4/3}(0,T;W_{0,\sigma}^{1,2}(\Omega)')} \\ \le c(n)\|u\|_{2,\infty;\Omega;T}^{1/4}\|v\|_{2,\infty;\Omega;T}^{1/4}\left(\|u\|_{L^2(0,T;H^1(\Omega)^n)}^{3/2} + \|v\|_{L^2(0,T;H^1(\Omega)^n)}^{3/2}\right) \qquad (4.43)$$

is satisfied.

(ii) Let

$$u \in L^\infty(0,T;L_\sigma^2(\Omega)) \cap L^2(0,T;W_{0,\sigma}^{1,2}(\Omega))\,, \theta \in L^\infty(0,T;L^2(\Omega)) \cap L^2(0,T;H^1(\Omega)).$$

Then for a.a. $t \in]0,T[$ the term $D(u,\theta)(t) \in H^1(\Omega)'$ is well defined by

$$\langle D(u,\theta)(t),\phi\rangle_{H^1(\Omega)';H^1(\Omega)} := d(u(t),\theta(t),\phi)\,, \quad \phi \in H^1(\Omega)\,, \qquad (4.44)$$

there holds $t \mapsto D(u,\theta)(t) \in L^{4/3}(0,T;H^1(\Omega)')$ and the estimate

$$\|D(u,\theta)\|_{L^{4/3}(0,T;(H^1(\Omega)')} \\ \le c(\Omega)\|u\|_{2,\infty;\Omega;T}^{1/4}\|\theta\|_{2,\infty;\Omega;T}^{1/4}\left(\|u\|_{L^2(0,T;H^1(\Omega)^n)}^{3/2} + \|\theta\|_{L^2(0,T;H^1(\Omega))}^{3/2}\right) \qquad (4.45)$$

is satisfied.

Proof.   (i) Combining (4.28), estimates (4.38), (4.39) and $n \in \{2,3\}$ we get

$$|b(u(t), v(t), w)| = |b(u(t), w, v(t))|$$
$$\leq c(n)\|u(t)\|_2^{1/4}\|u(t)\|_{H^1(\Omega)^n}^{3/4}\|\nabla w\|_2\|v(t)\|_2^{1/4}\|v(t)\|_{H^1(\Omega)^n}^{3/4}$$
$$\tag{4.46}$$

for almost all $t \in [0, T[$ and all $w \in W_{0,\sigma}^{1,2}(\Omega)$. Consequently

$$\|B(u,v)(t)\|_{W_{0,\sigma}^{1,2}(\Omega)'} \leq c(n)\|u(t)\|_2^{1/4}\|u(t)\|_{H^1(\Omega)^n}^{3/4}\|v(t)\|_2^{1/4}\|v(t)\|_{H^1(\Omega)^n}^{3/4}$$

for a.a. $t \in [0, T[$. It follows

$$\int_0^T \|B(u,v)(t)\|_{W_{0,\sigma}^{1,2}(\Omega)'}^{4/3}$$
$$\leq c(n)\|u\|_{2,\infty;\Omega;T}^{1/3}\|v\|_{2,\infty;\Omega;T}^{1/3}\int_0^T \|u(t)\|_{H^1(\Omega)^n}\|v(t)\|_{H^1(\Omega)^n}\,dt$$
$$\leq c(n)\|u\|_{2,\infty;\Omega;T}^{1/3}\|v\|_{2,\infty;\Omega;T}^{1/3}\int_0^T \left(\|u(t)\|_{H^1(\Omega)^n}^2 + \|v(t)\|_{H^1(\Omega)^n}^2\right)dt.$$

From the estimate above we get (4.43).

(ii) From (4.29), estimates (4.40), (4.41) and $n \in \{2,3\}$ it follows

$$\|D(u,\theta)(t)\|_{H^1(\Omega)'} \leq c(\Omega)\|u(t)\|_2^{1/4}\|u(t)\|_{H^1(\Omega)^n}^{3/4}\|\theta(t)\|_2^{1/4}\|\theta(t)\|_{H^1(\Omega)}^{3/4}$$

for a.a. $t \in [0, T[$. Having this estimate at hand the proof of (4.45) can be shown analogously to (i). $\square$

The next lemma is needed for the formulation of (4.25).

Lemma 4.15. *Let $\Omega \subseteq \mathbb{R}^n, n \in \{2,3\}$, be a domain satisfying the uniform Lipschitz condition, let $0 < T < \infty$. Let $u \in L^2(0,T;W_{0,\sigma}^{1,2}(\Omega))$ and $\theta \in L^2(0,T;H^1(\Omega))$.*

(i) We define $\Delta u \in L^2(0,T;W_{0,\sigma}^{1,2}(\Omega)')$ by

$$[\Delta u(t), w]_{W_{0,\sigma}^{1,2}(\Omega)';W_{0,\sigma}^{1,2}(\Omega)} := -\langle \nabla u(t), \nabla w\rangle_\Omega, \quad w \in W_{0,\sigma}^{1,2}(\Omega),$$

for a.a. $t \in]0, T[$. There holds

$$\|\Delta u\|_{L^2(0,T;W_{0,\sigma}^{1,2}(\Omega)')} \leq \|\nabla u\|_{2,2;\Omega;T}.$$

(ii) Define $\Delta\theta \in L^2(0,T;H^1(\Omega)')$ by

$$[\Delta\theta(t), \phi]_{H^1(\Omega)';H^1(\Omega)} := \langle \nabla\theta(t), \nabla\phi\rangle_\Omega, \quad \phi \in H^1(\Omega),$$

for a.a. $t \in]0, T[$. The estimate

$$\|\Delta\theta\|_{L^2(0,T;H^1(\Omega)')} \leq \|\nabla\theta\|_{2,2;\Omega;T}$$

holds.

(iii) Let $h \in L^2(0, T; L^2(\partial\Omega))$. Define the functional $Sh \in L^2(0, T; H^1(\Omega)')$ by

$$[Sh(t), \phi]_{H^1(\Omega)';H^1(\Omega)} := \langle h(t), \phi \rangle_{\partial\Omega}, \quad \phi \in H^1(\Omega)$$

for a.a. $t \in [0, T[$. There holds

$$\|Sh\|_{L^2((0,T;H^1(\Omega)')} \le c(\Omega)\|h\|_{2,2;\partial\Omega;T}. \tag{4.47}$$

Especially for $\Lambda \in L^\infty(\partial\Omega)$, $h_0 \in L^2(0, T; L^2(\partial\Omega))$ the functional $S(\Lambda(\theta - h_0)) \in L^2(0, T; H^1(\Omega)')$ is well defined by

$$[S(\Lambda(\theta - h_0))(t), \phi]_{H^1(\Omega)';H^1(\Omega)} := \langle \Lambda(\theta - h_0)(t), \phi \rangle_{\partial\Omega}, \quad \phi \in H^1(\Omega)$$

and there holds

$$\|S(\Lambda(\theta - h_0))\|_{L^2(0,T;H^1(\Omega)')} \le c(\Omega)\|\Lambda\|_{L^\infty(\partial\Omega)}\|\theta - h_0\|_{2,2;\partial\Omega;T}. \tag{4.48}$$

Proof. The proof of (i), (ii) is a direct application of Hölder's inequality. From the continuity of the trace operator (see Theorem 4.6) we get the estimate

$$|\langle h(t), \phi \rangle_{\partial\Omega}| \le \|h(t)\|_{2,\partial\Omega}\|\phi\|_{2,\partial\Omega} \le c(\Omega)\|h(t)\|_{2,\partial\Omega}\|\phi\|_{H^1(\Omega)}$$

for a.a. $t \in [0, T[$ and all $\phi \in H^1(\Omega)$. Consequently

$$\|Sh(t)\|_{H^1(\Omega)'} \le c\|h(t)\|_{L^2(\partial\Omega)}$$

for a.a. $t \in [0, T[$. Integrating the estimate above completes the proof of (4.47). Finally, Hölder's inequality shows that (4.48) is valid. $\qquad\square$

Consider the Gelfand triple $(W_{0,\sigma}^{1,2}(\Omega), L_\sigma^2(\Omega), W_{0,\sigma}^{1,2}(\Omega)')$. (For details about Gelfand triples we refer to the preliminaries.) Due to the identification in

$$W_{0,\sigma}^{1,2}(\Omega) \hookrightarrow L_\sigma^2(\Omega) \cong L_\sigma^2(\Omega)' \hookrightarrow W_{0,\sigma}^{1,2}(\Omega)' \tag{4.49}$$

we have that for all $f \in L_\sigma^2(\Omega)$ and all $w \in W_{0,\sigma}^{1,2}(\Omega)$ there holds

$$[f, w]_{W_{0,\sigma}^{1,2}(\Omega)';W_{0,\sigma}^{1,2}(\Omega)} = \langle f, w \rangle_\Omega. \tag{4.50}$$

Further, we also need the Gelfand triple $(H^1(\Omega), L^2(\Omega), H^1(\Omega)')$. Due to the identification in

$$H^1(\Omega) \hookrightarrow L^2(\Omega) \cong L^2(\Omega)' \hookrightarrow H^1(\Omega)' \tag{4.51}$$

we have that for all $f \in L^2(\Omega)$ and all $\phi \in H^1(\Omega)$ there holds

$$[f, \phi]_{H^1(\Omega)';H^1(\Omega)} = \langle f, \phi \rangle_\Omega \tag{4.52}$$

Now we can prove that the following formulation of the Boussinesq equations as nonlinear operator equation is equivalent to Definition 4.7.

Theorem 4.16. *Let $\Omega \subseteq \mathbb{R}^n, n \in \{2, 3\}$, be a domain satisfying the uniform Lipschitz condition, let $0 < T < \infty, g \in L^\infty(]0, T[\times\Omega)^n$, let $h_0 \in L^2(0, T; L^2(\partial\Omega)), \Lambda \in L^\infty(\partial\Omega)$ with $\Lambda(x) \geq 0$ for almost all $x \in \partial\Omega$. Further assume $f_1 \in L^2(0, T; L^2(\Omega)^n)$, assume $f_2 \in L^2(0, T; L^2(\Omega))$, assume $u_0 \in L^2_\sigma(\Omega), \theta_0 \in L^2(\Omega)$, and assume that ν, β, κ are positive, real constants. A pair (u, θ) satisfying*

$$
\begin{aligned}
&u \in L^\infty(0, T; L^2_\sigma(\Omega)) \cap L^2(0, T; W^{1,2}_{0,\sigma}(\Omega)), \quad u_t \in L^{4/3}(0, T; W^{1,2}_{0,\sigma}(\Omega)'), \\
&\theta \in L^\infty(0, T; L^2(\Omega)) \cap L^2(0, T; H^1(\Omega)), \quad \theta_t \in L^{4/3}(0, T; H^1(\Omega)')
\end{aligned}
\tag{4.53}
$$

is a weak solution of the Boussinesq equations (4.1) with boundary conditions (4.2) and (4.3) if and only if the following equations are fulfilled:

(i) $u_t - \nu\Delta u + B(u, u) = f_1 + \beta\theta g$ *in* $L^{4/3}(0, T; W^{1,2}_{0,\sigma}(\Omega)')$,

(ii) $\theta_t - \kappa\Delta\theta + D(u, \theta) + \kappa S(\Lambda(\theta - h_0)) = f_2$ *in* $L^{4/3}(0, T; H^1(\Omega)')$,

(iii) $u(0) = u_0$,

(iv) $\theta(0) = \theta_0$.

Remark. Consider an arbitrary pair (u, θ) satisfying (4.53). Then, after a redefinition on a null set, $u : [0, T[\to W^{1,2}_{0,\sigma}(\Omega)'$ and $\theta : [0, T[\to H^1(\Omega)'$ are (strongly) continuous. Moreover, by Lemma 2.32, there holds that $u : [0, T[\to L^2_\sigma(\Omega)$ and $\theta : [0, T[\to L^2(\Omega)$ are weakly continuous. Consequently, $u(0) \in L^2_\sigma(\Omega)$ and $\theta(0) \in L^2(\Omega)$ are well defined.

Proof. Throughout this proof (see the remark above) we assume that $u : [0, T[\to L^2_\sigma(\Omega)$ and $\theta : [0, T[\to L^2(\Omega)$ are weakly continuous for every (u, θ) satisfying (4.53). Furthermore, by Lemma 2.28 and (4.49), (4.51) there holds

$$
\frac{d}{dt}\langle u, w\rangle_\Omega = [u', w]_{W^{1,2}_{0,\sigma}(\Omega)';W^{1,2}_{0,\sigma}(\Omega)} ,
\tag{4.54}
$$

$$
\frac{d}{dt}\langle \theta, \phi\rangle_\Omega = [\theta', \phi]_{H^1(\Omega)';H^1(\Omega)}
\tag{4.55}
$$

in the weak sense of real-valued functions on $]0, T[$ for all $w \in W^{1,2}_{0,\sigma}(\Omega)$ and all $\phi \in H^1(\Omega)$.

Part 1. First, assume that (u, θ) satisfy (4.21), (4.22). Fix $w \in C^\infty_{0,\sigma}(\Omega)$ and $\eta \in C^\infty_0([0, T[)$. Setting $v(t, x) := \eta(t)w(x), (t, x) \in [0, T[\times\Omega$, in (4.21) it follows

$$
\begin{aligned}
&-\int_0^T \langle u, w\rangle_\Omega \eta'(t)\, dt + \int_0^T \left(\nu\langle\nabla u, \nabla w\rangle_\Omega + \langle u \cdot \nabla u, w\rangle_\Omega\right) \eta(t)\, dt \\
&= \int_0^T \left(\beta\langle\theta g, w\rangle_\Omega + \langle f_1, w\rangle_\Omega\right) \eta(t)\, dt + \langle u_0, w\rangle_\Omega \eta(0).
\end{aligned}
\tag{4.56}
$$

Due to our identifications, this means

$$-\int_0^T [u, w]_{W_{0,\sigma}^{1,2}(\Omega)', W_{0,\sigma}^{1,2}(\Omega)} \eta'(t)\, dt$$
$$= \int_0^T \left([\nu\Delta u - B(u,u) + \beta\theta g + f_1, w]_{W_{0,\sigma}^{1,2}(\Omega)'; W_{0,\sigma}^{1,2}(\Omega)} \eta(t) \right) dt + \langle u_0, w\rangle_\Omega \eta(0)$$
(4.57)

for all $w \in C_{0,\sigma}^\infty(\Omega)$ and $\eta \in C_0^\infty([0,T[)$. By taking $\eta \in C_0^\infty(]0,T[)$ in the identity above it follows (see Lemma 2.16).

$$u_t - \nu\Delta u + B(u,u) = f_1 + \beta\theta g \quad \text{in } L^{4/3}(0,T; W_{0,\sigma}^{1,2}(\Omega)').$$
(4.58)

Integration by parts (see Lemma 2.17) applied to the absolutely continuous functions $t \mapsto \langle u(t), w\rangle_\Omega$ and $t \mapsto \eta(t)$ gives us

$$-\int_0^T \langle u, w\rangle_\Omega \eta'(t)\, dt + \int_0^T \left(\nu\langle\nabla u, \nabla w\rangle_\Omega + \langle u\cdot\nabla u, w\rangle_\Omega \right) \eta(t)\, dt$$
$$= \int_0^T \left(\beta\langle\theta g, w\rangle_\Omega + \langle f_1, w\rangle_\Omega \right) \eta(t)\, dt + \langle u(0), w\rangle_\Omega \eta(0)$$
(4.59)

for all $w \in C_{0,\sigma}^\infty(\Omega)$ and $\eta \in C_0^\infty([0,T[)$. Comparing (4.58) and (4.59) implies

$$\langle u(0) - u_0, w\rangle_\Omega \eta(0) = 0$$
(4.60)

for all $\eta \in C_0^\infty([0,T[)$ and $w \in W_{0,\sigma}^{1,2}(\Omega)$. Choosing $\eta(0) \neq 0$ in (4.60) yields $u(0) = u_0$. Consequently, (i), (iii) are satisfied. In the same way as above we prove that (ii), (iv) are fulfilled.

Part 2. Let us conversely assume that (u, θ) satisfies (i)-(iv). Using (i), (iii), integration by parts and (4.54) it follows

$$-\int_0^T \langle u, w\rangle_\Omega \eta'(t)\, dt + \int_0^T \left(\nu\langle\nabla u, \nabla w\rangle_\Omega + \langle u\cdot\nabla u, w\rangle_\Omega \right) \eta(t)\, dt$$
$$= \int_0^T \left(\beta\langle\theta g, w\rangle_\Omega + \langle f_1, w\rangle_\Omega \right) \eta(t)\, dt + \langle u_0, w\rangle_\Omega \eta(0)$$
(4.61)

for all $w \in C_{0,\sigma}^\infty(\Omega)$ and $\eta \in C_0^\infty([0,T[)$. Looking at [Ga00, Lemma 2.3] we see that for every $v \in C_0^\infty([0,T[; C_{0,\sigma}^\infty(\Omega))$ and $\epsilon > 0$ there exists a function v_ϵ of the form

$$v_\epsilon(t,x) = \sum_{k=1}^n \eta_k(t) w_k(x), \quad (t,x) \in [0,T[\times\Omega,$$
(4.62)

with $N \in \mathbb{N}$ and $\eta_k \in C_0^\infty([0,T[)$, $w_k \in C_{0,\sigma}^\infty(\Omega)$ for all $k = 1, \ldots, N$, such that

$$\max_{t\in[0,T[} \left\| v_\epsilon(t) - v(t) \right\|_{C^1(\Omega)} + \max_{t\in[0,T[} \left\| \frac{\partial v_\epsilon}{\partial t}(t) - \frac{\partial v}{\partial t}(t) \right\|_{C^0(\Omega)} \leq \epsilon.$$

Combining (4.61) and (4.62) it follows that (4.21) holds for all needed test functions $v \in C_0^\infty([0, T[; C_{0,\sigma}^\infty(\Omega))$. Using (ii), (iv) and a similar density argument as in (4.62) we prove that (4.22) holds for all needed test functions $\phi \in C_0^\infty([0, T[\times\overline{\Omega})$. $\qquad\square$

The corollary below shows that every weak solution (u, θ) of (4.1) with (4.2), (4.3) is, after a redefinition on a null set, weakly continuous.

Corollary 4.17. *Let $\Omega \subseteq \mathbb{R}^n, n \in \{2, 3\}$, be a domain satisfying the uniform Lipschitz condition, let $0 < T < \infty, g \in L^\infty(]0, T[\times\Omega)^n$, let $h_0 \in L^2(0, T; L^2(\partial\Omega)), \Lambda \in L^\infty(\partial\Omega)$ with $\Lambda(x) \geq 0$ for almost all $x \in \partial\Omega$. Further assume $f_1 \in L^2(0, T; L^2(\Omega)^n)$, $f_2 \in L^2(0, T; L^2(\Omega))$, assume $u_0 \in L_\sigma^2(\Omega), \theta_0 \in L^2(\Omega)$, and assume that ν, β, κ are positive, real constants. Consider a weak solution (u, θ) of (4.1) with (4.2) and (4.3). Then, after a redefinition on a null set, $u : [0, T[\to L_\sigma^2(\Omega)$ and $\theta : [0, T[\to L^2(\Omega)$ are weakly continuous. The initial values u_0, θ_0 are taken in the following sense:*

$$\lim_{t \to 0+} \langle u(t), w \rangle_\Omega = \langle u_0, w \rangle_\Omega ,$$

$$\lim_{t \to 0+} \langle \theta(t), \phi \rangle_\Omega = \langle \theta_0, \phi \rangle_\Omega$$

for all $w \in L_\sigma^2(\Omega)$ and for all $\phi \in L^2(\Omega)$.

4.4 Existence of weak solutions

The following theorem states the existence of a global weak solution (u, θ) of the Boussinesq equations (4.1) with boundary conditions (4.2), (4.3). The special construction of the weak solution enables us to prove that this weak solution (u, θ) fulfils the energy inequalities. In general it is not known if every weak solution of the Boussinesq equations fulfils these energy inequalities. Actually, the corresponding problem for the Navier-Stokes equations is not solved.

Theorem 4.18. *Let $\Omega \subseteq \mathbb{R}^n, n \in \{2, 3\}$, be a domain satisfying the uniform Lipschitz condition, let $0 < T < \infty, g \in L^\infty(]0, T[\times\Omega)^n$, let $h_0 \in L^2(0, T; L^2(\partial\Omega)), \Lambda \in L^\infty(\partial\Omega)$ with $\Lambda(x) \geq 0$ for almost all $x \in \partial\Omega$. Further assume $f_1 \in L^2(0, T; L^2(\Omega)^n)$, assume $f_2 \in L^2(0, T; L^2(\Omega))$, assume $u_0 \in L_\sigma^2(\Omega), \theta_0 \in L^2(\Omega)$, and assume that ν, β, κ are positive, real constants. Then the following statements are satisfied.*

(i) *There exists a weak solution the Boussinesq equations of (4.1) with boundary conditions (4.2), (4.3).*

(ii) *If $n = 3$ and if additionally $g \in L^4(0, T; L^2(\Omega)^3)$ holds then there exists a weak solution of (4.1) with (4.2), (4.3) satisfying the energy inequalities (4.23), (4.24) for almost all $t \in [0, T[$.*

Remark. In the case $n = 2$ it follows from Theorem 4.23 that every weak solution (u, θ) of (4.1) with (4.2), (4.3) is uniquely determined and, after a redefinition on a null set of $]0, T[$, there holds that (u, θ) satisfies the energy equalities (4.131), (4.132).

The proof uses the **Galerkin method**. It is based on the construction of approximate solutions (u_m, θ_m), $m \in \mathbb{N}$, of certain finite dimensional problems. Energy estimates will be used to prove that $(u_m, \theta_m)_{m \in \mathbb{N}}$ contains a weakly convergent subsequence in a suitable function space with weak limit (u, θ). Then a passage to the limit will prove that (u, θ) is a weak solution; in this argument the compactness argument in Theorem 2.26 will be employed.

For the construction of the finite dimensional spaces choose a sequence $(w_m)_{m \in \mathbb{N}}$ of linear independent vectors $w_m \in C_{0,\sigma}^{\infty}(\Omega), m \in \mathbb{N}$, such that the linear span of $\{w_m; m \in \mathbb{N}\}$ is dense in $W_{0,\sigma}^{1,2}(\Omega)$ and, additionally, $\langle w_k, w_l \rangle = \delta_{k,l}$ for all $k, l \in \mathbb{N}$. Furthermore, we choose linear independent vectors $\psi_m \in C_0^{\infty}(\overline{\Omega})$, $m \in \mathbb{N}$, such that the linear span of $\{\psi_m; m \in \mathbb{N}\}$ is dense in $H^1(\Omega)$ and, additionally, $\langle \psi_k, \psi_l \rangle_\Omega = \delta_{k,l}$ for all $k, l \in \mathbb{N}$.

Fix $m \in \mathbb{N}$. We look for absolutely continuous functions $\alpha_{k,m}, \gamma_{k,m} : [0, T[\to \mathbb{R}, k = 1, \ldots, m$, such that the functions u_m, θ_m defined by

$$
\begin{aligned}
u_m(t) &:= \sum_{k=1}^{m} \alpha_{k,m}(t) w_k, \quad t \in [0, T[, \\
\theta_m(t) &:= \sum_{k=1}^{m} \gamma_{k,m}(t) \psi_k, \quad t \in [0, T[,
\end{aligned}
\tag{4.63}
$$

satisfy the finite dimensional **Galerkin approximation system**

$$
\frac{d}{dt} \langle u_m, w_k \rangle_\Omega + \nu \langle \nabla u_m, \nabla w_k \rangle_\Omega + \langle u_m \cdot \nabla u_m, w_k \rangle_\Omega = \beta \langle \theta_m g, w_k \rangle_\Omega + \langle f_1, w_k \rangle_\Omega \quad \text{(Gal1)}
$$

$$
\text{for a.a. } t \in]0, T[, \quad \forall k = 1, \ldots, m,
$$

$$
\frac{d}{dt} \langle \theta_m, \psi_l \rangle_\Omega + \kappa \langle \nabla \theta_m, \nabla \psi_l \rangle_\Omega + \langle u_m \cdot \nabla \theta_m, \psi_l \rangle_\Omega + \kappa \langle \Lambda(\theta_m - h_0), \psi_l \rangle_{\partial\Omega} = \langle f_2, \psi_l \rangle_\Omega
$$
$$
\text{(Gal2)}
$$

$$
\text{for a.a. } t \in]0, T[, \quad \forall l = 1, \ldots, m,
$$

$$
u_m(0) = \sum_{k=1}^{m} \langle u_0, w_k \rangle_\Omega \, w_k, \tag{Gal3}
$$

$$
\theta_m(0) = \sum_{k=1}^{m} \langle \theta_0, \psi_k \rangle_\Omega \, \psi_k. \tag{Gal4}
$$

In (Gal1) we denote by $\frac{d}{dt} \langle u_m(t), w_k \rangle_\Omega$ the derivative of the absolutely continuous function $t \mapsto \langle u_m(t), w_k \rangle_\Omega, t \in]0, T[$. Analogously for (Gal2).

Existence and needed properties of the approximate solutions

In Lemma 4.19 it will be proved that the Galerkin system can be solved using the Carathéodory existence theorem for ordinary differential equations.

Lemma 4.19. *For every $m \in \mathbb{N}$ the Galerkin system (Gal1)-(Gal4) has a unique solution (u_m, θ_m) of the form (4.63) defined on the whole interval $[0, T[$. This solution satisfies the energy equalities*

$$
\frac{1}{2}\|u_m(t)\|_2^2 + \nu \int_0^t \|\nabla u_m\|_2^2 \, d\tau
$$

$$
= \frac{1}{2}\|u_0\|_2^2 + \beta \int_0^t \langle \theta_m g, u_m \rangle_\Omega \, d\tau + \int_0^t \langle f_1, u_m \rangle_\Omega \, d\tau \,, \tag{4.64}
$$

$$
\frac{1}{2}\|\theta_m(t)\|_2^2 + \kappa \int_0^t \|\nabla \theta_m\|_2^2 \, d\tau + \kappa \int_0^t \|\sqrt{\Lambda}\,\theta_m\|_{2,\partial\Omega}^2 \, d\tau
$$

$$
= \frac{1}{2}\|\theta_0\|_2^2 + \kappa \int_0^t \langle \Lambda h_0, \theta_m \rangle_{\partial\Omega} \, d\tau + \int_0^t \langle f_2, \theta_m \rangle_\Omega \, d\tau \tag{4.65}
$$

for all $t \in [0, T]$. There exists a subsequence $(u_{m_k}, \theta_{m_k})_{k \in \mathbb{N}}$ and a pair (u, θ) with

$$
u \in L^\infty(0, T; L_\sigma^2(\Omega)) \cap L^2(0, T; W_{0,\sigma}^{1,2}(\Omega))\,, \quad \theta \in L^\infty(0, T; L^2(\Omega)) \cap L^2(0, T; H^1(\Omega))
$$

such that

$$
u_{m_k} \xrightarrow[k\to\infty]{*} u \quad \text{in } L^\infty(0, T; L_\sigma^2(\Omega))\,, \quad u_{m_k} \xrightarrow[k\to\infty]{} u \quad \text{in } L^2(0, T; W_{0,\sigma}^{1,2}(\Omega))\,, \tag{4.66}
$$

$$
\theta_{m_k} \xrightarrow[k\to\infty]{*} \theta \quad \text{in } L^\infty(0, T; L^2(\Omega))\,, \quad \theta_{m_k} \xrightarrow[k\to\infty]{} \theta \quad \text{in } L^2(0, T; H^1(\Omega))\,. \tag{4.67}
$$

Moreover

$$
(\sqrt{\Lambda}\,\theta_m)_{m \in \mathbb{N}} \text{ is bounded in } L^2(0, T; L^2(\partial\Omega)). \tag{4.68}
$$

Proof. We consider the following system of differential equations for $\alpha_{k,m}$ and $\gamma_{l,m}$ which is equivalent to (Gal1)-(Gal4):

$$
\alpha'_{k,m}(t) + \nu \sum_{i=1}^m \langle \nabla w_i, \nabla w_k \rangle_\Omega \alpha_{i,m}(t) + \sum_{i,j=1}^m \alpha_{i,m}(t) b(w_i, w_j, w_k) \alpha_{j,m}(t)
$$

$$
- \beta \sum_{i=1}^m \langle \psi_i g(t), w_k \rangle_\Omega \gamma_{i,m}(t) = \langle f_1(t), w_k \rangle_\Omega \tag{Gal5}
$$

$$
\text{for a.a. } t \in]0, T[\,, \quad \forall k = 1, \ldots, m\,,
$$

$$
\gamma'_{l,m}(t) + \kappa \sum_{i=1}^m \langle \nabla \psi_i, \nabla \psi_l \rangle_\Omega \gamma_{i,m}(t) + \sum_{i,j=1}^m \alpha_{i,m}(t) d(w_i, \psi_j, \psi_l) \gamma_{j,m}(t)
$$

$$
+ \kappa \sum_{i=1}^m \langle \Lambda \psi_i, \psi_l \rangle_{\partial\Omega} \gamma_{i,m}(t) = \kappa \langle \Lambda h_0(t), \psi_l \rangle_{\partial\Omega} + \langle f_2(t), \psi_l \rangle_\Omega \tag{Gal6}
$$

$$
\text{for a.a. } t \in]0, T[\,, \quad \forall l = 1, \ldots, m\,,
$$

$$
\alpha_{k,m}(0) = \langle w_k, u_0 \rangle_\Omega \quad \forall k = 1, \ldots, m\,, \tag{Gal7}
$$

$$
\gamma_{l,m}(0) = \langle \psi_l, \theta_0 \rangle_\Omega \quad \forall l = 1, \ldots, m\,. \tag{Gal8}
$$

By Carathéodory's existence theorem for differential equations (see [Wa00, § 10]) we obtain for every $m \in \mathbb{N}$ a maximal interval of existence $I_m \subseteq [0, T[$ and unique absolutely continuous solutions $\alpha_{k,m}, \gamma_{l,m}$ $(k, l = 1, \ldots m)$ on I_m of the system (Gal5)-(Gal8). From the assumptions on the data it follows that all functions $\alpha_{k,m}, \gamma_{l,m}$ $(k, l = 1, \ldots m)$, lie in the space $H^1(I_m)$. Consequently (u_m, θ_m) defined by (4.63) is the unique solution of the system (Gal1)-(Gal4) on I_m. To prove that $I_m = [0, T[$ holds we show that $\|u_m(t)\|_2$ and $\|\theta_m(t)\|_2$ remain bounded on I_m. By Lemma 4.14 there holds

$$b(u_m(t), u_m(t), u_m(t)) = 0, \; d(u_m(t), \theta_m(t), \theta_m(t)) = 0 \text{ for a.a. } t \in I_m. \tag{4.69}$$

We multiply (Gal1) by $\alpha_{k,m}$ and (Gal2) by $\gamma_{l,m}$ and sum over $k = 1, \ldots, m$ and $l = 1, \ldots, m$ to obtain with Lemma 2.30 and (4.69) that the equations

$$\frac{1}{2} \frac{d}{d\tau} \|u_m(\tau)\|_2^2 \Big|_{\tau=t} + \nu \|\nabla u_m(t)\|_2^2 \tag{4.70}$$
$$= \beta \langle \theta_m(t) g(t), u_m(t) \rangle_\Omega + \langle f_1(t), u_m(t) \rangle_\Omega,$$

$$\frac{1}{2} \frac{d}{d\tau} \|\theta_m(\tau)\|_2^2 \Big|_{\tau=t} + \kappa \|\nabla \theta_m(t)\|_2^2 + \kappa \|\sqrt{\Lambda} \theta_m(t)\|_{2,\partial\Omega}^2 \tag{4.71}$$
$$= \kappa \langle \Lambda h_0(t), \theta_m(t) \rangle_{\partial\Omega} + \langle f_2(t), \theta_m(t) \rangle_\Omega,$$

are satisfied for a.a. $t \in I_m$. Integrating (4.70) and (4.71) over I_m we get (4.64), (4.65) for a.a. $t \in I_m$. By continuity these inequalities are satisfied for all $t \in I_m$.

We get with Young's inequality and the trace estimate (4.17) that

$$\kappa |\langle \Lambda h_0(t), \theta_m(t) \rangle_{\partial\Omega}| \leq \kappa \|\Lambda\|_{\infty,\partial\Omega} \|h_0(t)\|_{2,\partial\Omega} \|\theta_m(t)\|_{2,\partial\Omega}$$
$$\leq c\kappa \|\Lambda\|_{\infty,\partial\Omega} \|h_0(t)\|_{2,\partial\Omega} (\|\theta_m(t)\|_2 + \|\nabla \theta_m(t)\|_2)$$
$$\leq c\|\theta_m(t)\|_2^2 + \frac{\kappa}{2} \|\nabla \theta_m(t)\|_2^2 + c\|h_0(t)\|_{2,\partial\Omega}^2$$

holds for a.a. $t \in [0, T[$ with a constant $c > 0$ independent of $m \in \mathbb{N}$. Consequently, from (4.71) it follows

$$\frac{1}{2} \frac{d}{d\tau} \|\theta_m(\tau)\|_2^2 \Big|_{\tau=t} + \frac{\kappa}{2} \|\nabla \theta_m(t)\|_2^2 + \kappa \|\sqrt{\Lambda} \theta_m(t)\|_{2,\partial\Omega}^2 \tag{4.72}$$
$$\leq c_1 \|\theta_m(t)\|_2^2 + c_2 (\|f_2(t)\|_2^2 + \|h_0(t)\|_{2,\partial\Omega}^2)$$

for almost all $t \in I_m$ with constants $c_1, c_2 > 0$ independent of $m \in \mathbb{N}$. The inequality of Gronwall yields

$$\|\theta_m(t)\|_2^2 \leq e^{\int_0^t 2c_1 \, ds} \left(\|\theta_m(0)\|_2^2 + 2c_2 \int_0^t (\|f_2(\tau)\|_2^2 + \|h_0(\tau)\|_{2,\partial\Omega}^2 \, d\tau) \right) \tag{4.73}$$

for almost all $t \in I_m$. Since $\theta_m(0) \to \theta_0$ as $m \to \infty$ in $L^2(\Omega)$ (see (Gal4)) and (4.73) we conclude that $(\theta_m)_{m \in \mathbb{N}}$ is bounded in $L^\infty(I_m; L^2(\Omega))$. Integrating (4.72) over I_m

implies that $(\nabla\theta_m)_{m\in\mathbb{N}}$ is bounded in $L^2(I_m; L^2(\Omega)^n)$ and $(\sqrt{\Lambda}\theta_m)_{m\in\mathbb{N}}$ is bounded in $L^2(I_m; L^2(\partial\Omega))$. Since the interval I_m is bounded (use $0 < T < \infty$), we obtain that

$$(\theta_m)_{m\in\mathbb{N}} \text{ is bounded in } L^\infty(I_m; L^2(\Omega)) \cap L^2(I_m; H^1(\Omega)). \tag{4.74}$$

Equality (4.70) can be estimated in the following way:

$$\frac{1}{2}\frac{d}{d\tau}\|u_m(\tau)\|_2^2\Big|_{\tau=t} + \nu\|\nabla u_m(t)\|_2^2 \le c_1\|u_m(t)\|_2^2 + c_2\|\theta_m(t)\|_2^2 + c_3\|f_1(t)\|_2^2 \tag{4.75}$$

for almost all $t \in I_m$ with constants $c_1, c_2, c_3 > 0$ independent of $m \in \mathbb{N}$. We employ in (4.75) Gronwall's inequality and integration by parts to show that

$$(u_m)_{m\in\mathbb{N}} \text{ is bounded in } L^\infty(I_m; L_\sigma^2(\Omega)) \cap L^2(I_m; W_{0,\sigma}^{1,2}(\Omega)). \tag{4.76}$$

Due to the proven boundedness of $\|u_m(t)\|_2$ and $\|\theta_m(t)\|_2$, we get $I_m = [0, T[$ for all $m \in \mathbb{N}$. We use (4.74), (4.76) to find a subsequence $(u_{m_k}, \theta_{m_k})_{k\in\mathbb{N}}$ and a pair (u, θ) satisfying (4.66), (4.67). □

Theorem 4.20. *Let $(u_{m_k}, \theta_{m_k})_{k\in\mathbb{N}}$ and (u, θ) be as in Lemma 4.19. Then*

$$u_{m_k} \underset{k\to\infty}{\to} u \quad \text{strongly in } L^2(0, T; L^2(G)^n), \tag{4.77}$$

$$\theta_{m_k} \underset{k\to\infty}{\to} \theta \quad \text{strongly in } L^2(0, T; L^2(G)), \quad \theta_{m_k} \underset{k\to\infty}{\to} \theta \quad \text{strongly in } L^2(0, T; L^2(\partial G)) \tag{4.78}$$

for all bounded Lipschitz domains $G \subseteq \Omega$. Furthermore

$$\sqrt{\Lambda}\theta_{m_k} \underset{k\to\infty}{\to} \sqrt{\Lambda}\theta \quad \text{in } L^2(0, T; L^2(\partial\Omega)). \tag{4.79}$$

Proof. Fix an arbitrary $0 < \gamma < \frac{1}{4}$. In this theorem we use the Fourier transform and therefore we have to work in the complexifications of the involved Sobolev spaces. We denote by $\widehat{u_{m_k}} : \mathbb{R} \to L_\sigma^2(\Omega)$ and by $\widehat{\theta_{m_k}} : \mathbb{R} \to L^2(\Omega)$ the Fourier transforms

$$\widehat{u_{m_k}}(\tau) := \int_{\mathbb{R}} e^{-2\pi i t\tau} 1_{[0,T[}(t)u_{m_k}(t)\, dt = \int_0^T e^{-2\pi i t\tau} u_{m_k}(t)\, dt, \quad \tau \in \mathbb{R},$$

$$\widehat{\theta_{m_k}}(\tau) := \int_{\mathbb{R}} e^{-2\pi i t\tau} 1_{[0,T[}(t)\theta_{m_k}(t)\, dt = \int_0^T e^{-2\pi i t\tau} \theta_{m_k}(t)\, dt, \quad \tau \in \mathbb{R}.$$

Our goal is to find a constant $M > 0$ such that

$$\int_{\mathbb{R}} |\tau|^{2\gamma}\|\widehat{u_{m_k}}(\tau)\|_{2,\Omega}^2\, d\tau \le M, \tag{4.80}$$

$$\int_{\mathbb{R}} |\tau|^{2\gamma}\|\widehat{\theta_{m_k}}(\tau)\|_{2,\Omega}^2\, d\tau \le M \tag{4.81}$$

for all $k \in \mathbb{N}$. Indeed, assume that (4.80), (4.81) are proven. Fix a bounded Lipschitz domain $G \subseteq \Omega$. Since $(u_{m_k})_{k \in \mathbb{N}}$ is bounded in $L^2(0, T; H^1(G)^n)$ and (4.80) holds, we obtain from Theorem 2.26 applied to the imbedding scheme

$$H^1(G)^n \underset{\text{compact}}{\hookrightarrow} L^2(G)^n \underset{\text{continuous}}{\hookrightarrow} L^2(G)^n \,, \qquad (4.82)$$

that $(u_{m_k})_{k \in \mathbb{N}}$ contains a strongly convergent subsequence in $L^2(0, T; L^2(G)^n)$. It follows from (4.66) that no additional subsequence is needed, i.e. that (4.77) is satisfied. Let s be any real number with $\frac{1}{2} < s < 1$. There holds that $(\theta_{m_k})_{k \in \mathbb{N}}$ is bounded in $L^2(0, T; H^1(G))$ and estimate (4.81) is satisfied. Consequently, applying Theorem 2.26 to the imbedding scheme

$$H^1(G) \underset{\text{compact}}{\hookrightarrow} W^{s,2}(G) \underset{\text{continuous}}{\hookrightarrow} L^2(G) \,, \qquad (4.83)$$

we get that $(\theta_{m_k})_{k \in \mathbb{N}}$ contains a strongly convergent subsequence in $L^2(0, T; W^{s,2}(G))$. Using (4.67) we conclude

$$\theta_{m_k} \underset{k \to \infty}{\to} \theta \quad \text{strongly in } L^2(0, T; W^{s,2}(G)). \qquad (4.84)$$

This implies the strong convergence $\theta_{m_k} \to \theta$ as $k \to \infty$ in $L^2(0, T; L^2(G))$. The continuous map $W^{s,2}(G) \mapsto L^2(\partial G)$ (see Theorem 2.8) applied to (4.84) yields $\theta_{m_k} \to \theta$ as $k \to \infty$ strongly in $L^2(0, T; L^2(\partial G))$. Altogether (4.78) holds. From (4.68) we know that $(\sqrt{\Lambda}\theta_{m_k})_{k \in \mathbb{N}}$ is bounded in $L^2(0, T; L^2(\partial \Omega))$ and with the help of (4.78) we can prove that (4.79) is satisfied.

To finish the proof of this theorem it remains to show that estimates (4.80), (4.81) can be fulfilled. We restrict ourselves to show (4.81) since this proof will present all arguments needed for the estimate (4.80). To simplify notations let $m_k = k$ for all $k \in \mathbb{N}$. Fix $m \in \mathbb{N}$. From (Gal2) we know

$$\frac{d}{dt} \langle \theta_m, \psi_l \rangle_\Omega + \kappa \langle \nabla \theta_m, \nabla \psi_l \rangle_\Omega + \langle u_m \cdot \nabla \theta_m, \psi_l \rangle_\Omega + \kappa \langle \Lambda(\theta_m - h_0), \psi_l \rangle_{\partial\Omega} = \langle f_2, \psi_l \rangle_\Omega \quad (4.85)$$

for a.a. $t \in]0, T[$ and all $l = 1, \ldots, m$. Fix $l \in \{1, \ldots, m\}$. Take the Fourier transform in (4.85), integrate by parts (see (2.23)) to obtain

$$
\begin{aligned}
\int_0^T & e^{-2\pi i t \tau} \frac{d}{dt} \langle \theta_m, \psi_l \rangle_\Omega \, dt \\
&= -\int_0^T \left(\frac{d}{dt} e^{-2\pi i t \tau} \right) \langle \theta_m, \psi_l \rangle_\Omega \, dt + e^{-2\pi i t \tau} \langle \theta_m, \psi_l \rangle_\Omega \Big|_{t=0}^{t=T} \\
&= 2\pi i \tau \int_0^T e^{-2\pi i t \tau} \langle \theta_m, \psi_l \rangle_\Omega \, dt + e^{-2\pi i T \tau} \langle \theta_m(T), \psi_l \rangle_\Omega - \langle \theta_m(0), \psi_l \rangle_\Omega \\
&= 2\pi i \tau \langle \widehat{\theta_m}(\tau), \psi_l \rangle_\Omega + e^{-2\pi i T \tau} \langle \theta_m(T), \psi_l \rangle_\Omega - \langle \theta_m(0), \psi_l \rangle_\Omega
\end{aligned}
\qquad (4.86)
$$

for all $\tau \in \mathbb{R}$. On the other hand

$$
\int_0^T e^{-2\pi i t \tau} \frac{d}{dt} \langle \theta_m, \psi_l \rangle_\Omega \, dt
$$
$$
= \int_0^T e^{-2\pi i t \tau} \Big(\langle f_2, \psi_l \rangle_\Omega - \kappa \langle \nabla \theta_m, \nabla \psi_l \rangle_\Omega - \langle u_m \cdot \nabla \theta_m, \psi_l \rangle_\Omega \tag{4.87}
$$
$$
- \kappa \langle \Lambda(\theta_m - h_0), \psi_l \rangle_{\partial\Omega} \Big) \, dt
$$
$$
= \langle \widehat{y_m}(\tau), \psi_l \rangle_{H^1(\Omega)'; H^1(\Omega)}
$$

for all $\tau \in \mathbb{R}$ where the function $y_m \in L^1(\mathbb{R}; H^1(\Omega)')$ is defined by

$$
y_m(t) := \begin{cases} f_2(t) + \kappa \Delta \theta_m(t) - D(u_m, \theta_m)(t) - \kappa S\big(\Lambda(\theta_m(t) - h_0(t))\big), & \text{a.a. } t \in [0, T[, \\ 0, & t \in \mathbb{R} \setminus [0, T[. \end{cases}
$$
$$
\tag{4.88}
$$

Combining (4.86), (4.87) yields

$$
2\pi i \tau \langle \widehat{\theta_m}(\tau), \psi_l \rangle_\Omega = \langle \widehat{y_m}(\tau), \psi_l \rangle_{H^1(\Omega)'; H^1(\Omega)} + \langle \theta_m(0), \psi_l \rangle_\Omega - e^{-2\pi i T \tau} \langle \theta_m(T), \psi_l \rangle_\Omega \tag{4.89}
$$

for all $\tau \in \mathbb{R}$. Multiplying (4.89) by $\overline{\gamma_{l,m}(\tau)}$ and adding these equations for $l = 1, \dots, m$, we see that

$$
2\pi i \tau \|\widehat{\theta_m}(\tau)\|_2^2 = \langle \widehat{y_m}(\tau), \widehat{\theta_m}(\tau) \rangle_{H^1(\Omega)'; H^1(\Omega)} + \langle \theta_m(0), \widehat{\theta_m}(\tau) \rangle_\Omega - e^{-2\pi i T \tau} \langle \theta_m(T), \widehat{\theta_m}(\tau) \rangle_\Omega \tag{4.90}
$$

for all $\tau \in \mathbb{R}$. We obtain with (4.17)

$$
\int_0^T \|y_m(\tau)\|_{H^1(\Omega)'} \, d\tau
$$
$$
\leq \int_0^T \Big(\|f_2(\tau)\|_2 + \kappa \|\nabla \theta_m(\tau)\|_2 + c\|u_m(\tau)\|_{H^1(\Omega)^n} \|\theta_m(\tau)\|_{H^1(\Omega)}
$$
$$
+ c\kappa \|\Lambda\|_{\infty, \partial\Omega} (\|\theta_m(\tau)\|_{2, \partial\Omega} + \|h_0(\tau)\|_{2, \partial\Omega}) \Big) \, d\tau \tag{4.91}
$$
$$
\leq \int_0^T \Big(\|f_2(\tau)\|_2 + \kappa \|\nabla \theta_m(\tau)\|_2 + c\big(\|u_m(\tau)\|_{H^1(\Omega)^n}^2 + \|\theta_m(\tau)\|_{H^1(\Omega)}^2\big)
$$
$$
+ \kappa c \|\Lambda\|_{\infty, \partial\Omega} (\|\theta_m(\tau)\|_{H^1(\Omega)} + \|h_0(\tau)\|_{2, \partial\Omega}) \Big) \, d\tau
$$

with a constant $c > 0$ independent of $m \in \mathbb{N}$. The boundedness of $(u_{m_k}, \theta_{m_k})_{k \in \mathbb{N}}$ in (4.66), (4.67) imply that there exists a constant $c > 0$ satisfying

$$
\|\widehat{y_m}(\tau)\|_{H^1(\Omega)'} \leq c \tag{4.92}
$$

for all $m \in \mathbb{N}$ and all $\tau \in \mathbb{R}$. We employ (4.90) and the fact that the sequences $(\|\theta_m(0)\|_2)_{m \in \mathbb{N}}$ and $(\|\theta_m(T)\|_2)_{m \in \mathbb{N}}$ are bounded, real sequences to obtain the existence of a constant $c > 0$ such that

$$
|\tau| \|\widehat{\theta_m}(\tau)\|_2^2 \leq c \|\widehat{\theta_m}(\tau)\|_{H^1(\Omega)} \tag{4.93}
$$

for all $\tau \in \mathbb{R}$ and $m \in \mathbb{N}$. Since $0 < \gamma < \frac{1}{4}$ there holds

$$|\tau|^{2\gamma} \le c(\gamma) \frac{1 + |\tau|}{1 + |\tau|^{1-2\gamma}} \tag{4.94}$$

for all $\tau \in \mathbb{R}$. Using (4.93), (4.94) we get

$$\int_{\mathbb{R}} |\tau|^{2\gamma} \|\widehat{\theta_m}(\tau)\|_2^2 \, d\tau \le c \int_{\mathbb{R}} \frac{1 + |\tau|}{1 + |\tau|^{1-2\gamma}} \|\widehat{\theta_m}(\tau)\|_2^2 \, d\tau$$

$$\le c \int_{\mathbb{R}} \|\widehat{\theta_m}(\tau)\|_2^2 \, d\tau + c \int_{\mathbb{R}} \frac{\|\widehat{\theta_m}(\tau)\|_{H^1(\Omega)}}{1 + |\tau|^{1-2\gamma}} \, d\tau \tag{4.95}$$

A consequence of our choice of γ is $\frac{1}{2} < 1 - 2\gamma$ and therefore

$$\int_{\mathbb{R}} \frac{1}{(1 + |\tau|^{1-2\gamma})^2} \, d\tau < \infty. \tag{4.96}$$

Application of the Parseval identity (2.32) gives us

$$\int_{\mathbb{R}} \|\widehat{\theta_m}(\tau)\|_{H^1(\Omega)}^2 \, d\tau = \int_{\mathbb{R}} \|\theta_m(\tau)\|_{H^1(\Omega)}^2 \, d\tau \le c \tag{4.97}$$

for all $m \in \mathbb{N}$ with a constant $c > 0$ independent of m. Consequently, by (4.96), (4.97) the Hölder inequality yields

$$\int_{\mathbb{R}} \frac{\|\widehat{\theta_m}(\tau)\|_{H^1(\Omega)}}{1 + |\tau|^{1-2\gamma}} \, d\tau \le \left(\int_{\mathbb{R}} \|\widehat{\theta_m}(\tau)\|_{H^1(\Omega)}^2 \, d\tau \right)^{1/2} \left(\int_{\mathbb{R}} \frac{1}{(1 + |\tau|^{1-2\gamma})^2} \, d\tau \right)^{1/2} \le c \tag{4.98}$$

with a constant $c > 0$ independent of $m \in \mathbb{N}$. Finally we use (4.98) and the Parseval identity to get

$$\int_{\mathbb{R}} |\tau|^{2\gamma} \|\widehat{\theta_m}(\tau)\|_2^2 \, d\tau \le c$$

for all $m \in \mathbb{N}$ with a constant $c > 0$ independent of m. $\qquad\square$

Proof of Theorem 4.18.

Let $(u_{m_k}, \theta_{m_k})_{k \in \mathbb{N}}$ and (u, θ) be as in Lemma 4.19. To simplify notation assume $m_k = k$ for $k \in \mathbb{N}$.

Proof of statement (i) of Theorem 4.18. Integrating (Gal1) yields

$$-\int_0^T \langle u_m, w \rangle_\Omega \eta'(t) \, dt + \int_0^T \left(\nu \langle \nabla u_m, \nabla w \rangle_\Omega + \langle u_m \cdot \nabla u_m, w \rangle_\Omega \right) \eta(t) \, dt$$

$$= \int_0^T \left(\beta \langle \theta_m g, w \rangle_\Omega + \langle f_1, w \rangle_\Omega \right) \eta(t) \, dt + \langle u_m(0), w \rangle_\Omega \eta(0) \tag{4.99}$$

for all $m \in \mathbb{N}$, all $\eta \in C_0^\infty([0,T[)$ and all $w \in \text{span}\{w_1, \ldots, w_m\}$. By (4.66), (4.77) we have

$$\lim_{m \to \infty} \int_0^T \langle u_m \cdot \nabla u_m, w \rangle_\Omega \eta(t)\, dt = \int_0^T \langle u \cdot \nabla u, w \rangle_\Omega \eta(t)\, dt \qquad (4.100)$$

for all $\eta \in C_0^\infty([0,T[)$ and all $w \in \text{span}\{w_k; k \in \mathbb{N}\}$. Now let $m \to \infty$ in (4.99), use $u_m(0) \to u_0$ in $L^2(\Omega)$ as $m \to \infty$, and (4.66), (4.67), (4.100) to obtain

$$-\int_0^T \langle u, w \rangle_\Omega \eta'(t)\, dt + \int_0^T \Big(\nu \langle \nabla u, \nabla w \rangle_\Omega + \langle u \cdot \nabla u, w \rangle_\Omega \Big) \eta(t)\, dt$$
$$= \int_0^T \Big(\beta \langle \theta g, w \rangle_\Omega + \langle f_1, w \rangle_\Omega \Big) \eta(t)\, dt + \langle u_0, w \rangle_\Omega \eta(0) \qquad (4.101)$$

for all $\eta \in C_0^\infty([0,T[)$ and all $w \in \text{span}\{w_k; k \in \mathbb{N}\}$. By construction $\text{span}\{w_k; k \in \mathbb{N}\}$ is dense in $W_{0,\sigma}^{1,2}(\Omega)$ and consequently (4.101) remains true for all $\eta \in C_0^\infty([0,T[)$ and all $w \in W_{0,\sigma}^{1,2}(\Omega)$. As a consequence of our identifications and (4.101), the identity

$$-\int_0^T [u, w]_{W_{0,\sigma}^{1,2}(\Omega)', W_{0,\sigma}^{1,2}(\Omega)} \eta'(t)\, dt$$
$$= \int_0^T \Big([\nu \Delta u - B(u,u) + \beta \theta g + f_1, w]_{W_{0,\sigma}^{1,2}(\Omega)'; W_{0,\sigma}^{1,2}(\Omega)} \eta(t) \Big)\, dt \qquad (4.102)$$

holds for all $w \in C_{0,\sigma}^\infty(\Omega)$ and $\eta \in C_0^\infty(]0,T[)$. Thus

$$u_t - \nu \Delta u + B(u,u) = f_1 + \beta \theta g \quad \text{in } L^{4/3}(0,T; W_{0,\sigma}^{1,2}(\Omega)'). \qquad (4.103)$$

In the next step we prove $u(0) = u_0$. After a redefinition on an null set of $[0,T[$, we have that $u : [0,T[\to W_{0,\sigma}^{1,2}(\Omega)'$ is (strongly) continuous. Moreover, by Lemma 2.32, there holds that $u : [0,T[\to L_\sigma^2(\Omega)$ is weakly continuous. Consequently, $u(0) \in L_\sigma^2(\Omega)$ is well defined. Multiplying (4.103) with $\eta \in C_0^\infty([0,T[)$ and integrating this identity it follows

$$-\int_0^T \langle u, w \rangle_\Omega \eta'(t)\, dt + \int_0^T \Big(\nu \langle \nabla u, \nabla w \rangle_\Omega + \langle u \cdot \nabla u, w \rangle_\Omega \Big) \eta(t)\, dt$$
$$= \int_0^T \Big(\beta \langle \theta g, w \rangle_\Omega + \langle f_1, w \rangle_\Omega \Big) \eta(t)\, dt + \langle u(0), w \rangle_\Omega \eta(0) \qquad (4.104)$$

for all $\eta \in C_0^\infty([0,T[)$ and $w \in W_{0,\sigma}^{1,2}(\Omega)$. Comparing (4.101) and (4.104) implies

$$\langle u(0) - u_0, w \rangle_\Omega \eta(0) = 0 \qquad (4.105)$$

for all $\eta \in C_0^\infty([0,T[)$ and $w \in W_{0,\sigma}^{1,2}(\Omega)$. Choosing $\eta(0) \neq 0$ in (4.105) implies $u(0) = u_0$. Thus conditions (i), (iii) of Theorem 4.16 are satisfied.

Integrating (Gal2) yields

$$-\int_0^T \langle \theta_m, \psi \rangle_\Omega \eta'(t)\, dt + \int_0^T \Big(\kappa \langle \nabla \theta_m, \nabla \psi \rangle_\Omega + \langle u_m \cdot \nabla \theta_m, \psi \rangle_\Omega + \kappa \langle \Lambda \theta_m, \psi \rangle_{\partial\Omega} \Big) \eta(t)\, dt$$
$$= \int_0^T \Big(\langle f_2, \psi \rangle_\Omega + \kappa \langle \Lambda h_0, \psi \rangle_{\partial\Omega} \Big) \eta(t)\, dt + \langle \theta_m(0), \psi \rangle_\Omega \eta(0) \qquad (4.106)$$

for all $m \in \mathbb{N}$, all $\eta \in C_0^\infty([0,T[)$, and all $\psi \in \mathrm{span}\{\psi_1, \ldots, \psi_m\}$. We make use of (4.29), (4.66), (4.78) to obtain

$$
\lim_{m \to \infty} \int_0^T \langle u_m \cdot \nabla \theta_m, \psi \rangle_\Omega \eta(t)\, dt = - \lim_{m \to \infty} \int_0^T \langle \theta_m u_m, \nabla \psi \rangle_\Omega \eta(t)\, dt
$$

$$
= - \int_0^T \langle \theta u, \nabla \psi \rangle_\Omega \eta(t)\, dt \tag{4.107}
$$

$$
= \int_0^T \langle u \cdot \nabla \theta, \psi \rangle_\Omega \eta(t)\, dt
$$

for all $\eta \in C_0^\infty([0,T[)$ and all $\psi \in \mathrm{span}\{\psi_k; k \in \mathbb{N}\}$. Furthermore, by (4.78) there holds

$$
\lim_{m \to \infty} \int_0^T \langle \Lambda \theta_m, \psi \rangle_{\partial\Omega} \eta(t)\, dt = \int_0^T \langle \Lambda \theta, \psi \rangle_{\partial\Omega} \eta(t)\, dt \tag{4.108}
$$

for all $\eta \in C_0^\infty([0,T[)$ and all $\psi \in \mathrm{span}\{\psi_k; k \in \mathbb{N}\}$. Let $m \to \infty$ in (4.106), use $\theta_m(0) \to \theta_0$ in $L^2(\Omega)$, and (4.67), (4.107), (4.108) to obtain

$$
-\int_0^T \langle \theta, \psi \rangle_\Omega \eta'(t)\, dt + \int_0^T \left(\kappa \langle \nabla \theta, \nabla \psi \rangle_\Omega + \langle u \cdot \nabla \theta, \psi \rangle_\Omega + \kappa \langle \Lambda \theta, \psi \rangle_{\partial\Omega} \right) \eta(t)\, dt
$$

$$
= \int_0^T \left(\langle f_2, \psi \rangle_\Omega + \kappa \langle \Lambda h_0, \psi \rangle_{\partial\Omega} \right) \eta(t)\, dt + \langle \theta_0, \psi \rangle_\Omega \eta(0) \tag{4.109}
$$

for all $\eta \in C_0^\infty([0,T[)$ and all $\psi \in \mathrm{span}\{\psi_k; k \in \mathbb{N}\}$. By density, (4.109) is still valid for all $\psi \in H^1(\Omega)$ and all $\eta \in C_0^\infty([0,T[)$. This means

$$
-\int_0^T [\theta, \psi]_{H^1(\Omega)', H^1(\Omega)}\, \eta'(t)\, dt
$$

$$
= \int_0^T \left([\kappa \Delta \theta - D(u,\theta) - \kappa S(\Lambda(\theta - h_0)) + f_2, \psi]_{H^1(\Omega)'; H^1(\Omega)}\, \eta(t) \right) dt \tag{4.110}
$$

for all $\psi \in H^1(\Omega)$ and $\eta \in C_0^\infty(]0,T[)$. Consequently

$$
\theta_t - \kappa \Delta \theta + D(u,\theta) + \kappa S(\Lambda(\theta - h_0)) = f_2 \quad \text{in } L^{4/3}(0,T; H^1(\Omega)'). \tag{4.111}
$$

After a redefinition on a null set of $[0,T[$, we have that $\theta : [0,T[\to H^1(\Omega)'$ is strongly continuous. Moreover, by Lemma 2.32, there holds that $\theta : [0,T[\to L^2(\Omega)$ is weakly continuous. Consequently, $\theta(0) \in L^2(\Omega)$ is well defined. Integrating (4.111) yields

$$
-\int_0^T \langle \theta, \psi \rangle_\Omega \eta'(t)\, dt + \int_0^T \left(\kappa \langle \nabla \theta, \nabla \psi \rangle_\Omega + \langle u \cdot \nabla \theta, \psi \rangle_\Omega + \kappa \langle \Lambda \theta, \psi \rangle_{\partial\Omega} \right) \eta(t)\, dt
$$

$$
= \int_0^T \left(\langle f_2, \psi \rangle_\Omega + \kappa \langle \Lambda h_0, \psi \rangle_{\partial\Omega} \right) \eta(t)\, dt + \langle \theta(0), \psi \rangle_\Omega \eta(0) \tag{4.112}
$$

for all $\psi \in H^1(\Omega)$ and $\eta \in C_0^\infty([0,T[)$. We combine (4.109) and (4.112) to obtain $\theta(0) = \theta_0$. Therefore, conditions (ii), (iv) of Theorem 4.16 are satisfied. Altogether, by this theorem, it follows that (u, θ) is a weak solution of (4.1) with (4.2), (4.3).

Proof of statement (ii) of Theorem 4.18. Assume $n = 3$ and that additionally $g \in L^4(0,T; L^2(\Omega)^3)$ is satisfied. Our goal is to prove the validity of the energy inequalities (4.23), (4.24). Let us first prove that (4.24) is fulfilled. Consider a bounded Lipschitz domain V with $\overline{V} \subseteq \Omega$. By (4.78) there is a null set $N = N(V)$ such that

$$\theta_m(t) \underset{m \to \infty}{\to} \theta(t) \quad \text{strongly in } L^2(V)$$

for all $t \in [0,T[\backslash N$. Writing

$$\Omega = \bigcup_{k \in \mathbb{N}} \Omega_k$$

with bounded Lipschitz domains $\Omega_k \subseteq \Omega$, $k \in \mathbb{N}$, we obtain the existence of a null set $N \subseteq [0,T[$ such that

$$\theta_m(t) \underset{m \to \infty}{\to} \theta(t) \quad \text{strongly in } L^2(G)$$

holds for all bounded Lipschitz domains $G \subseteq \Omega$ and for all $t \in [0,T[\backslash N$. Therefore we obtain the 'global' weak convergence

$$\theta_m(t) \underset{m \to \infty}{\rightharpoonup} \theta(t) \quad \text{in } L^2(\Omega) \tag{4.113}$$

for all $t \in [0,T[\backslash N$. The weak lower semicontinuity of the norm applied to (4.67), (4.79), (4.113), the energy equalities (4.64), (4.65) and (4.67), (4.78) yield

$$\frac{1}{2}\|\theta(t)\|_2^2 + \kappa \int_0^t \|\nabla\theta\|_2^2 \, d\tau + \kappa \int_0^t \|\sqrt{\Lambda}\theta\|_{2,\partial\Omega}^2 \, d\tau$$

$$\leq \frac{1}{2} \liminf_{m \to \infty} \|\theta_m(t)\|_2^2 + \kappa \liminf_{m \to \infty} \int_0^t \|\nabla\theta_m\|_2^2 \, d\tau + \kappa \liminf_{m \to \infty} \int_0^t \|\sqrt{\Lambda}\theta_m\|_{2,\partial\Omega}^2 \, d\tau$$

$$\leq \liminf_{m \to \infty} \left(\frac{1}{2}\|\theta_m(t)\|_2^2 + \kappa \int_0^t \|\nabla\theta_m\|_2^2 \, d\tau + \kappa \int_0^t \|\sqrt{\Lambda}\theta_m\|_{2,\partial\Omega}^2 \, d\tau \right)$$

$$= \liminf_{m \to \infty} \left(\frac{1}{2}\|\theta_0\|_2^2 + \kappa \int_0^t \langle\Lambda h_0, \theta_m\rangle_{\partial\Omega} + \int_0^t \langle f_2, \theta_m\rangle_\Omega \, d\tau \right)$$

$$= \frac{1}{2}\|\theta_0\|_2^2 + \kappa \int_0^t \langle\Lambda h_0, \theta\rangle_{\partial\Omega} \, d\tau + \int_0^t \langle f_2, \theta\rangle_\Omega \, d\tau \tag{4.114}$$

for almost all $t \in [0,T[$.

The crucial point in the proof that the energy inequality (4.23) holds is to show

$$\lim_{m \to \infty} \int_0^t \langle\theta_m u_m, g\rangle_\Omega \, d\tau = \int_0^t \langle\theta u, g\rangle_\Omega \, d\tau \tag{4.115}$$

for all $t \in [0, T[$. We obtain from with Hölder's inequality and Lemma 4.12

$$
\begin{aligned}
\|\theta_m(t)u_m(t)\|_2 &\leq \|\theta_m(t)\|_4 \|u_m(t)\|_4 \\
&\leq c\|\theta_m(t)\|_2^{1/4}\|\theta_m(t)\|_{H^1(\Omega)}^{3/4}\|u_m(t)\|_2^{1/4}\|u_m(t)\|_{H^1(\Omega)^3}^{3/4} \qquad (4.116)\\
&\leq c\|\theta_m\|_{2,\infty;\Omega;T}^{1/4}\|u_m\|_{2,\infty;\Omega;T}^{1/4}\left(\|\theta_m(t)\|_{H^1(\Omega)}^{3/2} + \|u_m(t)\|_{H^1(\Omega)^3}^{3/2}\right)
\end{aligned}
$$

for a.a. $t \in]0, T[$ and all $m \in \mathbb{N}$ with a constant $c > 0$. Integrating (4.116) with respect to $t \in]0, T[$ and making use of (4.66), (4.67) implies that the sequence $(\theta_m u_m)_{m \in \mathbb{N}}$ is bounded in $L^{4/3}(0, T; L^2(\Omega)^3)$. Consequently, there exists a subsequence $(u_{m_l}, \theta_{m_l})_{l \in \mathbb{N}}$ and $\overline{\theta u} \in L^{4/3}(0, T; L^2(\Omega)^3)$ such that

$$
\theta_{m_l} u_{m_l} \underset{l \to \infty}{\rightharpoonup} \overline{\theta u} \quad \text{in } L^{4/3}(0, T; L^2(\Omega)^3). \qquad (4.117)
$$

Fix $v \in C_0^\infty(]0, T[\times \Omega)^3$. Choose a bounded Lipschitz domain G with $G \subseteq \Omega$ such that $\operatorname{supp}(v) \subseteq]0, T[\times G$. By (4.77), (4.78)

$$
\lim_{l \to \infty} \int_0^T \langle \theta_{m_l} u_{m_l}, v \rangle_\Omega \, d\tau = \int_0^T \langle \theta u, v \rangle_\Omega \, d\tau. \qquad (4.118)
$$

Combining (4.117), (4.118) yields

$$
\int_0^T \langle \theta u, v \rangle_\Omega \, d\tau = \int_0^T \langle \overline{\theta u}, v \rangle_\Omega \, d\tau \qquad (4.119)
$$

for all $v \in C_0^\infty(]0, T[\times \Omega)^3$. By the fundamental theorem of calculus of variations we obtain $(\theta u)(t) = (\overline{\theta u})(t)$ for almost all $t \in]0, T[$. Furthermore it follows that it is possible to choose $m_l = l, l \in \mathbb{N}$. Altogether

$$
\theta_m u_m \underset{m \to \infty}{\rightharpoonup} \theta u \quad \text{in } L^{4/3}(0, T; L^2(\Omega)^3). \qquad (4.120)
$$

Thus, the assumption $g \in L^4(0, T; L^2(\Omega)^3)$ in combination with (4.120) yields (4.115). From (4.115) it follows immediately

$$
\lim_{m \to \infty} \int_0^t \langle \theta_m g, u_m \rangle_\Omega \, d\tau = \int_0^t \langle \theta g, u \rangle_\Omega \, d\tau \qquad (4.121)
$$

for all $t \in [0, T[$.

Using (4.77) we show analogously to (4.113) that

$$
u_m(t) \underset{m \to \infty}{\rightharpoonup} u(t) \quad \text{in } L^2(\Omega) \qquad (4.122)
$$

is satisfied for almost all $t \in [0, T[$. We obtain with (4.66), (4.121), (4.122)

$$\frac{1}{2}\|u(t)\|_2^2 + \nu \int_0^t \|\nabla u\|_2^2 \, d\tau \leq \frac{1}{2} \liminf_{m \to \infty} \|u_m(t)\|_2^2 + \nu \liminf_{m \to \infty} \int_0^t \|\nabla u_m\|_2^2 \, d\tau$$

$$\leq \liminf_{m \to \infty} \left(\frac{1}{2}\|u_m(t)\|_2^2 + \nu \int_0^t \|\nabla u_m\|_2^2 \, d\tau \right)$$

$$= \liminf_{m \to \infty} \left(\frac{1}{2}\|u_0\|_2^2 + \int_0^t \langle f_1, u_m \rangle_\Omega + \beta \int_0^t \langle \theta_m g, u_m \rangle_\Omega \, d\tau \right)$$

$$= \frac{1}{2}\|u_0\|_2^2 + \int_0^t \langle f_1, u \rangle_\Omega + \beta \int_0^t \langle \theta g, u \rangle_\Omega \, d\tau$$

for almost all $t \in [0, T[$. Altogether, the energy inequalities (4.23), (4.24) are satisfied for almost all $t \in [0, T[$. The proof of Theorem 4.18 is complete. $\qquad \square$

4.5 Further properties of weak solutions.

In this subsection we will present properties of weak solutions of the Boussinesq equations (4.1) with boundary conditions (4.2), (4.3) which will be used in the sequel. Especially, uniqueness will be proven in the two-dimensional case. Let us start with the following Lemma.

Lemma 4.21. *Let $\Omega \subseteq \mathbb{R}^n$, $n \in \{2, 3\}$, be a domain satisfying the uniform Lipschitz condition with parameters (α, β, N, L), let $0 < T < \infty$, let $g \in L^\infty(]0, T[\times\Omega)^n$, $h_0 \in L^2(0, T; L^2(\partial\Omega))$, and let $\Lambda \in L^\infty(\partial\Omega)$ with $\Lambda(x) \geq 0$ for a.a. $x \in \partial\Omega$. Further assume $f_1 \in L^2(0, T; L^2(\Omega)^n)$, $f_2 \in L^2(0, T; L^2(\Omega))$, assume $u_0 \in L^2_\sigma(\Omega)$, $\theta_0 \in L^2(\Omega)$, and assume that β, ν, κ are positive, real constants. Consider*

$$u \in L^\infty(0, T; L^2_\sigma(\Omega)) \cap L^2(0, T; W^{1,2}_{0,\sigma}(\Omega)),$$
$$\theta \in L^\infty(0, T; L^2(\Omega)) \cap L^2(0, T; H^1(\Omega)) \tag{4.123}$$

satisfying

$$\frac{1}{2}\|u(t)\|_2^2 + \nu \int_0^t \|\nabla u\|_2^2 \, d\tau$$
$$\leq \frac{1}{2}\|u_0\|_2^2 + \beta \int_0^t \langle \theta g, u \rangle_\Omega \, d\tau + \int_0^t \langle f_1, u \rangle_\Omega \, d\tau, \tag{4.124}$$

$$\frac{1}{2}\|\theta(t)\|_2^2 + \kappa \int_0^t \|\nabla\theta\|_2^2 \, d\tau + \kappa \int_0^t \|\sqrt{\Lambda}\,\theta\|_{2,\partial\Omega}^2 \, d\tau$$
$$\leq \frac{1}{2}\|\theta_0\|_2^2 + \kappa \int_0^t \langle \Lambda h_0, \theta \rangle_{\partial\Omega} \, d\tau + \int_0^t \langle f_2, \theta \rangle_\Omega \, d\tau \tag{4.125}$$

for almost all $t \in [0, T[$. Then the following estimates are satisfied:

$$\frac{1}{4}\|\theta\|_{2,\infty;\Omega;T}^2 + \frac{\kappa}{2}\|\nabla\theta\|_{2,2;\Omega;T}^2 + \kappa\|\sqrt{\Lambda}\theta\|_{2,2;\partial\Omega;T}^2$$
$$\leq \|\theta_0\|_2^2 + c\kappa^2\|\Lambda\|_{\infty,\partial\Omega}\|h_0\|_{2,1;\partial\Omega;T}^2 + c\kappa\|\Lambda\|_{\infty,\partial\Omega}\|h_0\|_{2,2;\partial\Omega;T}^2 + 4\|f_2\|_{2,1;\Omega;T}^2 \tag{4.126}$$

with a constant $c = c(\beta, N, L) > 0$ and

$$\frac{1}{4}\|u\|^2_{2,\infty;\Omega;T} + \nu\|\nabla u\|^2_{2,2;\Omega;T} \leq \|u_0\|^2_2 + 4\beta^2\|g\|^2_{L^\infty(]0,T[\times\Omega)^n}\|\theta\|^2_{2,1;\Omega;T} + 4\|f_1\|^2_{2,1;\Omega;T}.$$

(4.127)

Remark. Let $h : \mathbb{R}^{n-1} \to \mathbb{R}$ be a Lipschitz continuous function. Assume that the domain Ω is given by

$$\Omega = \{ (x', x_n) \in \mathbb{R}^n; x_n > h(x'), x' \in \mathbb{R}^{n-1} \}.$$

(4.128)

Then the constant c in (4.126) depends only on $\mathrm{Lip}(h)$.

Proof. We obtain with the trace estimate (4.17), Hölder's and Young's inequality

$$\kappa\left|\int_0^t \langle \Lambda h_0, \theta \rangle_{\partial\Omega}\, dt\right| \leq \kappa\|\Lambda\|_{\infty,\partial\Omega} \int_0^T \|h_0\|_{2,\partial\Omega}\|\theta\|_{2,\partial\Omega}\, d\tau$$

$$\leq c\kappa\|\Lambda\|_{\infty,\partial\Omega} \int_0^T \|h_0\|_{2,\partial\Omega}\big(\|\theta\|_2 + \|\nabla\theta\|_2\big)\, d\tau$$

$$\leq c\kappa\|\Lambda\|_{\infty,\partial\Omega}\|\theta\|_{2,\infty;\Omega;T}\|h_0\|_{2,1;\partial\Omega;T}$$
$$+ c\kappa\|\Lambda\|_{\infty,\partial\Omega}\|\nabla\theta\|_{2,2;\Omega;T}\|h_0\|_{2,2;\partial\Omega;T}$$

$$\leq \frac{1}{8}\|\theta\|^2_{2,\infty;\Omega;T} + 2\big(c\kappa\|\Lambda\|_{\infty,\partial\Omega}\|h_0\|_{2,1;\partial\Omega;T}\big)^2$$

$$+ \frac{\kappa}{2}\|\nabla\theta\|^2_{2,2;\Omega;T} + \frac{\big(c\kappa\|\Lambda\|_{\infty,\partial\Omega}\|h_0\|_{2,2;\partial\Omega;T}\big)^2}{2\kappa}$$

for all $t \in [0, T[$ with a constant $c = c(\beta, N, L) > 0$. Consequently, we get from (4.125) the estimate

$$\frac{1}{2}\|\theta(t)\|^2_2 + \kappa\int_0^t \|\nabla\theta\|^2_2\, d\tau + \kappa\int_0^t \|\sqrt{\Lambda}\theta\|^2_{2,\partial\Omega}\, d\tau$$

$$\leq \frac{1}{2}\|\theta_0\|^2_2 + 2\cdot\frac{1}{8}\|\theta\|^2_{2,\infty;\Omega;T} + 2\|f_2\|^2_{2,1;\Omega;T} + \frac{\kappa}{2}\|\nabla\theta\|^2_{2,2;\Omega;T}$$

(4.129)

$$+ c\big(\kappa\|\Lambda\|_{\infty,\partial\Omega}\|h_0\|_{2,1;\partial\Omega;T}\big)^2 + c\kappa\big(\|\Lambda\|_{\infty,\partial\Omega}\|h_0\|_{2,2;\partial\Omega;T}\big)^2$$

for a.a. $t \in [0, T[$ with a constant $c = c(\beta, N, L) > 0$. Taking the supremum in (4.129) with respect to $t \in]0, T[$ implies (4.126). There holds

$$\beta\left|\int_0^t \langle \theta g, u \rangle_\Omega\, d\tau\right| \leq \beta\|g\|_{L^\infty(]0,T[\times\Omega)^n} \int_0^T \|\theta\|_2\|u\|_2\, d\tau$$

$$\leq \beta\|g\|_{L^\infty(]0,T[\times\Omega)^n}\|u\|_{2,\infty;\Omega;T}\|\theta\|_{2,1;\Omega;T}$$

$$\leq \frac{1}{8}\|u\|^2_{2,\infty;\Omega;T} + 2\big(\beta\|g\|_{L^\infty(]0,T[\times\Omega)^n}\|\theta\|_{2,1;\Omega;T}\big)^2$$

for all $t \in [0, T[$. Therefore (4.124) can be estimated to obtain

$$
\begin{aligned}
&\frac{1}{2}\|u(t)\|_2^2 + \nu \int_0^t \|\nabla u\|_2^2 \, d\tau \\
&\leq \frac{1}{2}\|u_0\|_2^2 + 2 \cdot \frac{1}{8}\|u\|_{2,\infty;\Omega;T}^2 + 2\beta^2 \|g\|_{L^\infty(]0,T[\times\Omega)^n}^2 \|\theta\|_{2,1;\Omega;T}^2 + 2\|f_1\|_{2,1;\Omega;T}^2
\end{aligned}
\tag{4.130}
$$

for a.a. $t \in [0, T[$. By taking the supremum in (4.130) over $t \in]0, T[$ we obtain (4.127).

In the case that Ω is given by (4.128), it follows from (4.19) that the constant c in (4.126) depends only on L. $\qquad\square$

The auxiliary lemma below is needed for the proof of Theorem 4.23.

Lemma 4.22. (i) *Let* $\Omega \subseteq \mathbb{R}^2$ *be an arbitrary domain, let* $0 < T < \infty$, *and let* $u \in L^\infty(0, T; L_\sigma^2(\Omega)) \cap L^2(0, T; W_{0,\sigma}^{1,2}(\Omega))$. *Then* $B(u, u) \in L^2(0, T; W_{0,\sigma}^{1,2}(\Omega)')$ *and the estimate*

$$
\|B(u, u)\|_{L^2(0,T;W_{0,\sigma}^{1,2}(\Omega)')} \leq c\|u\|_{2,\infty;\Omega;T}\|\nabla u\|_{2,2;\Omega;T}
$$

is satisfied with an absolute constant $c > 0$ *(independent of* Ω*).*

(ii) *Let* $\Omega \subseteq \mathbb{R}^2$ *be a domain satisfying the uniform Lipschitz condition, and let* $0 < T < \infty$. *Consider*

$$
\begin{aligned}
&u \in L^\infty(0, T; L_\sigma^2(\Omega)) \cap L^2(0, T; W_{0,\sigma}^{1,2}(\Omega)), \\
&\theta \in L^\infty(0, T; L^2(\Omega)) \cap L^2(0, T; H^1(\Omega)).
\end{aligned}
$$

Then the estimate

$$
\|D(u, \theta)\|_{L^2(0,T;H^1(\Omega)')} \leq c\|u\|_{2,\infty;\Omega;T}^{1/2}\|\theta\|_{2,\infty;\Omega;T}^{1/2}\left(\|\nabla u\|_{2,2;\Omega;T} + \|\theta\|_{L^2(0,T;H^1(\Omega))}\right)
$$

holds with a constant $c = c(\Omega) > 0$.

Proof. (i) From (4.28), (4.38) we get, analogously to the proof of Lemma 4.14, that

$$
\|B(u, u)(t)\|_{W_{0,\sigma}^{1,2}(\Omega)'} \leq c\|u(t)\|_2\|\nabla u(t)\|_2
$$

is satisfied for almost all $t \in [0, T[$ with a constant $c > 0$ independent of Ω. Integrating the estimate above completes the proof.

(ii) From (4.29), (4.40) we get the estimate

$$
\|D(u, \theta)(t)\|_{H^1(\Omega)'} \leq c(\Omega)\|u(t)\|_2^{1/2}\|\nabla u(t)\|_2^{1/2}\|\theta(t)\|_2^{1/2}\|\theta(t)\|_{H^1(\Omega)}^{1/2}
$$

for almost all $t \in [0, T[$ with a constant $c = c(\Omega)$. Integrate the estimate above to finish the proof.

$\qquad\square$

Now we have all ingredients at hand for the following important theorem for weak solutions in the two-dimensional case.

Theorem 4.23. *Let $\Omega \subseteq \mathbb{R}^2$ be a domain satisfying the uniform Lipschitz condition, let $0 < T < \infty$, let $g \in L^\infty(]0, T[\times\Omega)^2$, let $h_0 \in L^2(0, T; L^2(\partial\Omega))$, $\Lambda \in L^\infty(\partial\Omega)$ with $\Lambda(x) \geq 0$ for a.a. $x \in \partial\Omega$, and let β, ν, κ be positive, real constants. Further assume $u_0 \in L^2_\sigma(\Omega)$, $\theta_0 \in L^2(\Omega)$ and $f_1 \in L^2(0, T; L^2(\Omega)^2)$, $f_2 \in L^2(0, T; L^2(\Omega))$.*

(i) *There exists exactly one weak solution*

$$u \in L^\infty(0, T; L^2_\sigma(\Omega)) \cap L^2(0, T; W^{1,2}_{0,\sigma}(\Omega)),$$
$$\theta \in L^\infty(0, T; L^2(\Omega)) \cap L^2(0, T; H^1(\Omega))$$

of the Boussinesq equations (4.1) with boundary conditions (4.2), (4.3).

(ii) *Let (u, θ) be the unique weak solution of (4.1) with (4.2), (4.3). Then*

$$u_t \in L^2(0, T; W^{1,2}_{0,\sigma}(\Omega)'), \quad \theta_t \in L^2(0, T; H^1(\Omega)').$$

After a redefinition on a null set of $[0, T[$ we have that $u : [0, T[\to L^2_\sigma(\Omega)$ is (strongly) continuous and $\theta : [0, T[\to L^2(\Omega)$ is (strongly) continuous. Furthermore the energy equalities

$$
\begin{aligned}
\frac{1}{2}\|u(t)\|_2^2 + \nu &\int_0^t \|\nabla u\|_2^2 \, d\tau \\
&= \frac{1}{2}\|u_0\|_2^2 + \beta \int_0^t \langle \theta g, u \rangle_\Omega \, d\tau + \int_0^t \langle f_1, u \rangle_\Omega \, d\tau,
\end{aligned}
\tag{4.131}
$$

$$
\begin{aligned}
\frac{1}{2}\|\theta(t)\|_2^2 + \kappa &\int_0^t \|\nabla\theta\|_2^2 \, d\tau + \kappa \int_0^t \|\sqrt{\Lambda}\,\theta\|_{2,\partial\Omega}^2 \, d\tau \\
&= \frac{1}{2}\|\theta_0\|_2^2 + \kappa \int_0^t \langle \Lambda h_0, \theta \rangle_{\partial\Omega} \, d\tau + \int_0^t \langle f_2, \theta \rangle_\Omega \, d\tau
\end{aligned}
\tag{4.132}
$$

are satisfied for all $t \in [0, T[$.

Proof. Let (u, θ) be a weak solution of (4.1) with (4.2), (4.3). From Lemma 4.22 it follows $B(u, u) \in L^2(0, T; W^{1,2}_{0,\sigma}(\Omega)')$ and $D(u, \theta) \in L^2(0, T; H^1(\Omega)')$. Consequently, (see Lemma 4.15 and Theorem 4.16) we get $u_t \in L^2(0, T; W^{1,2}_{0,\sigma}(\Omega)')$, $\theta_t \in L^2(0, T; H^1(\Omega)')$ and

$$u_t - \nu\Delta u = f_1 + \beta\theta g - B(u, u) \qquad \text{in } L^2(0, T; W^{1,2}_{0,\sigma}(\Omega)'), \tag{4.133}$$
$$\theta_t - \kappa\Delta\theta + \kappa S(\Lambda\theta) = f_2 + \kappa S(\Lambda h_0) - D(u, \theta) \quad \text{in } L^2(0, T; H^1(\Omega)'). \tag{4.134}$$

Applying Lemma 2.30 to the imbedding scheme

$$W^{1,2}_{0,\sigma}(\Omega) \hookrightarrow L^2_\sigma(\Omega) \cong L^2_\sigma(\Omega)' \hookrightarrow W^{1,2}_{0,\sigma}(\Omega)' \tag{4.135}$$

implies that, after a redefinition on a null set, $u : [0, T[\to L_\sigma^2(\Omega)$ is (strongly) continuous. The same Lemma applied to the imbedding scheme

$$H^1(\Omega) \hookrightarrow L^2(\Omega) \cong L^2(\Omega)' \hookrightarrow H^1(\Omega)' \tag{4.136}$$

implies that, after a redefinition on a null set, $\theta : [0, T[\to L^2(\Omega)$ is (strongly) continuous. Testing (4.133) with $u \in L^2(0, T; W_{0,\sigma}^{1,2}(\Omega))$ and (4.134) with $\theta \in L^2(0, T; H^1(\Omega))$, using Lemma 4.11 and Lemma 2.30 (with (4.135), (4.136)) we get

$$\frac{1}{2}\frac{d}{d\tau}\|u(\tau)\|_2^2\Big|_{\tau=t} + \nu\|\nabla u(t)\|_2^2 = \langle f_1(t), u(t)\rangle_\Omega + \beta\langle\theta(t)g(t), u(t)\rangle_\Omega, \tag{4.137}$$

$$\frac{1}{2}\frac{d}{d\tau}\|\theta(\tau)\|_2^2\Big|_{\tau=t} + \kappa\|\nabla\theta(t)\|_2^2 + \kappa\|\sqrt{\Lambda}\theta(t)\|_{2,\partial\Omega}^2 = \langle f_2(t), \theta(t)\rangle_\Omega + \langle\Lambda h_0(t), \theta(t)\rangle_{\partial\Omega} \tag{4.138}$$

for almost all $t \in [0, T[$. Integrating (4.137), (4.138) over $[0, t[$ we obtain the energy equalities (4.131), (4.132) for all $t \in [0, T[$. To finish the proof it remains to prove the uniqueness of weak solutions of (4.1) with (4.2), (4.3).

Proof of uniqueness. Assume that (u_1, θ_1) and (u_2, θ_2) are two weak solutions of (4.1) with (4.2), (4.3). Define $u := u_1 - u_2$ and $\theta := \theta_1 - \theta_2$. We subtract equation (4.133) fulfilled by (u_2, θ_2) from (4.133) fulfilled by (u_1, θ_1) and do the same with (4.134). We get

$$u_t - \nu\Delta u = \beta\theta g - B(u_1, u_1) + B(u_2, u_2) \quad \text{in } L^2(0, T; W_{0,\sigma}^{1,2}(\Omega)'), \tag{4.139}$$

$$\theta_t - \kappa\Delta\theta + \kappa S(\Lambda\theta) = -D(u_1, \theta_1) + D(u_2, \theta_2) \quad \text{in } L^2(0, T; H^1(\Omega)'). \tag{4.140}$$

A short calculation shows that the identities

$$\begin{aligned} -b(u_1, u_1, u) + b(u_2, u_2, u) &= -b(u, u_2, u), \\ -d(u_1, \theta_1, \theta) + d(u_2, \theta_2, \theta) &= -d(u, \theta_2, \theta) \end{aligned} \tag{4.141}$$

are satisfied. Testing (4.139) with $u \in L^2(0, T; W_{0,\sigma}^{1,2}(\Omega)')$ and (4.140) with $\theta \in L^2(0, T; H^1(\Omega)')$, using Lemma 2.30 and (4.141) we obtain

$$\frac{1}{2}\frac{d}{d\tau}\|u(\tau)\|_2^2\Big|_{\tau=t} + \nu\|\nabla u(t)\|_2^2 = \beta\langle\theta(t)g(t), u(t)\rangle_\Omega - b(u(t), u_2(t), u(t)), \tag{4.142}$$

$$\frac{1}{2}\frac{d}{d\tau}\|\theta(\tau)\|_2^2\Big|_{\tau=t} + \kappa\|\nabla\theta(t)\|_2^2 + \kappa\|\sqrt{\Lambda}\theta(t)\|_{2,\partial\Omega}^2 = -d(u(t), \theta_2(t), \theta(t)) \tag{4.143}$$

for a.a. $t \in [0, T[$. The next step is to estimate the right hand sides of (4.142), (4.143). We make use of (4.28), (4.31), Hölder's inequality and Young's inequality to obtain

$$\begin{aligned} |b(u, u_2, u)| + \beta|\langle\theta g, u\rangle_\Omega| &= |b(u, u, u_2)| + \beta|\langle\theta g, u\rangle_\Omega| \\ &\leq c\|u\|_2^{1/2}\|\nabla u\|_2^{3/2}\|u_2\|_4 + \beta\|g\|_{L^\infty(]0,T[\times\Omega)^n}\|\theta\|_2\|u\|_2 \\ &\leq \frac{\nu}{2}\|\nabla u\|_2^2 + c\|u\|_2^2\|u_2\|_4^4 + \frac{\beta\|g\|_{L^\infty(]0,T[\times\Omega)^n}}{2}(\|u\|_2^2 + \|\theta\|_2^2). \end{aligned} \tag{4.144}$$

for a.a. $t \in [0, T[$ with a constant $c > 0$ independent of t. Hence

$$\frac{1}{2}\frac{d}{d\tau}\|u(\tau)\|_2^2\Big|_{\tau=t} + \frac{\nu}{2}\|\nabla u(t)\|_2^2 \leq c\|u_2(t)\|_4^4\|u(t)\|_2^2 + \frac{\beta\|g\|_{L^\infty(]0,T[\times\Omega)^n}}{2}\left(\|u(t)\|_2^2 + \|\theta(t)\|_2^2\right)$$

$$(4.145)$$

for a.a. $t \in [0, T[$ with a constant $c > 0$ independent of t. We employ (4.29), (4.33) to obtain

$$|d(u, \theta_2, \theta)| = |d(u, \theta, \theta_2)|$$
$$\leq c\|u\|_2^{1/2}\|\nabla u\|_2^{1/2}\|\nabla\theta\|_2\|\theta_2\|_4 \qquad (4.146)$$
$$\leq \frac{\nu}{2}\|\nabla u\|_2^2 + \kappa\|\nabla\theta\|_2^2 + c\|u\|_2^2\|\theta_2\|_4^4$$

for a.a. $t \in [0, T[$ with a constant $c > 0$ independent of $t \in [0, T[$. Hence (4.143) can be written as

$$\frac{1}{2}\frac{d}{d\tau}\|\theta(\tau)\|_2^2\Big|_{\tau=t} + \kappa\|\sqrt{\Lambda}\theta(t)\|_{2,\partial\Omega}^2 \leq \frac{\nu}{2}\|\nabla u(t)\|_2^2 + c\|u(t)\|_2^2\|\theta_2(t)\|_4^4 \qquad (4.147)$$

for a.a. $t \in [0, T[$ with a constant $c > 0$ independent of t. Adding (4.145), (4.147) yields

$$\frac{1}{2}\frac{d}{d\tau}\left(\|u(\tau)\|_2^2 + \|\theta(\tau)\|_2^2\right)\Big|_{\tau=t}$$
$$\leq c\left(\|u_2(t)\|_4^4 + \|\theta_2(t)\|_4^4 + \beta\|g\|_{L^\infty(]0,T[\times\Omega)^n}\right)\left(\|u(t)\|_2^2 + \|\theta(t)\|_2^2\right)$$

$$(4.148)$$

for almost all $t \in [0, T[$ with a constant $c > 0$ independent of t. Gronwall's inequality and $u(0) = 0$, $\theta(0) = 0$ imply $u(t) = 0$ and $\theta(t) = 0$ for almost all $t \in [0, T[$. $\qquad\square$

5 Strong solutions of the Boussinesq equations in general domains

Let $\Omega \subseteq \mathbb{R}^3$ be a domain, and let $0 < T \leq \infty$. Throughout this chapter we consider the Boussinesq system with Dirichlet boundary conditions

$$
\begin{aligned}
u_t - \Delta u + (u \cdot \nabla)u + \nabla p = \theta g + f_1 & \quad \text{in }]0, T[\times\Omega\,, \\
\operatorname{div} u = 0 & \quad \text{in }]0, T[\times\Omega\,, \\
\theta_t - \Delta\theta + (u \cdot \nabla)\theta = f_2 & \quad \text{in }]0, T[\times\Omega\,, \\
u = 0 & \quad \text{on }]0, T[\times\partial\Omega\,, \\
\theta = 0 & \quad \text{on }]0, T[\times\partial\Omega\,, \\
u(0) = u_0 & \quad \text{in } \Omega\,, \\
\theta(0) = \theta_0 & \quad \text{in } \Omega.
\end{aligned}
\tag{5.1}
$$

As already motivated in Section 1.2, we have defined a strong solution of (5.1) in $[0, T[\times\Omega$ where $\Omega \subseteq \mathbb{R}^3$ is a general domain, which has not necessarily any boundary regularity, as a weak solution satisfying the additional condition $u \in L^8(0, T; L^4(\Omega)^3)$. (For a precise definition we refer to Definition 5.1.)

As already outlined in the introduction, this chapter deals with existence, uniqueness and regularity of strong solutions of the Boussinesq system (5.1). Furthermore we will investigate regularity properties of weak solutions of (5.1) satisfying the strong energy inequalities (5.144), (5.145).

Before we return to rigorous mathematics, we motivate our *functional analytic approach* to construct strong solutions of the Boussinesq equations. The guiding idea of this approach is to consider the partial differential equations (5.1) as an ordinary differential equation in an infinite dimensional Banach space. Formally applying the Helmholtz projection $P = P_\Omega$ to the first equation in (5.1) and using $P(\nabla p) = 0$, $A = -P\Delta$ yields

$$
u_t + Au + P(u \cdot \nabla u) = Pf_1 + P\theta g. \tag{5.2}
$$

Taking also the initial values into account the system (5.2) can be (formally) reformulated as

$$
\begin{aligned}
u_t + Au = Pf_1 + P(\theta g) - P(u \cdot \nabla u)\,, & \quad u(0) = u_0\,, \\
\theta_t + (-\Delta)\theta = f_2 - u \cdot \nabla\theta\,, & \quad \theta(0) = \theta_0.
\end{aligned}
\tag{5.3}
$$

Formally applying the variation of constants formula to (5.3) it follows that (u, θ) solves the system of coupled integral equations

$$u(t) = e^{-tA}u_0 + \int_0^t e^{-(t-\tau)A}\big(Pf_1 + P(\theta g) - P(u \cdot \nabla u)\big) \, d\tau \,,$$

$$\theta(t) = e^{t\Delta}\theta_0 + \int_0^t e^{(t-\tau)\Delta}\big(f_2 - u \cdot \nabla\theta\big) \, d\tau$$

(5.4)

for a.a. $t \in [0, T[$.

In general a weak solution (u, θ) of (5.1) does not have enough regularity so that every term in (5.4) is well defined. Therefore, we rewrite the system (5.4) in the form

$$u(t) = e^{-tA}u_0 + \int_0^t e^{-(t-\tau)A}Pf_1(\tau) \, d\tau$$

$$- A^{1/2}\int_0^t e^{-(t-\tau)A}A^{-1/2}P\text{div}\big(u(\tau) \otimes u(\tau)\big) \, d\tau + \int_0^t e^{-(t-\tau)A}P\big(\theta(\tau)g(\tau)\big) \, d\tau \,,$$

$$\theta(t) = e^{t\Delta}\theta_0 + \int_0^t e^{(t-\tau)\Delta}f_2(\tau) \, d\tau - (-\Delta)^{1/2}\int_0^t e^{(t-\tau)\Delta}(-\Delta)^{-1/2}\text{div}\big(\theta(\tau)u(\tau)\big) \, d\tau.$$

using the operators $A^{-1/2}P\text{div}$ and $(-\Delta)^{-1/2}\text{div}$ (see Lemma 2.39).

In the following section we give a precise definition of weak and strong solutions of (5.1).

5.1 Weak solutions and strong solutions

Multiplying the Boussinesq system (5.1) with test functions w, ϕ, using integration by parts we obtain the following definition of weak solutions. In the following definition $T = \infty$ is allowed.

Definition 5.1. Let $\Omega \subseteq \mathbb{R}^3$ be a general domain, let $u_0 \in L^2_\sigma(\Omega), \theta_0 \in L^2(\Omega)$. Further assume $0 < T \leq \infty, g \in L^{8/5}_{\text{loc}}([0, T[; L^4(\Omega)^3)$ and $f_1 \in L^1_{\text{loc}}([0, T[; L^2(\Omega)^3)$, $f_2 \in L^1_{\text{loc}}([0, T[; L^2(\Omega))$. A pair

$$u \in L^\infty_{\text{loc}}([0, T[; L^2_\sigma(\Omega)) \cap L^2_{\text{loc}}([0, T[; W^{1,2}_{0,\sigma}(\Omega)) \,,$$

$$\theta \in L^\infty_{\text{loc}}([0, T[; L^2(\Omega)) \cap L^2_{\text{loc}}([0, T[; H^1_0(\Omega)) \,,$$

(5.5)

is called a weak solution of the Boussinesq system (5.1) if there holds

$$-\int_0^T \langle u, w_t \rangle_\Omega \, dt + \int_0^T \langle \nabla u, \nabla w \rangle_\Omega \, dt + \int_0^T \langle u \cdot \nabla u, w \rangle_\Omega \, dt$$

$$= \int_0^T \langle f_1, w \rangle_\Omega \, dt + \int_0^T \langle \theta g, w \rangle_\Omega \, dt + \langle u_0, w \rangle_\Omega$$

(5.6)

for all $w \in C_0^\infty([0,T[; C_{0,\sigma}^\infty(\Omega))$ and

$$
\begin{aligned}
-\int_0^T \langle \theta, \phi_t \rangle_\Omega \, dt &+ \int_0^T \langle \nabla\theta, \nabla\phi \rangle_\Omega \, dt + \int_0^T \langle u \cdot \nabla\theta, \phi \rangle_\Omega \, dt \\
&= \int_0^T \langle f_2, \phi \rangle_\Omega \, dt + \langle \theta_0, \phi(0) \rangle_\Omega
\end{aligned}
\tag{5.7}
$$

for all $\phi \in C_0^\infty([0,T[; C_0^\infty(\Omega))$.

To proceed we formulate integral equations which characterize weak solutions of the Boussinesq system (5.1).

Theorem 5.2. *Let $\Omega \subseteq \mathbb{R}^3$ be a general domain, and let $u_0 \in L_\sigma^2(\Omega)$, $\theta_0 \in L^2(\Omega)$. Further assume $0 < T \leq \infty$, $g \in L_{loc}^{8/5}([0,T[; L^4(\Omega)^3)$ and $f_1 \in L_{loc}^1([0,T[; L^2(\Omega)^3)$, $f_2 \in L_{loc}^1([0,T[; L^2(\Omega))$.*

(i) Then (u, θ) satisfying (5.5) is a weak solution of (5.1) if and only if the integral equations

$$
\begin{aligned}
u(t) = e^{-tA} u_0 &+ \int_0^t e^{-(t-\tau)A} P f_1(\tau) \, d\tau + \int_0^t e^{-(t-\tau)A} P\big(\theta(\tau)g(\tau)\big) \, d\tau \\
&- A^{1/2} \int_0^t e^{-(t-\tau)A} A^{-1/2} P \mathrm{div}\big(u(\tau) \otimes u(\tau)\big) \, d\tau,
\end{aligned}
\tag{5.8}
$$

$$
\begin{aligned}
\theta(t) = e^{t\Delta} \theta_0 &+ \int_0^t e^{(t-\tau)\Delta} f_2(\tau) \, d\tau \\
&- (-\Delta)^{1/2} \int_0^t e^{(t-\tau)\Delta} (-\Delta)^{-1/2} \mathrm{div}\big(\theta(\tau)u(\tau)\big) \, d\tau
\end{aligned}
\tag{5.9}
$$

are satisfied for almost all $t \in [0,T[$.

(ii) Let (u, θ) be a weak solution of (5.1). Then, after a redefinition on a null set of $[0,T[$, we have that $u : [0,T[\to L_\sigma^2(\Omega)$ and $\theta : [0,T[\to L^2(\Omega)$ are weakly continuous. This means, by definition, that the functions

$$
\begin{aligned}
t &\mapsto \langle u(t), w \rangle_\Omega, \quad t \in [0,T[, \\
t &\mapsto \langle \theta(t), \phi \rangle_\Omega, \quad t \in [0,T[
\end{aligned}
$$

are real-valued continuous functions for all $w \in L_\sigma^2(\Omega)$ and all $\phi \in L^2(\Omega)$. Especially $u(0) = u_0$ in $L_\sigma^2(\Omega)$ and $\theta(0) = \theta_0$ in $L^2(\Omega)$.

Proof. (i) The representation formula (5.8) follows directly from [So01, Chapter IV, Section 2.4]. To prove (5.9) we replace $-A$ by Δ and use the same argumentation as in the proof of (5.8).

(ii) From [So01, Chapter IV, Lemma 2.4.2] we get, after a redefinition on a null set of $[0,T[$, that $u : [0,T[\to L_\sigma^2(\Omega)$ is weakly continuous. Furthermore, that

$\theta : [0, T[\to L^2(\Omega)$ is, after a redefinition on a null set of $[0, T[$, weakly continuous follows from an analogous version of [So01, Chapter IV, Lemma 2.4.2] with A replaced by $-\Delta$.

$\square$

We want to construct a weak solution (u, θ) of (5.1) which additionally satisfies the Serrin condition $u \in L^8(0, T; L^4(\Omega)^3)$. For this reason we introduce the following notation:

Definition 5.3. Let $\Omega \subseteq \mathbb{R}^3$ be a general domain, let $u_0 \in L^2_\sigma(\Omega), \theta_0 \in L^2(\Omega)$. Further assume $0 < T \leq \infty, g \in L^{8/3}(0, T; L^4(\Omega)^3)$, and $f_1 \in L^{4/3}(0, T; L^2(\Omega)^3)$, $f_2 \in L^1(0, T; L^2(\Omega))$. We say that (u, θ) is a *strong solution* of (5.1) if (u, θ) is a weak solution of (5.1) and if the condition $u \in L^8(0, T; L^4(\Omega)^3)$ is satisfied.

5.2 Construction of strong solutions in general domains

The goal of this section is to give necessary and sufficient conditions for the existence of strong solutions of (5.1).

To construct a strong solution we will proceed as follows.

Step 1. Prove the existence of $u \in L^8(0, T; L^4_\sigma(\Omega))$ and $\theta \in L^{8/3}(0, T; L^4(\Omega))$ which satisfy the equations (5.8), (5.9). Define

$$E_1(t) := e^{-tA} u_0 + \int_0^t e^{-(t-\tau)A} P f_1(\tau) \, d\tau \, ,$$
$$E_2(t) := e^{t\Delta} \theta_0 + \int_0^t e^{(t-\tau)\Delta} f_2(\tau) \, d\tau \tag{5.10}$$

for almost all $t \in [0, T[$. The main idea to construct (u, θ) as in the first step is to rewrite the integral equations (5.8), (5.9) in the form

$$u = E_1 + \mathcal{F}_1(u, u) + \mathcal{L}\theta \, ,$$
$$\theta = E_2 + \mathcal{F}_2(u, \theta) \, , \tag{5.11}$$

where $\mathcal{F}_1, \mathcal{F}_2, \mathcal{L}$ are defined as in Lemma 5.5 below, and to obtain (u, θ) as a fixed point of this system. This fixed point argument, which is based on Banach's fixed point theorem, will be formulated in Lemma 5.4.

Step 2. Show that (u, θ) satisfy (5.5).

The next lemma gives a sufficient criterion how to solve the system of equations (5.16) with the help of Banach's fixed point theorem.

Lemma 5.4. *Let X, Y be Banach spaces, let $\mathcal{F}_1 : X \times X \to X$ and $\mathcal{F}_2 : X \times Y \to Y$ be bilinear maps, and let $\mathcal{L} : Y \to X$ be a linear map. Assume that $\alpha_1, \alpha_2, \alpha_3 > 0$ are constants such that*

$$\|\mathcal{F}_1(u, v)\|_X \le \alpha_1 \|u\|_X \|v\|_X, \tag{5.12}$$
$$\|\mathcal{F}_2(u, \theta)\|_Y \le \alpha_2 \|u\|_X \|\theta\|_Y, \tag{5.13}$$
$$\|\mathcal{L}\theta\|_X \le \alpha_3 \|\theta\|_Y \tag{5.14}$$

for all $u, v \in X$ and all $\theta \in Y$. Assume that $E_1 \in X$, $E_2 \in Y$ satisfy the condition

$$\frac{1}{2}\|E_1\|_X + \alpha_3\|E_2\|_Y \le \frac{1}{8\,(2\alpha_1 + \alpha_2)^2}. \tag{5.15}$$

Then the system

$$\begin{aligned} u &= E_1 + \mathcal{F}_1(u, u) + \mathcal{L}\theta, \\ \theta &= E_2 + \mathcal{F}_2(u, \theta) \end{aligned} \tag{5.16}$$

has a solution $(\tilde{u}, \tilde{\theta}) \in X \times Y$. This solution $(\tilde{u}, \tilde{\theta})$ is the only solution in $X \times Y$ of (5.16) which satisfies the condition

$$\frac{1}{2}\|\tilde{u}\|_X + \alpha_3\|\tilde{\theta}\|_Y \le \frac{1}{4}\,(2\alpha_1 + \alpha_2). \tag{5.17}$$

Proof. This is [BrSc12, Lemma 5.1] with $\eta = \frac{1}{2}$. □

We have to define the terms appearing in (5.10). The next lemma provides us with estimates of the terms $\mathcal{F}_1, \mathcal{F}_2$ and $\mathcal{L}$ appearing in the previous lemma.

Lemma 5.5. *Let $\Omega \subseteq \mathbb{R}^3$ be a general domain, let $0 < T \le \infty$, $g \in L^{8/3}(0, T; L^4(\Omega)^3)$.*

1. *Define the bilinear form*

   $$\mathcal{F}_1 : L^8(0, T; L^4_\sigma(\Omega)) \times L^8(0, T; L^4_\sigma(\Omega)) \to L^8(0, T; L^4_\sigma(\Omega)),$$
   $$\left(\mathcal{F}_1(u, v)\right)(t) := -A^{1/2} \int_0^t e^{-(t-\tau)A} A^{-1/2} P\mathrm{div}\left(u(\tau) \otimes v(\tau)\right) d\tau \tag{5.18}$$
   for a.a. $t \in [0, T[$.

   Then
   $$\|\mathcal{F}_1(u, v)\|_{4,8;T} \le K\|u \otimes v\|_{2,4;T} \le K\|u\|_{4,8;T}\|v\|_{4,8;T} \tag{5.19}$$
   for all $u, v \in L^8(0, T; L^4_\sigma(\Omega))$ where $K > 0$ is an absolute constant (independent of Ω, T, g, u, v).

2. *Define the bilinear form*

   $$\mathcal{F}_2 : L^8(0, T; L^4_\sigma(\Omega)) \times L^{8/3}(0, T; L^4(\Omega)) \to L^{8/3}(0, T; L^4(\Omega)),$$
   $$\left(\mathcal{F}_2(u, \theta)\right)(t) := -(-\Delta^{1/2}) \int_0^t e^{(t-\tau)\Delta} (-\Delta)^{-1/2} \mathrm{div}\left(\theta(\tau) u(\tau)\right) d\tau \tag{5.20}$$
   for a.a. $t \in [0, T[$.

Then

$$\|\mathcal{F}_2(u,\theta)\|_{4,\frac{8}{3};T} \leq K\|\theta u\|_{2,2;T} \leq K\|u\|_{4,8;T}\|\theta\|_{4,\frac{8}{3};T} \tag{5.21}$$

for all $u \in L^8(0,T;L^4_\sigma(\Omega))$, $\theta \in L^{8/3}(0,T;L^4(\Omega))$ *with an absolute constant* $K > 0$.

3. *Define the linear map*

$$\mathcal{L}: L^{8/3}(0,T;L^4(\Omega)) \rightarrow L^8(0,T;L^4_\sigma(\Omega)),$$

$$(\mathcal{L}\theta)(t) := \int_0^t e^{-(t-\tau)A} P\big(\theta(\tau)g(\tau)\big)\,d\tau \quad \textit{for a.a. } t \in [0,T[. \tag{5.22}$$

Then

$$\|\mathcal{L}\theta\|_{4,8;T} \leq K\|\theta g\|_{2,\frac{4}{3};T} \leq K\|g\|_{4,\frac{8}{3};T}\|\theta\|_{4,\frac{8}{3};T} \tag{5.23}$$

for all $\theta \in L^{8/3}(0,T;L^4(\Omega))$ *with an absolute constant* $K > 0$.

Proof. Fix $u,v \in L^8(0,T;L^4_\sigma(\Omega))$ and $\theta \in L^{8/3}(0,T;L^4(\Omega))$.

1. Estimates (2.56), (2.58), (2.62) yield

$$\|(\mathcal{F}_1(u,v))(t)\|_4 \leq K\left\|A^{7/8}\int_0^t e^{-(t-\tau)A} A^{-1/2} P\mathrm{div}\big(u(\tau)\otimes v(\tau)\big)\,d\tau\right\|_2$$

$$\leq K\int_0^t |t-\tau|^{-7/8}\|A^{-1/2}P\mathrm{div}\big(u(\tau)\otimes v(\tau)\big)\|_2\,d\tau \tag{5.24}$$

$$\leq K\int_0^T |t-\tau|^{-7/8}\|u(\tau)\otimes v(\tau)\|_2\,d\tau$$

for almost $t \in [0,T[$ with an absolute constant $K > 0$. Since $(1 - \frac{7}{8}) + \frac{1}{8} = \frac{1}{4}$ we can apply the Hardy-Littlewood inequality and Hölder's inequality to (5.24) and obtain

$$\|\mathcal{F}_1(u,v)\|_{4,8;T} \leq K\|u\otimes v\|_{2,4;T}$$
$$\leq K\|u\|_{4,8;T}\|v\|_{4,8;T} \tag{5.25}$$

with an absolute constant $K > 0$.

2. A consequence of estimates (2.57), (2.56), (2.64) is

$$\|(\mathcal{F}_2(u,\theta))(t)\|_4 \leq K\left\|(-\Delta)^{7/8}\int_0^t e^{(t-\tau)\Delta}(-\Delta)^{-1/2}\mathrm{div}\big(u(\tau)\otimes\theta(\tau)\big)\,d\tau\right\|_2$$

$$\leq K\int_0^t |t-\tau|^{-7/8}\|(-\Delta)^{-1/2}\mathrm{div}\big(u(\tau)\otimes\theta(\tau)\big)\|_2\,d\tau$$

$$\leq K\int_0^T |t-\tau|^{-7/8}\|u(\tau)\otimes\theta(\tau)\|_2\,d\tau$$

for almost $t \in [0, T[$ with an absolute constant $K > 0$. The Hardy-Littlewood inequality with $(1 - \frac{7}{8}) + \frac{1}{\left(\frac{8}{3}\right)} = \frac{1}{2}$, combined with Hölder's inequality, yields

$$
\begin{aligned}
\|\mathcal{F}_2(u, \theta)\|_{4, \frac{8}{3}; T} &\leq K \|\theta u\|_{2, 2; T} \\
&\leq K \|u\|_{4, 8; T} \|\theta\|_{4, \frac{8}{3}; T}
\end{aligned}
\tag{5.26}
$$

with an absolute constant $K > 0$.

3. From (2.58) it follows

$$
\|(\mathcal{L}\theta)(t)\|_4 \leq K \int_0^T |t - \tau|^{-3/8} \|\theta(\tau) g(\tau)\|_2 \, d\tau
\tag{5.27}
$$

for almost all $t \in [0, T[$ with an absolute constant $K > 0$. Since $(1 - \frac{3}{8}) + \frac{1}{8} = \frac{1}{\left(\frac{4}{3}\right)}$ we can apply the Hardy-Littlewood estimate to (5.27) and get

$$
\begin{aligned}
\|\mathcal{L}\theta\|_{4, 8; T} &\leq K \|\theta g\|_{2, \frac{4}{3}; T} \\
&\leq K \|g\|_{4, \frac{8}{3}; T} \|\theta\|_{4, \frac{8}{3}; T}
\end{aligned}
\tag{5.28}
$$

with an absolute constant $K > 0$.

$\square$

Lemma 5.6. *Let $\Omega \subseteq \mathbb{R}^3$ be a general domain, let $0 < T \leq \infty$. The following statements are fulfilled:*

(i) For $f_1 \in L^{4/3}(0, T; L^2(\Omega)^3)$ define

$$
(\mathcal{G} f_1)(t) := \int_0^t e^{-(t-\tau)A} P f_1(\tau) \, d\tau \quad \text{for a.a. } t \in [0, T[.
$$

Then

$$
\|\mathcal{G} f_1\|_{4, 8; T} \leq K \|f_1\|_{2, \frac{4}{3}; T}.
\tag{5.29}
$$

with an absolute constant $K > 0$ (independent of Ω, T, f_1).

(ii) For $f_2 \in L^1(0, T; L^2(\Omega))$ define

$$
(\mathcal{H} f_2)(t) := \int_0^t e^{(t-\tau)\Delta} f_2(\tau) \, d\tau \quad \text{for a.a. } t \in [0, T[.
$$

Then

$$
\|\mathcal{H} f_2\|_{4, \frac{8}{3}; T} \leq K \|f_2\|_{2, 1; T}
\tag{5.30}
$$

with an absolute constant $K > 0$.

(iii) We have

$$
\left(\int_0^T \|e^{t\Delta} \theta_0\|_4^{8/3} \, dt \right)^{3/8} \leq K \|\theta_0\|_2
\tag{5.31}
$$

for all $\theta_0 \in L^2(\Omega)$ with an absolute constant $K > 0$.

Proof. In this proof we will make frequent use of the following interpolation inequality. For $E \in L^\infty(0, T; L^2(\Omega)) \cap L^2(0, T; H_0^1(\Omega))$ there holds

$$
\begin{aligned}
\|E\|_{4,\frac{8}{3};T} &\le \|E\|_{2,\infty;T}^{1/4} \|E\|_{6,2;T}^{3/4} \\
&\le K \|E\|_{2,\infty;T}^{1/4} \|(-\Delta)^{1/2} E\|_{2,2;T}^{3/4} \\
&\le K \big(\|E\|_{2,\infty;T} + \|(-\Delta)^{1/2} E\|_{2,2;T} \big)
\end{aligned}
\tag{5.32}
$$

with an absolute constant $K > 0$.

(i) Replace θg by f_1 in the proof of (5.23).

(ii) An analogous energy estimate as in ([So01, Chapter IV, Lemma 2.3.1]) yields

$$
\|\mathcal{H} f_2\|_{2,\infty;T}^2 + \|(-\Delta)^{1/2} \mathcal{H} f_2\|_{2,2;T}^2 \le K \|f\|_{2,1;T}^2 .
$$

Therefore, by monotonicity of the square root

$$
\|\mathcal{H} f_2\|_{2,\infty;T} + \|(-\Delta)^{1/2} \mathcal{H} f_2\|_{2,2;T} \le K \|f\|_{2,1;T} .
$$

Consequently, by (5.32) it follows that (5.30) is satisfied.

(iii) From the estimates $\|e^{t\Delta}\theta_0\|_{2,\infty;T} \le \|\theta_0\|_2$ and $\|(-\Delta)^{1/2} e^{t\Delta}\theta_0\|_{2,2;T} \le \|\theta_0\|_2$ (analogous to [So01, Chapter IV, (1.5.3)]) applied to $E(t) = e^{t\Delta}\theta_0$ in (5.32) we get (5.31).

$\square$

The following uniqueness theorem on weak solution of (5.1) needs no additional assumption on θ. Let us introduce the following notation: For $0 < T' \le T \le \infty$, and $f \in L^s(0, T; L^q(\Omega)^d)$ where $d \in \mathbb{N}$, we define

$$
\|f\|_{q,s;T',T} := \left(\int_{T'}^T \|f(t)\|_q^s \, dt \right)^{\frac{1}{s}} .
$$

Theorem 5.7. *Let $\Omega \subseteq \mathbb{R}^3$ be a general domain, let $u_0 \in L_\sigma^2(\Omega), \theta_0 \in L^2(\Omega)$. Consider $0 < T \le \infty, g \in L^{8/3}(0, T; L^4(\Omega)^3)$, consider $f_1 \in L^{4/3}(0, T; L^2(\Omega)^3)$, $f_2 \in L^1(0, T; L^2(\Omega))$. Assume that (u_1, θ_1) and (u_2, θ_2) are weak solutions of (5.1) in $\Omega \times [0, T[$ such that additionally $u_1, u_2 \in L^8(0, T; L^4(\Omega)^3)$. Then $u_1(t) = u_2(t)$ and $\theta_1(t) = \theta_2(t)$ for almost all $t \in [0, T[$.*

Proof. Throughout this proof we will make frequently use of Lemma 5.5 without referring to it at every every single occurrence. By Theorem 5.2 there holds

$$
u_1 - u_2 = \mathcal{F}_1(u_1, u_1 - u_2) + \mathcal{F}_1(u_1 - u_2, u_2) + \mathcal{L}(\theta_1 - \theta_2) ,
\tag{5.33}
$$

$$
\theta_1 - \theta_2 = \mathcal{F}_2(u_1 - u_2, \theta_1) + \mathcal{F}_2(u_2, \theta_1 - \theta_2) .
\tag{5.34}
$$

Define

$$
T_{max} := \sup \{ T' \in [0, T[; \, u_1(t) = u_2(t), \theta_1(t) = \theta_2(t) \quad \text{for a.a. } t \in [0, T'[\} .
\tag{5.35}
$$

Suppose by contradiction $T_{max} < T$. Consider any $T' \in [T_{max}, T[$. Then

$$\|u_1 - u_2\|_{4,8;T'}$$
$$\leq K\big(\|u_1 \otimes (u_1 - u_2)\|_{2,4;T'} + \|(u_1 - u_2) \otimes u_2\|_{2,4;T'} + \|(\theta_1 - \theta_2)g\|_{2,\frac{4}{3};T'} \big)$$
$$= K\big(\|u_1 \otimes (u_1 - u_2)\|_{2,4;T_{max},T'} + \|(u_1 - u_2) \otimes u_2\|_{2,4;T_{max},T'}$$
$$+ \|(\theta_1 - \theta_2)g\|_{2,\frac{4}{3};T_{max},T'} \big) \tag{5.36}$$
$$\leq K\big(\|u_1\|_{4,8;T_{max},T'} + \|u_2\|_{4,8;T_{max},T'} \big)\|u_1 - u_2\|_{4,8;T_{max},T'}$$
$$+ K\|g\|_{4,\frac{8}{3};T_{max},T'}\|\theta_1 - \theta_2\|_{4,\frac{8}{3};T_{max},T'}$$

with an absolute constant $K > 0$. The difference $\|\theta_1 - \theta_2\|_{4,\frac{8}{3};T'}$ can be estimated with an analogous argumentation. We get

$$\|\theta_1 - \theta_2\|_{4,\frac{8}{3};T'} \leq K\big(\|u_1 - u_2\|_{4,8;T_{max},T'}\|\theta_1\|_{4,\frac{8}{3};T_{max},T'} + \|u_2\|_{4,8;T_{max},T'}\|\theta_1 - \theta_2\|_{4,\frac{8}{3};T_{max},T'} \big).$$

Altogether

$$\|u_1 - u_2\|_{4,8;T'} + \|\theta_1 - \theta_2\|_{4,\frac{8}{3};T'}$$
$$\leq K\big(\|u_1\|_{4,8;T_{max},T'} + \|u_2\|_{4,8;T_{max},T'} + \|\theta_1\|_{4,\frac{8}{3};T_{max},T'} \big)\|u_1 - u_2\|_{4,8;T_{max},T'} \tag{5.37}$$
$$+ K\big(\|g\|_{4,\frac{8}{3};T_{max},T'} + \|u_2\|_{4,8;T_{max},T'} \big)\|\theta_1 - \theta_2\|_{4,\frac{8}{3};T_{max},T'}$$

with an absolute constant $K > 0$. Now $T' > T_{max}$ can be chosen such that the two conditions

$$K\big(\|u_1\|_{4,8;T_{max},T'} + \|u_2\|_{4,8;T_{max},T'} + \|\theta\|_{4,\frac{8}{3};T_{max},T'} \big) \leq \frac{1}{4}, \tag{5.38}$$
$$K\big(\|g\|_{4,\frac{8}{3};T_{max},T'} + \|u_2\|_{4,8;T_{max},T'} \big) \leq \frac{1}{4}$$

are satisfied. From (5.37) we get with (5.38) that

$$\|u_1 - u_2\|_{4,8;T'} + \|\theta_1 - \theta_2\|_{4,\frac{8}{3};T'} = 0. \tag{5.39}$$

But since $T' > T_{max}$ this is a contradiction to (5.35). Therefore $T_{max} = T$.

$\square$

Our first main theorem gives a sufficient criterion for the existence of strong solutions of the Boussinesq equations (5.1).

Theorem 5.8. *Let $\Omega \subseteq \mathbb{R}^3$ be a general domain, let $u_0 \in L^2_\sigma(\Omega)$, $\theta_0 \in L^2(\Omega)$. Further assume $0 < T \leq \infty$, $g \in L^{8/3}(0, T; L^4(\Omega)^3)$, and $f_1 \in L^{4/3}(0, T; L^2(\Omega)^3)$, $f_2 \in L^1(0, T; L^2(\Omega))$. Then there exist absolute constants $\epsilon_*, K_* > 0$ (which are independent of $\Omega, T, g, f_1, f_2, g, u_0, \theta_0$) with the following property: If the conditions*

$$\left(\int_0^T \|e^{-tA}u_0\|_4^8 \, dt \right)^{1/8} \leq \epsilon_*, \tag{5.40}$$

$$\left(\int_0^T \|e^{t\Delta}\theta_0\|_4^{8/3} \, dt \right)^{3/8} \leq \frac{\epsilon_*}{\|g\|_{4,\frac{8}{3};T}}, \tag{5.41}$$

$$\|f_1\|_{2,\frac{4}{3};T} + \|g\|_{4,\frac{8}{3};T}\|f_2\|_{2,1;T} \leq \epsilon_* \tag{5.42}$$

are satisfied, then there exists a strong solution (u, θ) of (5.1) in $[0, T[\times \Omega$ satisfying the estimates

$$\|u\|_{4,8;T} \le \frac{3}{2} K_* , \tag{5.43}$$

$$\|g\|_{4,\frac{8}{3};T} \|\theta\|_{4,\frac{8}{3};T} \le \frac{3}{4} . \tag{5.44}$$

After a possible redefinition on a null set, $u : [0, T[\to L^2_\sigma(\Omega)$ and $\theta : [0, T[\to L^2(\Omega)$ are strongly continuous and the energy equalities

$$\frac{1}{2}\|u(t)\|_2^2 + \int_0^t \|\nabla u(\tau)\|_2^2 \, d\tau = \frac{1}{2}\|u_0\|_2^2 + \int_0^t \langle f_1, u \rangle_\Omega \, d\tau + \int_0^t \langle \theta g, u \rangle_\Omega \, d\tau , \tag{5.45}$$

$$\frac{1}{2}\|\theta(t)\|_2^2 + \int_0^t \|\nabla \theta(\tau)\|_2^2 \, d\tau = \frac{1}{2}\|\theta_0\|_2^2 + \int_0^t \langle f_2, \theta \rangle_\Omega \, d\tau \tag{5.46}$$

are satisfied for all $t \in [0, T[$. Moreover (u, θ) is the only strong solution of (5.1) in $[0, T[\times \Omega$. (independent of the smallness conditions (5.43), (5.44).)

Remark. Combining (5.41) and (5.31) if follows that ϵ_* can be replaced by a smaller constant (if necessary), to be denoted again by ϵ_*, such that the following property is satisfied: If (5.40), (5.42) and if the condition

$$\|\theta_0\|_2 \le \frac{\epsilon_*}{\|g\|_{4,\frac{8}{3};T}} \tag{5.47}$$

is satisfied, then there exists a strong solution (u, θ) of (5.1) on $[0, T[$ satisfying the estimates (5.43), (5.44).

Proof. Step 1. Choose a constant $K_* > 0$ such that (5.19), (5.21), (5.23) are satisfied where K is replaced by K_*. By construction the constant K_* is an absolute constant independent of $\Omega, T, g, f_1, f_2, u_0, \theta$ appearing in these estimates.

Define $X := L^8(0, T; L^4_\sigma(\Omega)), Y := L^{8/3}(0, T; L^4(\Omega))$ and let $\mathcal{F}_1, \mathcal{F}_2, \mathcal{L}$ be defined as in Lemma 5.5. Furthermore define

$$E_1(t) := e^{-tA} u_0 + \int_0^t e^{-(t-\tau)A} P f_1(\tau) \, d\tau ,$$

$$E_2(t) := e^{t\Delta} \theta_0 + \int_0^t e^{(t-\tau)\Delta} f_2(\tau) \, d\tau \tag{5.48}$$

for almost all $t \in [0, T[$. Consider the system

$$u = E_1 + \mathcal{F}_1(u, u) + \mathcal{L}\theta ,$$

$$\theta = E_2 + \mathcal{F}_2(u, \theta). \tag{5.49}$$

From Lemma 5.5 we see that (5.12), (5.13), (5.14) are fulfilled with $\alpha_1 := \alpha_2 := K_*$ and $\alpha_3 := K_* \|g\|_{4,\frac{8}{3};T}$. Therefore we can apply Lemma 5.4 and obtain that the following statement holds:

If the condition

$$\frac{1}{2}\|E_1\|_{4,8;T} + K_*\|g\|_{4,\frac{8}{3};T}\|E_2\|_{4,\frac{8}{3};T} \leq \frac{1}{72K_*^2} \tag{5.50}$$

is satisfied, there exists a solution $u \in L^8(0,T;L^4_\sigma(\Omega))\,,\theta \in L^{8/3}(0,T;L^4(\Omega))$ of (5.49) satisfying the estimate

$$\frac{1}{2}\|u\|_{4,8;T} + K_*\|g\|_{4,\frac{8}{3};T}\|\theta\|_{4,\frac{8}{3};T} \leq \frac{3}{4}K_*. \tag{5.51}$$

Looking at (5.29), (5.30) we get the estimate

$$\frac{1}{2}\|E_1\|_{4,8;T} + K_*\|g\|_{4,\frac{8}{3};T}\|E_2\|_{4,\frac{8}{3};T}$$

$$\leq \left(\int_0^T \|e^{-tA}u_0\|_4^8\,dt\right)^{1/8} + K_*\|f_1\|_{2,\frac{4}{3};T} + K_*\|g\|_{4,\frac{8}{3};T}\left(\int_0^T \|e^{t\Delta}\theta_0\|_4^{8/3}\,dt\right)^{3/8} \tag{5.52}$$

$$+ K_*K\|g\|_{4,\frac{8}{3};T}\|f_2\|_{2,1;T}$$

with the absolute constant $K > 0$ appearing in the estimate (5.30).

Define $\epsilon_* := \min\left\{\dfrac{1}{288K_*^2}, \dfrac{1}{288K_*^3}, \dfrac{1}{288K_*^3K}\right\}$. Thus, if (5.40), (5.41), (5.42) are satisfied, then we get from (5.52) that

$$\frac{1}{2}\|E_1\|_{4,8;T} + K_*\|g\|_{4,\frac{8}{3};T}\|E_2\|_{4,\frac{8}{3};T} \leq \frac{1}{72K_*^2}.$$

Altogether there exists an absolute constant $\epsilon_* > 0$ with the following property: If (5.40), (5.41), (5.42) are satisfied, then condition (5.50) is fulfilled and consequently there is a solution $u \in L^8(0,T;L^4_\sigma(\Omega))\,,\theta \in L^{8/3}(0,T;L^4(\Omega))$ of (5.1) satisfying the estimates (5.43), (5.44).

Step 2. Let (u,θ) with $u \in L^8(0,T;L^4_\sigma(\Omega))$ and $\theta \in L^{8/3}(0,T;L^4(\Omega))$ be a solution of (5.49). Consider any $0 < T' \leq T$ with $T' < \infty$. It follows $u \otimes u \in L^2(0,T';L^2(\Omega)^{3\times3})$. Consequently, from [So01, Chapter IV, Lemma 2.4.2] (with $F = u \otimes u$) we get

$$\mathcal{F}_1(u,u) \in \mathcal{LH}_{T'} := L^\infty(0,T';L^2_\sigma(\Omega)) \cap L^2(0,T';W^{1,2}_{0,\sigma}(\Omega)). \tag{5.53}$$

From $P(\theta g) \in L^{4/3}(0,T;L^2(\Omega)^3)$ and [So01, Chapter IV, Lemma 2.4.2] (with $f = \theta g\,, F = 0$) it follows $\mathcal{L}\theta \in \mathcal{LH}_{T'}$. Thus $u \in \mathcal{LH}_{T'}$. Using $\theta u \in L^2(0,T;L^2(\Omega)^3)$ and an analogous version of [So01, Chapter IV, Lemma 2.4.2] with A replaced by $-\Delta$ we get

$$\theta \in L^\infty(0,T';L^2_\sigma(\Omega)) \cap L^2(0,T';W^{1,2}_{0,\sigma}(\Omega)).$$

Altogether

$$u \in L^\infty_{\text{loc}}(0,T;L^2_\sigma(\Omega)) \cap L^2_{\text{loc}}(0,T;W^{1,2}_{0,\sigma}(\Omega))\,,$$
$$\theta \in L^\infty_{\text{loc}}(0,T;L^2(\Omega)) \cap L^2_{\text{loc}}(0,T;H^1_0(\Omega)).$$

Thus, (u, θ) is a solution of the system (5.49) and fulfils (5.5). From Theorem 5.2 we get that (u, θ) is a weak solution of the Boussinesq equations (5.1) in $[0, T[\times\Omega$.

Step 3. The uniqueness of a strong solution of the Boussinesq equations (5.1) in $[0, T[\times\Omega$ is proven in Theorem 5.7. $\qquad\square$

Next we present a criterion when the conditions (5.40), (5.41), (5.42) are satisfied.

Corollary 5.9. *Let $\Omega \subseteq \mathbb{R}^3$ be a general domain, let $0 < T \leq \infty$, and let $g \in L^{8/3}(0, T; L^4(\Omega)^3)$. Assume $f_1 \in L^{4/3}(0, T; L^2(\Omega)^3)$, $f_2 \in L^1(0, T; L^2(\Omega))$. Then there exist absolute constants $\epsilon_*, K_* > 0$ (which are independent of Ω, T, g, f_1, f_2) such that the following two statements are fulfilled:*

(i) Let $u_0 \in \mathcal{D}(A^{1/4})$, $\theta_0 \in L^2(\Omega)$, and assume that the conditions

$$\|A^{1/4}u_0\|_2 \leq \epsilon_* ,$$

$$\|\theta_0\|_2 \leq \frac{\epsilon_*}{\|g\|_{4,\frac{8}{3};T}} , \tag{5.54}$$

$$\|f_1\|_{2,\frac{4}{3};T} + \|g\|_{4,\frac{8}{3};T}\|f_2\|_{2,1;T} \leq \epsilon_*$$

are satisfied. Then there exists a strong solution (u, θ) of (5.1) in $[0, T[\times\Omega$.

(ii) Let $u_0 \in W_{0,\sigma}^{1,2}(\Omega)$ and assume additionally $0 < T < \infty$. If the condition

$$T^{\frac{1}{8}}\|u_0\|_{H^1(\Omega)} + \|g\|_{4,\frac{8}{3};T}\|\theta_0\|_2 + \|f_1\|_{2,\frac{4}{3};T} + \|g\|_{4,\frac{8}{3};T}\|f_2\|_{2,1;T} \leq \epsilon_* \tag{5.55}$$

is satisfied, then there exists a strong solution (u, θ) of (5.1) in $[0, T[\times\Omega$.

In both cases (u, θ) satisfies the estimates

$$\|u\|_{4,8;T} \leq \frac{3}{2}K_* ,$$

$$\|g\|_{4,\frac{8}{3};T}\|\theta\|_{4,\frac{8}{3};T} \leq \frac{3}{4}.$$

Proof. Let $\epsilon_*, K_* > 0$ be the absolute constants of Theorem 5.8.

(i) The proof is based on the following observation. If $u_0 \in \mathcal{D}(A^{1/4})$, then (2.58) and [So01, Chapter IV, (1.5.24)] yield

$$\int_0^T \|e^{-tA}u_0\|_4^8 \, dt \leq K \int_0^T \|A^{3/8}e^{-tA}u_0\|_2^8 \, dt$$

$$= K \int_0^T \|A^{1/8}e^{-tA}(A^{1/4}u_0)\|_2^8 \, dt \tag{5.56}$$

$$\leq K\|A^{1/4}u_0\|_2$$

with an absolute constant $K > 0$. By Lemma 5.6 we get $\|e^{t\Delta}\theta_0\|_{4,\frac{8}{3};T} \leq K\|\theta_0\|_2$ with an absolute constant $K > 0$. Hence, after a possible reduction of ϵ_*, we see from (5.56) that if (5.54) holds, then the conditions (5.40), (5.41), (5.42) are satisfied.

(ii) Let $\epsilon_* > 0$ be as in (i). Let $u_0 \in W_{0,\sigma}^{1,2}(\Omega)$. From (2.48) and (2.54) we get that $\|A^{3/8} u_0\|_2 \leq \|u_0\|_{H^1(\Omega)}$ holds. We make use of (2.56), (2.58) and get

$$
\begin{aligned}
\int_0^T \|e^{-tA} u_0\|_4^8 \, dt &\leq K \int_0^T \|A^{3/8} e^{-tA} u_0\|_2^8 \, dt \\
&= K \int_0^T \|e^{-tA} A^{3/8} u_0\|_2^8 \, dt \\
&\leq K \int_0^T \|A^{3/8} u_0\|_2^8 \, dt \\
&= KT \|A^{3/8} u_0\|_2^8 \\
&\leq K_1 T \|u_0\|_{H^1(\Omega)}^8
\end{aligned}
\tag{5.57}
$$

with an absolute constant $K_1 > 0$ which will be fixed up to the end of the proof. Assume that

$$
K_1^{\frac{1}{8}}(T)^{\frac{1}{8}} \|u_0\|_{H^1(\Omega)} \leq \epsilon_* ,
$$
$$
\|\theta_0\|_2 \leq \frac{\epsilon_*}{\|g\|_{4,\frac{8}{3};T}} ,
\tag{5.58}
$$

then (5.41), (5.47) are satisfied. Looking at (5.58), (5.57) and the remark after Theorem 5.8 we see that it is possible to redefine ϵ_* (if necessary) such that if condition (5.55) holds then (5.40), (5.41), (5.42) are satisfied. Therefore, there exists a strong solution on $[0, T[$ of (5.1).

$\qquad\qquad\qquad\qquad\qquad\qquad\qquad\qquad\qquad\qquad\qquad\qquad\qquad\qquad\square$

Theorem 5.10. *Let I be an arbitrary index set, let $\Omega_i \subseteq \mathbb{R}^3 , i \in I$, be general domains, let $0 < T < \infty$. For $i \in I$ let $g_i \in L^{8/3}(0, T; L^4(\Omega_i)^3)$, $f_{1,i} \in L^{4/3}(0, T; L^2(\Omega_i)^3)$, let $f_{2,i} \in L^1(0, T; L^2(\Omega_i))$. Further consider $u_{0,i} \in W_{0,\sigma}^{1,2}(\Omega_i)$, $\theta_{0,i} \in L^2(\Omega_i)$ for every $i \in I$. Then there exists absolute constants $\epsilon_*, K_* > 0$ (which are independent of $\Omega_i, T, f_{1,i}, f_{2,i}, u_{0,i}, \theta_{0,i}$) with the following property: Assume that the condition*

$$
T^{\frac{1}{8}} \|u_{0,i}\|_{H^1(\Omega_i)} + \|g_i\|_{4,\frac{8}{3};\Omega_i;T} \|\theta_{0,i}\|_{L^2(\Omega_i)} + \|f_{1,i}\|_{2,\frac{4}{3};\Omega_i;T}
$$
$$
+ \|g_i\|_{4,\frac{8}{3};\Omega_i;T} \|f_{2,i}\|_{2,1;\Omega_i;T} \leq \epsilon_*
\tag{5.59}
$$

is satisfied for every $i \in I$. Then, for every $i \in I$, there exists a (uniquely determined) strong $L^8(L^4)$-solution (u_i, θ_i) of (5.1) in $\Omega_i \times [0, T[$. Moreover, the estimates

$$
\|u_i\|_{4,8;\Omega_i;T} \leq \frac{3}{2} K_* ,
$$
$$
\|g_i\|_{4,\frac{8}{3};\Omega_i;T} \|\theta_i\|_{4,\frac{8}{3};\Omega_i;T} \leq \frac{3}{4}
\tag{5.60}
$$

are satisfied for every $i \in I$.

Proof. Let ϵ_*, K_* be the absolute constants obtained in Corollary 5.9. Fix $i \in I$. Looking at (5.54) we see that if (5.59) is satisfied, then there exists a strong solution (u_i, θ_i) of (5.1) in $\Omega_i \times [0, T[$ satisfying the estimate (5.60). □

We obtain the following characterization of initial values u_0, θ_0 which allow strong solutions.

Corollary 5.11. *Let $\Omega \subseteq \mathbb{R}^3$ be a general domain, let $u_0 \in L^2_\sigma(\Omega)$, $\theta_0 \in L^2(\Omega)$. Further assume $0 < T \leq \infty$, $g \in L^{8/3}(0, T; L^4(\Omega)^3)$, and $f_1 \in L^{4/3}(0, T; L^2(\Omega)^3)$, $f_2 \in L^1(0, T; L^2(\Omega))$. Then the condition*

$$\int_0^\infty \|e^{-tA} u_0\|_4^8 \, dt < \infty \tag{5.61}$$

is necessary and sufficient for the existence of $0 < T' \leq T$ and a strong solution (u, θ) of (5.1) in $[0, T'[\times\Omega$.

Proof. Let $\epsilon_* > 0$ be the constant of Theorem 5.8 (we do not need the constant K_* in this proof). First assume that (5.61) is satisfied. Consequently, it is possible to choose $T' > 0$ with

$$\left(\int_0^{T'} \|e^{-tA} u_0\|_4^8 \, dt \right)^{1/8} \leq \epsilon_*. \tag{5.62}$$

By (5.31) we have $\int_0^\infty \|e^{t\Delta} \theta_0\|_4^{8/3} \, dt < \infty$. Therefore, after reducing T' (if necessary), we obtain

$$\left(\int_0^{T'} \|e^{t\Delta} \theta_0\|_4^{8/3} \, dt \right)^{3/8} \leq \frac{\epsilon_*}{\|g\|_{4, \frac{8}{3}; T'}}. \tag{5.63}$$

Moreover, after reducing T' again (if necessary), we obtain that (5.42) is also satisfied. By Theorem 5.8 we obtain the existence of a strong solution of (5.1) in $[0, T'[\times\Omega$.

For the proof of the converse direction assume that there is $T' > 0$ and a strong solution (u, θ) of (5.1) in $[0, T'[\times\Omega$. By Theorem 5.2

$$e^{-tA} u_0 = u(t) - \left(\mathcal{F}_1(u, u) \right)(t) - \left(\mathcal{L}\theta \right)(t) - \int_0^t e^{-(t-\tau)A} P f_1(\tau) \, d\tau \tag{5.64}$$

for almost all $t \in [0, T'[$. Since (u, θ) is a strong $L^8(L^4)$-solution in $[0, T'[\times\Omega$ we obtain $u \in L^8(0, T'; L^4(\Omega)^3)$ and $\theta \in L^{8/3}(0, T'; L^4(\Omega))$. Thus, we apply (5.19), (5.23), (5.29) to (5.64) and get $\int_0^{T'} \|e^{-tA} u_0\|_4^8 \, dt < \infty$. There holds (see (2.56), (2.58))

$$\int_{T'}^\infty \|e^{-tA} u_0\|_4^8 \, dt \leq K \int_{T'}^\infty (t^{-3/8} \|u_0\|_2)^8 \, dt < \infty.$$

Consequently, (5.61) is fulfilled. □

5.3 Regularity of strong solutions of the Boussinesq equations

One part of the motivation of defining the concept of a strong solution of (5.1) is the idea that such a solution (u, θ) is smooth if the data are sufficiently smooth. For this reason consider a strong solution (u, θ) of (5.1) in $[0, T[\times\Omega$. In the main result of this section, Theorem 5.16, we prove that $u \in C^\infty(]0, T[\times\Omega)^3$, $\theta \in C^\infty(]0, T[\times\Omega)$ holds under regularity assumptions on the data; especially we have to require that the boundary $\partial\Omega$ is smooth. Further a pressure $p \in C^\infty(]0, T[\times\Omega)$ associated to u will be constructed. For the proof of this theorem we will use results and techniques presented in [So01, Chapter IV, Section 2.7 and Chapter V, Section 1.8].

5.3.1 Outline of this section

Let (u, θ) be a strong solution of (5.1) in $[0, T[\times\Omega$ where Ω, f_1, f_2, g, u_0, θ_0 are sufficiently smooth. In this subsection we present the main ideas how to prove smoothness of (u, θ).

1. The first regularity step is the content of Theorem 5.15. Considering u as a weak solution of the Navier-Stokes equations with external force $f := f_1 + \theta g$ and initial value u_0 we get from the regularity theory of the Navier-Stokes equations in [So01, Chapter V, Section 1.8] that

$$u \in L^\infty(0, T; W^{1,2}_{0,\sigma}(\Omega)) \cap L^2(0, T; H^2(\Omega)^3) \tag{5.65}$$

and that the equation

$$u_t - \Delta u + u \cdot \nabla u + \nabla p = f_1 + \theta g \tag{5.66}$$

is satisfied in $L^2(0, T; L^2(\Omega)^3)$ with an associated pressure p. These better integrability properties of u can be used to increase the regularity of θ. Note that, in contrast to the condition $u \in L^8(0, T; L^4(\Omega)^3)$, we have imposed no additional integrability condition on θ in the definition of a strong solution of (5.1). By a bootstrapping argument we are able to prove

$$\theta \in L^\infty(0, T; H^1_0(\Omega)) \cap L^2(0, T; H^2(\Omega)). \tag{5.67}$$

Consequently

$$\theta_t - \Delta\theta + u \cdot \nabla\theta = f_2 \tag{5.68}$$

is satisfied in $L^2(0, T; L^2(\Omega))$.

2. Let $\eta \in C^\infty_0(]0, T[)$. In the proof of Theorem 5.16 we consider the equations

$$(\eta u)_t + A(\eta u) = P(\eta f_1) + P(\eta\theta g) + \eta_t u - P(\eta u \cdot \nabla u), \tag{5.69}$$
$$(\eta\theta)_t + (-\Delta)(\eta\theta) = \eta f_2 + \eta_t\theta - \eta u \cdot \nabla\theta \tag{5.70}$$

and improve the regularity of ηu and $\eta \theta$. The reason why we consider the functions $\eta u, \eta \theta$ instead of u, θ is, since $(\eta u)(0) = 0, (\eta \theta)(0) = 0$, we get that the time derivatives $(\eta u)^{(k)}, (\eta \theta)^{(k)}$ are integrable in a neighbourhood of 0. Differentiating (5.69), (5.70) in the time direction and using bootstrapping arguments we obtain

$$A\big((\eta u)^{(k)}\big) \in L^s\big(0, T; L^2_\sigma(\Omega)\big), \quad (-\Delta)\big((\eta \theta)^{(k)}\big) \in L^s(0, T; L^2(\Omega)) \tag{5.71}$$

for all $\eta \in C_0^\infty(]0, T[)$ and all $2 \le s < \infty$.

3. To improve the regularity of $\eta u, \eta \theta$ in the spatial direction we write (5.66), (5.68) in the form

$$(-\Delta)(\eta u) + \nabla(\eta p) = \widetilde{f}_1, \tag{5.72}$$

$$(-\Delta)(\eta \theta) = \widetilde{f}_2, \tag{5.73}$$

and use well known regularity theory for the stationary Stokes equation and elliptic regularity theory to increase the spatial integrability for ηu and $\eta \theta$, respectively. Differentiating these equations in time and repeating the procedure again and again leads to

$$\eta u \in W^{k,2}(]0, T[\times G)^3, \quad \eta \theta \in W^{k,2}(]0, T[\times G) \tag{5.74}$$

for all $\eta \in C_0^\infty(]0, T[)$, all bounded domains $G \subseteq \Omega$ and all $k \in \mathbb{N}, 2 \le s < \infty$. Considering $\eta \in C_0^\infty(]0, T[)$ with $\eta(t) = 1$ for $t \in [\epsilon, T']$, and using Sobolev's imbedding theorem shows

$$u \in C_{\mathrm{loc}}^\infty(\overline{]\epsilon, T'[\times \Omega})^3, \quad \theta \in C_{\mathrm{loc}}^\infty(\overline{]\epsilon, T'[\times \Omega}) \tag{5.75}$$

for all $0 < \epsilon, T' < T$. Finally $p \in C_{\mathrm{loc}}^\infty(\overline{]\epsilon, T'[\times \Omega})$ for all $0 < \epsilon, T' < T$ will be proven.

In the following subsection we collect several lemmata that will be used to prove Theorem 5.15 and Theorem 5.16.

5.3.2 Preparatory lemmata

In the case that Ω is sufficiently smooth we can determine explicitly the domains of $(-\Delta)$ and of A. Further the estimates (5.76), (5.77) below will be used frequently throughout this section without giving explicit reference to them.

Lemma 5.12. *Let* $\Omega \subseteq \mathbb{R}^3$ *be a uniform* C^2*-domain. Then*

$$\mathcal{D}(-\Delta) = H_0^1(\Omega) \cap H^2(\Omega), \tag{5.76}$$

$$\mathcal{D}(A) = W_{0,\sigma}^{1,2}(\Omega) \cap H^2(\Omega)^3, \quad Au = -P\Delta u. \tag{5.77}$$

Moreover the estimates

$$\|\phi\|_{H^2(\Omega)} \le c(\Omega)\big(\|\phi\|_2 + \|(-\Delta)\phi\|_2\big) \quad \forall \phi \in \mathcal{D}(-\Delta), \tag{5.78}$$

$$\|u\|_{H^2(\Omega)^3} \le c(\Omega)\big(\|u\|_2 + \|Au\|_2\big) \quad \forall u \in \mathcal{D}(A) \tag{5.79}$$

are satisfied.

Proof. From [So01, Chapter III, Theorem 2.1.1 d)] we get (5.77). By ([So01, Chapter III, Section 2.4] it follow that (5.79) holds. The proof of (5.76), (5.78) is analogous. $\square$

The next lemma deals with regularity of the linear part of a solution of the Boussinesq equations in a uniform C^2-domain. Its proof is based on the maximal regularity of $(-\Delta)$ and Lemma 5.12. Moreover Yosida's smoothing procedure will be used.

Lemma 5.13. *Let $\Omega \subseteq \mathbb{R}^3$ be a uniform C^2-domain, let $0 < T < \infty$, $\theta_0 \in H_0^1(\Omega)$, and let $f \in L^2(0, T; L^2(\Omega))$. Define*

$$E(t) := E_1(t) + E_2(t) := e^{t\Delta}\theta_0 + \int_0^t e^{(t-\tau)\Delta} f(\tau)\, d\tau \tag{5.80}$$

for all $t \in [0, T[$. Then

$$E \in L^\infty(0, T; H_0^1(\Omega)) \cap L^2(0, T; H^2(\Omega)). \tag{5.81}$$

Further $E \in C([0, T[; L^2(\Omega))$ and

$$\begin{aligned} E_t + (-\Delta)E &= f && in\ ]0, T[\,, \\ E(0) &= \theta_0. \end{aligned} \tag{5.82}$$

Proof. Case E_1. We use [So01, Chapter IV, Lemma 1.5.3] (also true for $(-\Delta)$) and (2.53) obtain:

$$\begin{aligned} \|(-\Delta)E_1\|_{2,2;T} &= \|(-\Delta)^{1/2}e^{t\Delta}(-\Delta)^{1/2}\theta_0\|_{2,2;T} \\ &\leq \|(-\Delta)^{1/2}\theta_0\|_2 \\ &= \|\nabla\theta_0\|_2. \end{aligned}$$

Since $E_1'(t) = \Delta e^{t\Delta}\theta_0$, $t > 0$, it follows $E_1' \in L^2(0, T; L^2(\Omega))$ and (see Theorem 2.35) that E_1 satisfies (5.82) where f is replaced by 0. We make use of (5.78) to obtain $E_1 \in L^2(0, T; H^2(\Omega))$. There holds

$$\begin{aligned} \|(-\Delta)^{1/2}e^{t\Delta}\theta_0\|_2 &= \|e^{t\Delta}(-\Delta)^{1/2}\theta_0\|_2 \\ &\leq \|(-\Delta)^{1/2}\theta_0\|_2 \end{aligned}$$

for all $t > 0$. Consequently, $E_1 \in L^\infty(0, T; H_0^1(\Omega))$.

Case E_2. The maximal regularity property of $(-\Delta)$ yields that $E_2', (-\Delta)E_2 \in L^2(0, T; L^2\Omega))$ and that (5.82) holds. By (5.78) it follows $E_2 \in L^2(0, T; H^2(\Omega))$. To prove $\nabla E_2 \in L^\infty(0, T; L^2(\Omega)^3)$ we introduce the *Yosida approximation* of $(-\Delta)^{1/2}$ by

$$J_k : L^2(\Omega) \to \mathcal{D}((-\Delta)^{1/2})\,, \quad J_k v := \left(I + \frac{1}{k}(-\Delta)^{1/2}\right)^{-1} v.$$

There holds $\|J_k\| \leq 1$ and $\|(-\Delta)^{1/2}J_k\| \leq k$ for all $k \in \mathbb{N}$. For further properties of J_k we refer to [So01, Chapter II, Section 3.4]. Let us define

$$v(t) := (-\Delta)^{1/2} \int_0^t e^{(t-\tau)\Delta} f(\tau)\, d\tau\,, \quad t \in [0, T[\,, \tag{5.83}$$

and

$$v_k(t) := J_k(-\Delta)^{1/2} \int_0^t e^{(t-\tau)\Delta} f(\tau) \, d\tau \,, \quad t \in [0, T[\,, \tag{5.84}$$

for all $k \in \mathbb{N}$. By [So01, Chapter II, (3.4.8)]

$$v_k(t) \underset{k \to \infty}{\to} v(t) \quad \text{strongly in } L^2(\Omega) \tag{5.85}$$

for all $t \in [0, T[$. Since $J_k, (-\Delta)^{1/2}$ and $e^{t\Delta}$ commute, it follows

$$v_k(t) = \int_0^t e^{(t-\tau)\Delta} (-\Delta)^{1/2} J_k f(\tau) \, d\tau$$

for all $t \in [0, T[$. Fix $k \in \mathbb{N}$. Since $(-\Delta)^{1/2} J_k f \in L^2(0, T; L^2(\Omega))$ we obtain by the maximal regularity of $(-\Delta)$ that the equation

$$\begin{aligned} (v_k)_t + (-\Delta) v_k &= (-\Delta)^{1/2} J_k f \,, \\ v_k(0) &= 0 \end{aligned} \tag{5.86}$$

is satisfied in $L^2(0, T; L^2(\Omega))$. Multiplying (5.86) by v_k and integration over $[0, t[$ yields

$$\int_0^t \langle (v_k)', v_k \rangle_\Omega \, d\tau + \int_0^t \langle (-\Delta) v_k, v_k \rangle_\Omega \, d\tau = \int_0^t \langle (-\Delta)^{1/2} J_k f, v_k \rangle_\Omega \, d\tau$$

for all $t \in [0, T[$ and $k \in \mathbb{N}$. We integrate by parts (see [So01, Lemma 1.3.2]), use the self-adjointness of $(-\Delta)^{1/2}$ and obtain

$$\frac{1}{2} \|v_k(t)\|_2^2 + \int_0^t \|\nabla v_k(\tau)\|_2^2 \, d\tau = \int_0^t \langle J_k f, (-\Delta)^{1/2} v_k \rangle_\Omega \, d\tau. \tag{5.87}$$

There holds

$$\int_0^t |\langle J_k f, (-\Delta)^{1/2} v_k \rangle_\Omega| \, d\tau \le \frac{1}{4} \|J_k f\|_{2,2;t}^2 + \|(-\Delta)^{1/2} v_k\|_{2,2;t}^2 \le \frac{1}{4} \|f\|_{2,2;t}^2 + \|\nabla v_k\|_{2,2;t}^2.$$

Therefore $\|v_k(t)\|_2^2 \le \frac{1}{2} \|f\|_{2,2;t}^2$ for all $t \in [0, T[$ and $k \in \mathbb{N}$. By (5.85) we have that the inequality

$$\|v(t)\|_2^2 \le \frac{1}{2} \|f\|_{2,2;T}^2$$

holds for all $t \in [0, T[$. Consequently $\nabla E_2 \in L^\infty(0, T; L^2(\Omega))$. Altogether (5.81) is satisfied. $\qquad \square$

The following auxiliary Lemma will be useful in the proofs of Theorem 5.15 and Theorem 5.16. It is based on imbedding properties of A^α.

Lemma 5.14. *Let $\Omega \subseteq \mathbb{R}^3$ be a uniform C^2-domain, let $0 < T < \infty$, and let $2 \leq s_1 < \infty$. Assume that*

$$u \in L^\infty(0, T; W^{1,2}_{0,\sigma}(\Omega)) \cap L^{s_1}(0, T; H^2(\Omega)^3).$$

Then

$$u \in L^s(0, T; L^\infty(\Omega)^3) \tag{5.88}$$

for all $2 \leq s < 2s_1$.

Proof. Fix $2 < s < \infty$ with $s_1 < s < 2s_1$. Define $\alpha := \frac{s_1}{2s}$. Then $\frac{1}{4} < \alpha < \frac{1}{2}$. Define $3 < q < 6$, such that $2\alpha + \frac{3}{q} = \frac{3}{2}$. Sobolev's imbedding theorem (see [AdFo03, Theorem 4.12]) implies that the imbedding

$$W^{1,q}(\Omega) \hookrightarrow L^\infty(\Omega)$$

is continuous. We get with [So01, Chapter III, (2.4.18)] and (2.48) that

$$\begin{aligned}
\|u(t)\|_\infty &\leq c\|u(t)\|_{W^{1,q}(\Omega)} \\
&\leq c\big(\|A^{1/2+\alpha}u(t)\|_2 + \|u(t)\|_2\big) \\
&\leq c\big(\|Au(t)\|_2^{2\alpha}\|A^{1/2}u(t)\|_2^{1-2\alpha} + \|u(t)\|_2\big)
\end{aligned}$$

for almost all $t \in [0, T[$ with a constant $c > 0$ independent of t. Therefore

$$\|u\|_{\infty,s;T} \leq c\big(\|Au\|_{2,s_1;T}^{2\alpha}\|A^{1/2}u\|_{2,\infty;T}^{1-2\alpha} + \|u\|_{2,s;T}\big) < \infty.$$

$\square$

5.3.3 Proof of the smoothness of a strong solution

Theorem 5.15. *Let $\Omega \subseteq \mathbb{R}^3$ be a uniform C^2-domain, let $0 < T < \infty$, and let $u_0 \in W^{1,2}_{0,\sigma}(\Omega)$, $\theta_0 \in H^1_0(\Omega)$. Consider $g \in L^8(0, T; L^4(\Omega)^3)$ and $f_1 \in L^2(0, T; L^2(\Omega)^3)$, $f_2 \in L^2(0, T; L^2(\Omega))$. Let (u, θ) be a weak solution of (5.1) in $[0, T[\times\Omega$ and assume $u \in L^8(0, T; L^4(\Omega)^3)$. Then the following statements are satisfied:*

(i) u fulfils

$$u \in L^\infty(0, T; W^{1,2}_{0,\sigma}(\Omega)) \cap L^2(0, T; H^2(\Omega)^3), \tag{5.89}$$

$$u_t \in L^2(0, T; L^2_\sigma(\Omega)), \tag{5.90}$$

and

$$u \cdot \nabla u \in L^2(0, T; L^2(\Omega)^3). \tag{5.91}$$

Moreover, there exists an associated pressure p of u with

$$p \in L^2(0, T; L^2_{loc}(\overline{\Omega})), \quad \nabla p \in L^2(0, T; L^2(\Omega)^3) \tag{5.92}$$

such that

$$u_t - \Delta u + u \cdot \nabla u + \nabla p = f_1 + \theta g \quad \text{in } L^2(0, T; L^2(\Omega)). \tag{5.93}$$

(ii) θ fulfils

$$\theta \in L^{\infty}(0,T;H^1_0(\Omega)) \cap L^2(0,T;H^2(\Omega)), \qquad (5.94)$$
$$\theta_t \in L^2(0,T;L^2(\Omega)), \qquad (5.95)$$

and

$$u \cdot \nabla\theta \in L^2(0,T;L^2(\Omega)). \qquad (5.96)$$

Moreover

$$\theta_t - \Delta\theta + u \cdot \nabla\theta = f_2 \quad in \ L^2(0,T;L^2(\Omega)). \qquad (5.97)$$

Proof. Step 1. Since $\theta \in L^{8/3}(0,T;L^4(\Omega)^3)$ it follows by Hölder's inequality $\theta g \in L^2(0,T;L^2(\Omega)^3)$. There holds that u is a weak solution of the instationary Navier-Stokes equations with initial value u_0 and external force $\tilde{f} := \theta g + f_1$. Since the Serrin condition $u \in L^8(0,T;L^4(\Omega)^3)$ is satisfied, we apply [So01, Chapter V, Theorem 1.8.1] and obtain (5.89), (5.90), (5.91). Moreover, by this Theorem there exists p satisfying (5.92), (5.93).

Step 2. The crucial point of this step is to prove $u \cdot \nabla\theta \in L^2(0,T;L^2(\Omega))$. We know that the representation formula (see (5.9))

$$\theta(t) = E(t) + \tilde{\theta}(t) := E(t) - (-\Delta)^{1/2} \int_0^t e^{(t-\tau)\Delta}(-\Delta)^{-1/2}\mathrm{div}\big(\theta u\big)(\tau)\,d\tau \qquad (5.98)$$

holds for almost all $t \in [0,T[$. By (5.88) and $\nabla\theta \in L^2(0,T;L^2(\Omega)^3)$ it follows $u \cdot \nabla\theta \in L^1(0,T;L^2(\Omega))$. Consequently, since $\mathrm{div}\big((\theta u)(t)\big) = (u \cdot \nabla\theta)(t)$ for a.a. $t \in [0,T[$ in the sense of distributions on Ω, we can apply Lemma 2.40 and obtain

$$\tilde{\theta}(t) = -\int_0^t e^{(t-\tau)\Delta}\big(u \cdot \nabla\theta\big)(\tau)\,d\tau \qquad (5.99)$$

for almost all $t \in [0,T[$. By Lemma 5.13

$$E \in L^{\infty}(0,T;H^1_0(\Omega)). \qquad (5.100)$$

The proof of $u \cdot \nabla\theta \in L^2(0,T;L^2(\Omega))$ is based on the following

Assertion. Let $2 \le s_1 < \infty$. Assume $\nabla\theta \in L^{s_1}(0,T;L^2(\Omega)^3)$. Let $2 < s_2 < \infty$ be defined by $\frac{1}{4} + \frac{1}{s_2} = \frac{1}{s_1}$. Then

$$\nabla\theta \in L^s(0,T;L^2(\Omega)^3) \qquad (5.101)$$

for all $2 \le s < s_2$.

Proof of the assertion. We get (see (2.23), (2.55))

$$\|(-\Delta)^{1/2}\tilde{\theta}(t)\|_2 = \left\| \int_0^t (-\Delta)^{1/2}e^{(t-\tau)\Delta}(u \cdot \nabla\theta)(\tau)\,d\tau \right\|_2$$
$$\le \int_0^T |t-\tau|^{-1/2}\|(u \cdot \nabla\theta)(\tau)\|_2\,d\tau$$

for a.a. $t \in [0, T[$. By (5.88) there holds $u \in L^s(0, T; L^\infty(\Omega)^3)$ for all $1 \le s <$ 4. Define $1 < s_3 < \infty$ by $\frac{1}{s_3} = \frac{1}{4} + \frac{1}{s_1}$. Consequently, the elementary estimate $\|u \cdot \nabla\theta\|_2 \le \|u\|_\infty \|\nabla\theta\|_2$, and Hölder's inequality imply $u \cdot \nabla\theta \in L^s(0, T; L^2(\Omega))$ for all $1 \le s < s_3$. Thus, the Hardy-Littlewood inequality with $\frac{1}{2} + \frac{1}{s_2} = \frac{1}{s_3}$ implies $(-\Delta)^{1/2}\tilde{\theta} \in L^s(0, T; L^2(\Omega))$ for all $1 \le s < s_2$. By (5.98), (5.100) and (2.53) the proof of the claim is finished. $\qquad\square$

The requirements of the assertion are fulfilled with $s_1 = 2$. Thus $\nabla\theta \in L^s(0, T; L^2(\Omega)^3)$ for all $1 \le s < 4$. Consequently, the requirements of the assertion are satisfied for all $2 \le s_1 < 4$. Thus $\nabla\theta \in L^s(0, T; L^2(\Omega)^3)$ for all $2 \le s < \infty$. Altogether $u \cdot \nabla\theta \in L^2(0, T, L^2(\Omega))$.

Step 3. We make use of Lemma 5.13 with $\tilde{f} := (f_2 - u \cdot \nabla\theta) \in L^2(0, T; L^2(\Omega))$ to obtain that (5.94), (5.95) are satisfied. Consequently, (5.97) is fulfilled, and therefore, the proof of this theorem is completed. $\qquad\square$

Theorem 5.16. *Let $\Omega \subseteq \mathbb{R}^3$ be a uniform C^2-domain such that Ω is also a C^∞-domain. Let $0 < T \le \infty$ let $f_1, g \in C_0^\infty(]0, T[\times\Omega)^3$, let $f_2 \in C_0^\infty(]0, T[\times\Omega)$. Consider $u_0 \in W_{0,\sigma}^{1,2}(\Omega)$, $\theta_0 \in H_0^1(\Omega)$. Let (u, θ) be a weak solution of (5.1) in $[0, T[\times\Omega$ and assume $u \in L^8(0, T; L^4(\Omega)^3)$. Then, after redefinition on a null set of $]0, T[\times\Omega$,*

$$u \in C^\infty(]0, T[\times\Omega)^3, \quad \theta \in C^\infty(]0, T[\times\Omega). \tag{5.102}$$

Moreover

$$u \in C_{loc}^\infty(]\epsilon, T'[\times\overline{\Omega})^3, \quad \theta \in C_{loc}^\infty(]\epsilon, T'[\times\overline{\Omega}) \tag{5.103}$$

for all ϵ, T' with $0 < \epsilon < T' < T$. There exists an associated pressure $p \in C^\infty(]0, T[\times\Omega)$ such that

$$u_t - \Delta u + u \cdot \nabla u + \nabla p = \theta g + f_1. \tag{5.104}$$

Furthermore $p \in C_{loc}^\infty(]\epsilon, T'[\times\overline{\Omega})$ for all ϵ, T' with $0 < \epsilon < T' < T$.

Proof. With no loss of generality assume $0 < T < \infty$. From Theorem 5.15 we get

$$u \in L^\infty(0, T; W_{0,\sigma}^{1,2}(\Omega)) \cap L^2(0, T; H^2(\Omega)^3), \quad u_t \in L^2(0, T; L_\sigma^2(\Omega)), \tag{5.105}$$

and

$$u_t + Au = P(\theta g) + Pf_1 - P(u \cdot \nabla u) \quad \text{in } L^2(0, T; L_\sigma^2(\Omega)). \tag{5.106}$$

Let p be an associated pressure of u satisfying (5.92) such that

$$u_t - \Delta u + u \cdot \nabla u + \nabla p = f_1 + \theta g \quad \text{in } L^2(0, T; L^2(\Omega)^3). \tag{5.107}$$

Moreover, by this theorem

$$\theta \in L^\infty(0, T; H_0^1(\Omega)) \cap L^2(0, T; H^2(\Omega)), \quad \theta_t \in L^2(0, T; L^2(\Omega)) \tag{5.108}$$

and

$$\theta_t - \Delta\theta = f_2 - u \cdot \nabla\theta \quad \text{in } L^2(0, T; L^2(\Omega)). \tag{5.109}$$

Fix a non-negative function $\eta \in C_0^\infty(]0, T[)$.

Step 1a. By (5.106)

$$(\eta u)_t + A(\eta u) = P(\eta f_1) + P(\eta \theta g) + \eta_t u - P(\eta u \cdot \nabla u) \quad \text{in } L^2(0, T; L_\sigma^2(\Omega)). \quad (5.110)$$

Therefore (see (2.52))

$$(\eta u)(t) = \int_0^t e^{-(t-\tau)A}\Big(P(\eta f_1) + P(\eta \theta g) + \eta_t u - P(\eta u \cdot \nabla u)\Big)(\tau)\, d\tau \quad (5.111)$$

for almost all $t \in [0, T[$. By (5.88), (5.105) and Hölder's inequality we get $\eta u \cdot \nabla u \in L^s(0, T; L^2(\Omega)^3)$ for all $2 \le s < 4$. It follows

$$P(\eta f_1) + P(\eta \theta g) + \eta_t u - P(\eta u \cdot \nabla u) \in L^s(0, T; L_\sigma^2(\Omega))$$

for all $2 \le s < 4$. Therefore the maximal regularity of A applied to (5.111) implies

$$(\eta u)_t \in L^s(0, T; L_\sigma^2(\Omega)), \quad A(\eta u) \in L^s(0, T; L_\sigma^2(\Omega)) \quad (5.112)$$

for all $2 \le s < 4$. From (5.88), (5.105), (5.112) we get $\eta u \cdot \nabla u \in L^s(0, T; L^2(\Omega)^3)$ for all $4 \le s < 8$. Consequently, the maximal regularity of A applied to (5.111) yields that (5.112) holds for all $4 \le s < 8$. Repeating this procedure again and again we finally get

$$(\eta u)_t \in L^s(0, T; L_\sigma^2(\Omega)), \quad A(\eta u) \in L^s(0, T; L_\sigma^2(\Omega)) \quad \text{for all } 2 \le s < \infty. \quad (5.113)$$

From (5.79), (5.113), and the continuous imbedding $H^2(\Omega)^3 \hookrightarrow L^\infty(\Omega)^3$ we get

$$\eta u \in L^s(0, T; L^\infty(\Omega)^3) \quad (5.114)$$

for all $2 \le s < \infty$.

Step 1b. By (2.52), (5.109) it follows

$$(\eta \theta)_t + (-\Delta)(\eta \theta) = \eta f_2 + \eta_t \theta - \eta u \cdot \nabla \theta \quad \text{in } L^2(0, T; L^2(\Omega)), \quad (5.115)$$

$$(\eta \theta)(t) = \int_0^t e^{(t-\tau)\Delta}\Big(\eta f_2 + \eta_t \theta - \eta u \cdot \nabla \theta\Big)(\tau)\, d\tau \quad \text{for a.a. } t \in [0, T[. \quad (5.116)$$

Due to (5.108), (5.114) and Hölder's inequality we get $\eta u \cdot \nabla \theta \in L^s(0, T; L^2(\Omega))$ for all $2 \le s < \infty$. Consequently $\eta f_2 + \eta_t \theta - \eta u \cdot \nabla u \in L^s(0, T; L^2(\Omega))$ for all $2 \le s < \infty$. Thus the maximal regularity of $(-\Delta)$ applied to (5.116) yields

$$(\eta \theta)_t \in L^s(0, T; L^2(\Omega)), \quad (-\Delta)(\eta \theta) \in L^s(0, T; L^2(\Omega)) \quad (5.117)$$

for all $2 \le s < \infty$. Therefore, from (5.78), (5.117) and the continuous imbedding $H^2(\Omega) \hookrightarrow L^\infty(\Omega)$ we get

$$\eta \theta \in L^s(0, T; L^\infty(\Omega)) \quad (5.118)$$

for all $2 \leq s < \infty$.

Step 2a. The goal of this step is to prove that

$$(-\Delta)^{1/2}\big((\eta\theta)_t\big) \in L^s(0,T;L^2(\Omega)) \tag{5.119}$$

for all $2 \leq s < \infty$. Fix $2 \leq s < \infty$. We use $\mathrm{div}(\eta\theta u) = \eta u \cdot \nabla\theta$ in combination with Lemma 2.40 to rewrite (5.116) as

$$
\begin{aligned}
(\eta\theta)(t) &= \int_0^t e^{(t-\tau)\Delta}\Big(\eta f_2 + \eta_t\theta\Big)(\tau)\,d\tau \\
&\quad - (-\Delta)^{1/2}\int_0^t e^{(t-\tau)\Delta}(-\Delta)^{-1/2}\mathrm{div}\big(\eta\theta u\big)\,d\tau \\
&=: E(t) + \tilde{\theta}(t)
\end{aligned}
\tag{5.120}
$$

for almost all $t \in [0,T[$. From (5.114), (5.117) it follows $\eta\theta_t u \in L^s(0,T;L^2(\Omega)^3)$. We make use of (5.113), (5.118) to obtain $\eta\theta u_t \in L^s(0,T;L^2(\Omega)^3)$. It follows $(\eta\theta u)_t \in L^s(0,T;L^2(\Omega)^3)$. Applying Lemma 2.20 to $f = (-\Delta)^{-1/2}\mathrm{div}(\eta\theta u)$ yields

$$f_t = (-\Delta)^{-1/2}\mathrm{div}\big((\eta\theta u)_t\big) \in L^s(0,T;L^2(\Omega)).$$

Consequently, by Lemma 2.41 (ii) we get

$$\tilde{\theta}_t(t) = (-\Delta)^{1/2}\int_0^t e^{(t-\tau)\Delta}(-\Delta)^{-1/2}\mathrm{div}\big((\eta\theta u)_t(\tau)\big)\,d\tau \ , \quad \text{a.a. } t \in [0,T[, \tag{5.121}$$

and $(-\Delta)^{1/2}(\tilde{\theta}_t) \in L^s(0,T,L^2(\Omega))$. Since $\big(\eta f_2 + \eta_t\theta\big)_t \in L^s(0,T;L^2(\Omega))$ (see (5.117)) we get from Lemma 2.41 (i) that $(-\Delta)(E_t) \in L^s(0,T;L^2(\Omega))$. Altogether (5.119) is fulfilled.

Step 2b. We use that Lemma 2.40 and Lemma 2.41 hold true where $(-\Delta)$ is replaced by A and $L^2(\Omega)$ is replaced by $L^2_\sigma(\Omega)$ (see the remark following the proof of Lemma 2.41.) We use $\mathrm{div}(\eta u \otimes u) = \eta u \cdot \nabla u$ and Lemma 2.40 to rewrite (5.111) as

$$
\begin{aligned}
(\eta u)(t) &= \int_0^t e^{-(t-\tau)A}\Big(P(\eta f_1) + P(\eta\theta g) + \eta_t u\Big)(\tau)\,d\tau \\
&\quad - A^{1/2}\int_0^t e^{-(t-\tau)A}A^{-1/2}P\mathrm{div}\big(\eta u \otimes u\big)(\tau)\,d\tau
\end{aligned}
$$

for almost all $t \in [0,T[$. From (5.113), (5.114) it follows that an analogous argumentation as in step 2a yields

$$A^{1/2}\big((\eta u)_t\big) \in L^s(0,T;L^2_\sigma(\Omega)) \tag{5.122}$$

for all $2 \leq s < \infty$.

Step 2c. Fix $2 \leq s < \infty$. We use Hölder's inequality, (5.114), (5.119) to obtain $\eta u \cdot (\nabla\theta)_t \in L^s(0,T;L^2(\Omega))$. From Hölder's inequality, (5.117), (5.122) we get $\eta u_t \cdot$

$\nabla\theta \in L^s(0,T;L^2(\Omega))$. It follows $\big(\eta f_2 + \eta_t\theta - \eta u \cdot \nabla\theta\big)_t \in L^s(0,T;L^2(\Omega))$. Therefore Lemma 2.41 (i) applied to (5.116) yields

$$(\eta\theta)_{tt}, \quad (-\Delta)\big((\eta\theta)_t\big) \in L^s(0,T;L^2(\Omega)) \tag{5.123}$$

for all $2 \le s < \infty$.

Step 2d. Fix $2 \le s < \infty$. We use Hölder's inequality and (5.113), (5.114), (5.122) to obtain

$$\eta u_t \cdot \nabla u, \eta u \cdot (\nabla u)_t \in L^s(0,T;L^2(\Omega)^3).$$

It follows $\big(P(\eta f_1) + P(\eta\theta g) + \eta_t u - P(\eta u \cdot \nabla u)\big)_t \in L^s(0,T;L^2_\sigma(\Omega))$. Therefore Lemma 2.41 (i) applied to (5.111) yields

$$(\eta u)_{tt} \in L^s(0,T;L^2_\sigma(\Omega)), \quad A\big((\eta u)_t\big) \in L^s(0,T;L^2_\sigma(\Omega)) \tag{5.124}$$

for all $2 \le s < \infty$.

Step 3. We can iterate the procedure in Step 2 to obtain

$$(\eta u)^{(k+1)} \in L^s(0,T;L^2_\sigma(\Omega)), \quad A\big((\eta u)^{(k)}\big) \in L^s(0,T;L^2_\sigma(\Omega)), \tag{5.125}$$

$$(\eta\theta)^{(k+1)} \in L^s(0,T;L^2(\Omega)), \quad (-\Delta)\big((\eta\theta)^{(k)}\big) \in L^s(0,T;L^2(\Omega)), \tag{5.126}$$

for all non-negative functions $\eta \in C_0^\infty(]0,T[)$ and all $k \in \mathbb{N}, 2 \le s < \infty$. Consequently we have

$$(\eta u)^{(k)} \in L^s(0,T;H^2(\Omega)^3), \tag{5.127}$$

$$(\eta\theta)^{(k)} \in L^s(0,T;H^2(\Omega)) \tag{5.128}$$

for all η, k, s as above.

Step 4a. Fix a non-negative function $\eta \in C_0^\infty(]0,T[)$. By (5.115) we have

$$(-\Delta)(\eta\theta) = \widetilde{f}_2 \tag{5.129}$$

where

$$\widetilde{f}_2 := -\eta\theta_t + \eta f_2 - \eta u \cdot \nabla\theta.$$

Using (5.114), (5.127), (5.128) and Hölder's inequality it follows $\widetilde{f}_2 \in L^s(0,T;H^1(\Omega))$ for all $2 \le s < \infty$. Fix a bounded domain $G \subseteq \Omega$. Applying well known elliptic regularity theory to (5.129) we get $\|(\eta\theta)(t)\|_{H^3(G)} \le c\|\widetilde{f}_2(t)\|_{H^1(\Omega)}$ for a.a. $t \in [0,T[$ with a constant $c > 0$ independent of t. Integration yields $\eta\theta \in L^s(0,T;H^3(G))$. Therefore, by the definition of $H^3_{\text{loc}}(\overline{\Omega})$, this means

$$\eta\theta \in L^s(0,T;H^3_{\text{loc}}(\overline{\Omega})), \quad 2 \le s < \infty. \tag{5.130}$$

Combining this spatial regularity of $\eta\theta$ with (5.113) and making use of (5.123), it can be proved that $\widetilde{f}_2 \in L^s(0,T;H^2_{\text{loc}}(\overline{\Omega}))$ is satisfied. With the help of elliptic regularity theory applied to (5.129), we prove

$$\eta\theta \in L^s(0,T;H^4_{\text{loc}}(\overline{\Omega})), \quad 2 \le s < \infty. \tag{5.131}$$

To proceed in this way, we need more regularity of θ_t and of u. To increase the regularity of $(\eta\theta)_t$ we use Lemma 2.41 to differentiate (5.129). Consequently

$$(-\Delta)\big((\eta\theta)_t\big) = (\widetilde{f}_2)_t. \tag{5.132}$$

We use (5.125), (5.126) and Hölder's inequality to obtain $(\eta u \cdot \nabla\theta)_t \in L^s(0,T;H^1(\Omega))$. Consequently $(\widetilde{f}_2)_t \in L^s(0,T;H^1(\Omega))$. Consequently, by regularity of the Dirichlet Problem (5.132), it follows

$$(\eta\theta)_t \in L^s(0,T;H^3_{\mathrm{loc}}(\overline{\Omega})), \quad 2 \le s < \infty. \tag{5.133}$$

Step 4b. Next we will increase the spatial regularity of u. There holds

$$(-\Delta)(\eta u) + \nabla(\eta p) = \widetilde{f}_1 \tag{5.134}$$

where

$$\widetilde{f}_1 := -\eta u_t + \eta f_1 + \eta\theta g - \eta u \cdot \nabla u.$$

Then $\widetilde{f}_1 \in L^s(0,T;H^1(\Omega)^3)$ for all $2 \le s < \infty$. Consequently for each bounded domain $G \subseteq \Omega$, we apply the stationary estimates of [So01, Chapter III, Theorem 1.5.1] to (5.134) and obtain $\|(\eta u)(t)\|_{H^3(G)^3} \le c\|\widetilde{f}_1(t)\|_{H^1(\Omega)^3}$ for a.a. $t \in [0,T[$ with a constant $c > 0$. Thus $\eta u \in L^s(0,T;H^3_{\mathrm{loc}}(\overline{\Omega})^3)$ for all $2 \le s < \infty$. Now, $\widetilde{f}_1 \in L^s(0,T;H^2_{\mathrm{loc}}(\overline{\Omega})^3)$, and so [So01, Chapter III, Theorem 1.5.1] applied to (5.134) yields

$$\eta u \in L^s(0,T;H^4_{\mathrm{loc}}(\overline{\Omega})^3), \quad 2 \le s < \infty. \tag{5.135}$$

Step 4c. At this point, we have enough spatial regularity of θ_t and u to show that $\widetilde{f}_2 \in L^s(0,T;H^3_{\mathrm{loc}}(\overline{\Omega}))$ holds for all $2 \le s < \infty$. Consequently, application of elliptic regularity theory to (5.129) yields

$$\eta\theta \in L^s(0,T;H^5_{\mathrm{loc}}(\overline{\Omega})), \quad 2 \le s < \infty. \tag{5.136}$$

Step 4d. It can be proved that $(\widetilde{f}_1)_t \in L^s(0,T;H^1(\Omega)^3), 2 \le s < \infty$, holds. Thus, differentiating (5.134) and using regularity theory of the stationary Stokes equations ([So01, Chapter III, Theorem 1.5.1]) yields $(\eta u)_t \in L^s(0,T;H^3_{\mathrm{loc}}(\overline{\Omega})^3)$. Thus, $\widetilde{f}_1 \in L^s(0,T;H^3_{\mathrm{loc}}(\overline{\Omega})^3)$ and so $\eta u \in L^s(0,T;H^5_{\mathrm{loc}}(\overline{\Omega})^3)$ for all $2 \le s < \infty$.

Step 5. Repeating the procedure presented above again and again we obtain that

$$(\eta u)^{(k)} \in L^s(0,T;H^j_{\mathrm{loc}}(\overline{\Omega})^3), \quad (\eta\theta)^{(k)} \in L^s(0,T;H^j_{\mathrm{loc}}(\overline{\Omega})) \tag{5.137}$$

for all $k,j \in \mathbb{N}$ and all $2 \le s < \infty$. Fix a bounded Lipschitz domain $G \subseteq \Omega$ and fix ϵ, T' with $0 < \epsilon < T' < T$. Let $\eta_\epsilon \in C_0^\infty(]0,T[)$ be a non-negative cut-off function with $\eta_\epsilon(t) = 1$ for all $t \in [\epsilon, T']$. Then it follows from (5.137) that

$$\eta_\epsilon u \in W^{j,2}(]0,T[\times G)^3, \quad \eta_\epsilon\theta \in W^{j,2}(]0,T[\times G) \tag{5.138}$$

97

for all $j \in \mathbb{N}$. The Sobolev imbedding theorem [AdFo03, Theorem 4.12, Part II] yields, after a redefinition on a null set, $\eta_\epsilon u \in C^j(\overline{]0, T[\times G})^3$ and $\eta_\epsilon \theta \in C^j(\overline{]0, T[\times G})$ for all $j \in \mathbb{N}$. By definition of η_ϵ we have

$$u \in C^j(\overline{]\epsilon, T'[\times G})^3, \quad \theta \in C^j(\overline{]\epsilon, T'[\times G}) \tag{5.139}$$

for all $j \in \mathbb{N}$. Since the statement above holds for all bounded Lipschitz domains $G \subseteq \Omega$, all ϵ, T' with $0 < \epsilon < T' < T$ and all $j \in \mathbb{N}$ we see that u and θ can be redefined on a null set of $]0, T[\times \Omega$ such that (5.102) holds.

It remains to prove the regularity properties of the pressure p. By (5.107), we have

$$\nabla p = f_1 + \theta g - u_t + \nabla u - u \cdot \nabla u. \tag{5.140}$$

Thus, by (5.102) we get $\nabla p \in C^\infty_{\mathrm{loc}}(\overline{]\epsilon, T'[\times \Omega})^3$ for all ϵ, T' with $0 < \epsilon < T' < T$. In the same way as in [So01, Chapter IV, Theorem 2.7.3] it can be proved that p can be redefined on a null set of $]0, T[\times \Omega$ such that $p \in C^\infty_{\mathrm{loc}}(\overline{]\epsilon, T'[\times \Omega})$ holds for all ϵ, T' with $0 < \epsilon < T' < T$. □

5.4 Regularity criteria for the Boussinesq equations in general domains

Consider the instationary Navier Stokes equations

$$\begin{aligned}
u_t - \Delta u + u \cdot \nabla u + \nabla p &= f &&\text{in }]0, T[\times \Omega, \\
\mathrm{div}\, u &= 0 &&\text{in }]0, T[\times \Omega, \\
u &= 0 &&\text{on }]0, T[\times \partial\Omega, \\
u &= u_0 &&\text{in } \Omega,
\end{aligned} \tag{5.141}$$

in the space time cylinder $[0, T[\times \Omega$ with external force f and initial value u_0.

From Theorem 5.8 we get the following result, c.f. [FSV09, Theorem 4.1] with a slightly different condition on f.

Theorem 5.17. *Let $\Omega \subseteq \mathbb{R}^3$ be a general domain, let $u_0 \in L^2_\sigma(\Omega)$, let $0 < T \leq \infty$, $f \in L^{4/3}(0, T; L^2(\Omega)^3)$. Then there exists an absolute constant $\epsilon_* > 0$ (independent of Ω, T, f, u_0) with the following property: Assume that the two conditions*

$$\left(\int_0^T \|e^{-\tau A} u_0\|_4^8 \, d\tau \right)^{1/8} \leq \epsilon_*, \tag{5.142}$$

$$\|f\|_{2, \frac{4}{3}; T} \leq \epsilon_* \tag{5.143}$$

are satisfied. Then there exists a strong solution $u \in L^8(0, T; L^4(\Omega)^3)$ of the instationary Navier-Stokes equations (5.141) with initial value u_0 and external force f.

Proof. Let $\epsilon_* > 0$ be the constant from Theorem 5.8 (we do not need the constant K_* here). Apply Theorem 5.8 with $f_1 := f$, $f_2 := 0$, gravitational force $g := 0$ and initial values u_0 and $\theta_0 := 0$ to obtain that the conditions (5.142), (5.143) are sufficient for the existence of a strong solution $u \in L^8(0, T; L^4(\Omega)^3)$ of (5.141).

$\square$

This section deals with regularity of weak solutions (u, θ) of the Boussinesq equations. Since we consider a general domain with not necessarily any boundary regularity we cannot expect that (u, θ) has improved differentiability conditions. Therefore the goal is to formulate criteria that (u, θ) is a strong solution locally in a neighbourhood of a point or even globally in an interval $I \subseteq \mathbb{R}$. We begin with the following definition.

Definition 5.18. Let $\Omega \subseteq \mathbb{R}^3$ be a general domain, let $u_0 \in L^2_\sigma(\Omega)$, $\theta_0 \in L^2(\Omega)$. Further assume $0 < T \leq \infty$, $g \in L^{8/3}(0, T; L^4(\Omega)^3)$, and $f_1 \in L^{4/3}(0, T; L^2(\Omega)^3)$, $f_2 \in L^1(0, T; L^2(\Omega))$. Consider a weak solution (u, θ) of the Boussinesq system (5.1) in $[0, T[\times\Omega$.

1. We say that (u, θ) is *regular at* $t \in]0, T[$ if there exists a $\delta > 0$ such that $u \in L^8(t - \delta, t + \delta; L^4(\Omega)^3)$.

2. We say that (u, θ) satisfies the *strong energy inequalities* if there is a null set $N \subseteq]0, T[$ such that

$$\frac{1}{2}\|u(t)\|_2^2 + \int_s^t \|\nabla u\|_2^2 \, d\tau \leq \frac{1}{2}\|u(s)\|_2^2 + \int_s^t \langle \theta g, u \rangle_\Omega \, d\tau + \int_s^t \langle f_1, u \rangle_\Omega \, d\tau,$$
(5.144)

$$\frac{1}{2}\|\theta(t)\|_2^2 + \int_s^t \|\nabla\theta\|_2^2 \, d\tau \leq \frac{1}{2}\|\theta(s)\|_2^2 + \int_s^t \langle f_2, \theta \rangle_\Omega \, d\tau$$
(5.145)

for all $s \in (\,]0, T[\backslash N\,) \cup \{0\}$ (with $u(0) := u_0$, $\theta(0) := \theta_0$) and all $t \in [s, T[$.

Especially, (5.144), (5.145) are fulfilled for $s = 0$ and all $t \in [0, T[$. Therefore the strong energy inequality implies the energy inequality.

Remark. (i) Up to now the existence of weak solutions of the Boussinesq equations in general domains satisfying the strong energy inequality is not known.

(ii) Since $\theta g \in L^{4/3}(0, T; L^2(\Omega)^3)$, $u \in L^4_{\text{loc}}(0, T; L^2(\Omega)^3)$ the integral $\int_s^t \langle \theta g, u \rangle_\Omega \, d\tau$ in (5.144) is well defined.

(iii) Let (u, θ) be a weak solution of (5.1). Then (see Theorem 5.2 (ii)), after a redefinition on a null set of $[0, T[$, there holds that $u : [0, T[\rightarrow L^2_\sigma(\Omega)$ and $\theta : [0, T[\rightarrow L^2(\Omega)$ are both weakly continuous functions. Especially, $u(t) \in L^2_\sigma(\Omega)$, $\theta(t) \in L^2(\Omega)$ are well defined for all $t \in [0, T[$ and $u(0) = u_0$, $\theta(0) = \theta_0$.

Notation. Throughout this section we will assume (see the remark above) that for every weak solution (u, θ) of (5.1) the functions $u : [0, T[\rightarrow L^2_\sigma(\Omega)$, $\theta : [0, T[\rightarrow L^2(\Omega)$ are weakly continuous.

The following auxiliary lemma is needed for the proof of Theorem 5.20 below.

Lemma 5.19. *Let $\Omega \subseteq \mathbb{R}^3$ be a general domain, let $u_0 \in L^2_\sigma(\Omega), \theta_0 \in L^2(\Omega)$. Further assume $0 < T \leq \infty, g \in L^{8/3}(0,T; L^4(\Omega)^3)$, assume $f_1 \in L^{4/3}(0,T; L^2(\Omega)^3)$, assume $f_2 \in L^1(0,T; L^2(\Omega))$. Consider a weak solution (u,θ) of the Boussinesq equations (5.1) in $[0,T[\times\Omega$. Assume θ fulfils (5.145), let $0 < t_0 < t_1 < T$, and let $t_0 \in (\,]0,T[\backslash N\,) \cup \{0\}$. (In the case $t_0 = 0$ define $\theta(0) := \theta_0$). Then*

$$\|\theta g\|_{2,\frac{4}{3};t_0,t_1} \leq K\|g\|_{4,\frac{8}{3};t_0,t_1}\left(\|\theta(t_0)\|_2 + \|f_2\|_{2,1;t_0,t_1}\right) \tag{5.146}$$

with an absolute constant $K > 0$.

Proof. We get from (5.145) that

$$\frac{1}{2}\|\theta\|^2_{2,\infty;t_0,t_1} \leq \frac{1}{2}\|\theta(t_0)\|^2_2 + \frac{1}{8}\|\theta\|^2_{2,\infty;t_0,t_1} + 2\|f_2\|^2_{2,1;t_0,t_1},$$

$$\int_{t_0}^{t_1} \|\nabla\theta\|^2_2\, d\tau \leq \frac{1}{2}\|\theta(t_0)\|^2_2 + \frac{1}{8}\|\theta\|^2_{2,\infty;t_0,t_1} + 2\|f_2\|^2_{2,1;t_0,t_1}$$

is satisfied. Adding these equations yields

$$\frac{1}{4}\|\theta\|^2_{2,\infty;t_0,t_1} + \int_{t_0}^{t_1} \|\nabla\theta\|^2_2\, d\tau \leq \|\theta(t_0)\|^2_2 + 4\|f_2\|^2_{2,1;t_0,t_1}. \tag{5.147}$$

It follows

$$\|\theta\|_{2,\infty;t_0,t_1} + \|\nabla\theta\|_{2,2;t_0,t_1} \leq K\left(\|\theta(t_0)\|_2 + \|f_2\|_{2,1;t_0,t_1}\right) \tag{5.148}$$

with an absolute constant $K > 0$. By interpolation, Sobolev imbedding and Young's inequality it follows

$$
\begin{aligned}
\|\theta\|_{4,\frac{8}{3};t_0,t_1} &\leq \|\theta\|^{1/4}_{2,\infty;t_0,t_1}\|\theta\|^{3/4}_{6,2;t_0,t_1} \\
&\leq K\|\theta\|^{1/4}_{2,\infty;t_0,t_1}\|\nabla\theta\|^{3/4}_{2,2;t_0,t_1} \\
&\leq K\left(\|\theta\|_{2,\infty;t_0,t_1} + \|\nabla\theta\|_{2,2;t_0,t_1}\right)
\end{aligned}
\tag{5.149}
$$

with an absolute constant $K > 0$. From (5.149) and Hölder's inequality we obtain (5.146). □

Assume we want to prove regularity of a weak solution (u,θ) of the Boussinesq system (5.1) in an interval $]t_1,t_2,[$. The idea is to construct a strong solution v of the Navier-Stokes equations (5.141) with initial value $v(t_1) = u(t_1)$ and external force $\tilde{f} := \theta g + f_1$. To prove $L^8(L^4)$- integrability of u we want to identify u and v. For this identification we need that (u,θ) satisfies the energy inequality starting in t_1. This is the reason why the strong energy inequality plays a central role in this argumentation. The next theorem will be a precise formulation of this method.

Theorem 5.20. *Let* $\Omega \subseteq \mathbb{R}^3$ *be a general domain, let* $u_0 \in L^2_\sigma(\Omega)$, $\theta_0 \in L^2(\Omega)$. *Moreover assume* $0 < T \le \infty$, $g \in L^{8/3}(0, T; L^4(\Omega)^3)$, *assume* $f_1 \in L^{4/3}(0, T; L^2(\Omega)^3)$, $f_2 \in L^1(0, T; L^2(\Omega))$. *Consider a weak solution* (u, θ) *of the Boussinesq equations (5.1) in* $[0, T[\times\Omega$. *Assume that* (u, θ) *satisfies (5.144), (5.145) for all* $s \in (\,]0, T[\backslash N) \cup \{0\}$ *(with* $u(0) := u_0$, $\theta(0) := \theta_0$) *and all* $t \in [s, T[$. *The following statements are valid:*

(i) *There exists an absolute constant* $\epsilon_* > 0$ *(independent on* $\Omega, T, g, f_1, f_2, u_0, \theta_0$) *with the following property: Let* $t \in]0, T[$ *be arbitrary. If there is* $t_0 \in [0, t[\backslash N$ *and a* $t_1 \in]t, T]$ *with*

$$\left(\int_0^{t_1 - t_0} \|e^{-\tau A} u(t_0)\|_4^8 \, d\tau \right)^{1/8} \le \epsilon_* \,,$$

$$\|\theta(t_0)\|_2 \le \frac{\epsilon_*}{\|g\|_{4, \frac{8}{3}; t_0, t_1}} \,, \qquad (5.150)$$

$$\|f_1\|_{2, \frac{4}{3}; t_0, t_1} + \|g\|_{4, \frac{8}{3}; t_0, t_1} \|f_2\|_{2, 1; t_0, t_1} \le \epsilon_* \,,$$

then $u \in L^8(t_0, t_1; L^4(\Omega)^3)$. *Thus* (u, θ) *is regular at* t. *In the case* $t_1 = \infty$ *define* $t_1 - t_0 := \infty$.

(ii) *Let* $t_0 \in [0, T[\backslash N$ *and assume* $u(t_0) \in \mathcal{D}(A^{1/4})$. *Then there is a* $\delta = \delta(t_0) > 0$ *such that* $u \in L^8(t_0, t_0 + \delta; L^4(\Omega)^3)$.

Remark.

Proof. (i) Let ϵ_* be an absolute constant such that the statements of Theorem 5.17 are satisfied. Define

$$\tilde{u}(t) := u(t + t_0) \quad \text{for almost all } t \in]0, t_1 - t_0[.$$

A standard argumentation, which uses the weak continuity of u, shows that $\tilde{u}$ is a weak solution of the Navier-Stokes equations (5.141) in $[0, t_1 - t_0[\times\Omega$ with initial value $u(t_0)$ and external force $\tilde{f}(t) := \theta(t)g(t) + f_1(t)$, a.a. $t \in [0, t_1 - t_0[$.

From (5.146) we get

$$\|\tilde{f}\|_{2, \frac{4}{3}; 0, t_1 - t_0} \le K \|g\|_{4, \frac{8}{3}; t_0, t_1} \left(\|\theta(t_0)\|_2 + \|f_2\|_{2, 1; t_0, t_1}) \right) + \|f_1\|_{2, \frac{4}{3}; t_0, t_1}$$

with an absolute constant $K > 0$. Therefore, replacing ϵ_* by $\min\{\epsilon_*, \frac{\epsilon_*}{K}\}$ we see that (5.142), (5.143) are fulfilled. Consequently, by Theorem 5.17 there exists a strong solution $v \in L^8(0, t_1 - t_0; L^4(\Omega)^3)$ of the instationary Navier-Stokes equations (5.141) with initial value $u(t_0)$ and external force $\tilde{f}$. By Serrin's uniqueness criterion (see [So01, Chapter V, Theorem 1.5.1]) we get $\tilde{u}(t) = v(t)$ for almost all $t \in [0, t_1 - t_0[$. Therefore $u \in L^8(t_0, t_1; L^4(\Omega)^3)$.

(ii) Let $\epsilon_* > 0$ be the constant obtained in (i) of this theorem. Since $u(t_0) \in \mathcal{D}(A^{1/4})$ it follows from (see (5.56)) that there is $\delta > 0$ such that

$$\left(\int_0^\delta \|e^{-tA} u(t_0)\|_4^8 \, dt \right)^{1/8} \le \epsilon_* \,.$$

By (5.31) and reducing δ, if necessary, we can assume that

$$\left(\int_0^\delta \|e^{t\Delta}\theta(t_0)\|_4^{8/3}\, dt \right)^{3/8} \leq \frac{\epsilon_*}{\|g\|_{4,\frac{8}{3};0,\delta}}\,.$$

Since the third condition in (5.150) can be fulfilled with $t_1 = \delta$ (after a possible reduction of δ if necessary), we can apply statement (i) of this theorem to obtain $u \in L^8(t_0, t_0 + \delta; L^4(\Omega)^3)$. $\qquad\square$

Lemma 5.21. *Let $h : [a,b] \to \mathbb{R}$ be an integrable function and let $N \subseteq [a,b]$ be a null set. Then there exists $x_0 \in\,]a,b[\,\backslash N$ such that*

$$h(x_0) \leq \frac{1}{b-a}\int_a^b h(x)\, dx.$$

Proof. Suppose that the claim of this Lemma is false. Then

$$h(t) - \frac{1}{b-a}\int_a^b h(x)\, dx > 0 \tag{5.151}$$

for all $t \in\,]a,b[\,\backslash N$. Integration of (5.151) yields

$$\int_a^b h(t)\, dt > \frac{1}{b-a}(b-a)\int_a^b h(x)\, dx.$$

This is a contradiction. $\qquad\square$

Lemma 5.22. *Let $\Omega \subseteq \mathbb{R}^3$ be a general domain, let $u_0 \in L^2_\sigma(\Omega)\,, \theta_0 \in L^2(\Omega)$. Further assume $0 < T \leq \infty\,, g \in L^{8/3}(0,T;L^4(\Omega)^3)$, and $f_1 \in L^{4/3}(0,T;L^2(\Omega)^3)$, $f_2 \in L^1(0,T;L^2(\Omega))$. Let (u,θ) be a weak solution of the Boussinesq equations (5.1) in $[0,T[\times\Omega$ satisfying the strong energy inequalities. Then there exists an absolute constant $\epsilon_* > 0$ with the following property: Let $0 < t_0 < t_1 < T$ and assume that the conditions*

$$\int_{t_0}^{t_1} \|A^{1/4}u(t)\|_2\, dt + \|g\|_{4,\frac{8}{3};t_0,T} \int_{t_0}^{t_1} \|\theta(t)\|_2\, dt \leq \epsilon_*(t_1 - t_0)\,, \tag{5.152}$$

$$\|f_1\|_{2,\frac{4}{3};t_0,T} + \|g\|_{4,\frac{8}{3};t_0,T}\|f_2\|_{2,1;t_0,T} \leq \epsilon_* \tag{5.153}$$

are satisfied. Then $u \in L^8(t_1,T;L^4(\Omega)^3)$.

Proof. Let ϵ_* be the constant from Theorem 5.20. Define $h : [0,T[\to \mathbb{R}$ by

$$h(t) := \begin{cases} \|A^{1/4}u(t)\|_2 + \|g\|_{4,\frac{8}{3};t_0,T}\|\theta(t)\|_2\,, & \text{if } u(t) \in \mathcal{D}(A^{1/4})\,, \\ 0\,, & \text{otherwise.} \end{cases}$$

By assumption

$$\int_{t_0}^{t_1} h(t)\, dt \leq \epsilon_*(t_1 - t_0).$$

Consequently, by Lemma 5.21 there is $s_0 \in]t_0, t_1[$ such that $u(s_0) \in \mathcal{D}(A^{1/4})$ and

$$\|A^{1/4}u(s_0)\|_2 + \|g\|_{4,\frac{8}{3};t_0,T}\|\theta(s_0)\|_2 \leq \epsilon_*. \tag{5.154}$$

We get (see (5.31), (5.56))

$$\left(\int_0^{T-s_0} \|e^{-\tau A}u(s_0)\|_4^8 \, d\tau\right)^{1/8} + \|g\|_{4,\frac{8}{3};s_0,T}\left(\int_0^{T-s_0} \|e^{\tau\Delta}\theta(s_0)\|_4^{8/3} \, d\tau\right)^{3/8}$$
$$\leq K\left(\|A^{1/4}u(s_0)\|_2 + \|g\|_{4,\frac{8}{3};t_0,T}\|\theta(s_0)\|_2\right) \tag{5.155}$$
$$\leq \epsilon_* K$$

with an absolute constant $K > 0$. Replace ϵ_* by $\min\{\frac{\epsilon_*}{2}, \frac{\epsilon_*}{K}\}$. We combine (5.153) and (5.155) to get from Theorem 5.20 (ii) with $\delta = T - s_0$ that $u \in L^8(s_0, T; L^4(\Omega)^3)$ is valid. $\qquad\square$

Lemma 5.23. *Let $\Omega \subseteq \mathbb{R}^3$ be a general domain, let $g \in L^{8/3}(0, \infty; L^4(\Omega)^3)$, let $f_1 \in L^{4/3}(0, \infty; L^2(\Omega)^3)$, $f_2 \in L^1(0, \infty; L^2(\Omega))$, and let $u_0 \in L^2_\sigma(\Omega), \theta_0 \in L^2(\Omega)$. Assume that (u, θ) is a weak solution of the Boussinesq equations (5.1) in $[0, \infty[\times\Omega$ satisfying the strong energy inequalities. Then there exists a $t_1 > 0$ such that $u \in L^8(t_1, \infty; L^4(\Omega)^3)$.*

Proof. Let ϵ_* be the constant from Lemma 5.22. Fix $t_0 > 0$ such that

$$\|f_1\|_{2,\frac{4}{3};t_0,\infty} + \|g\|_{4,\frac{8}{3};t_0,\infty}\|f_2\|_{2,1;t_0,\infty} \leq \epsilon_* \tag{5.156}$$

and

$$\|g\|_{4,\frac{8}{3};t_0,\infty}\|\theta\|_{L^\infty(0,\infty;L^2(\Omega))} \leq \frac{\epsilon_*}{2}. \tag{5.157}$$

Fix $t_1 > 0$ such that

$$t_1 - t_0 > \left(\frac{2}{\epsilon_*}\right)^4 \|u\|_{L^\infty(0,\infty;L^2(\Omega)^3)}^2 \int_0^\infty \|\nabla u\|_2^2 \, dt. \tag{5.158}$$

In the following we will prove that (5.152), (5.153) are fulfilled. We get with (2.48)

$$\|A^{1/4}u(t)\|_2 \leq \|A^{1/2}u(t)\|_2^{1/2}\|u(t)\|_2^{1/2} = \|\nabla u(t)\|_2^{1/2}\|u(t)\|_2^{1/2}$$

for almost all $t \in [0, T[$. Thus, Hölder's inequality implies

$$\int_{t_0}^{t_1} \|A^{1/4}u(t)\|_2 \, dt \leq (t_1 - t_0)^{3/4}\left(\int_{t_0}^{t_1} \|\nabla u(t)\|_2^2\|u(t)\|_2^2 \, dt\right)^{1/4}$$
$$\leq (t_1 - t_0)^{3/4}\left(\|u\|_{L^\infty(0,\infty;L^2(\Omega)^3)}^2 \int_{t_0}^{t_1} \|\nabla u(t)\|_2^2 \, dt\right)^{1/4}.$$

Therefore (see (5.157), (5.158))

$$
\int_{t_0}^{t_1} \|A^{1/4}u(t)\|\, dt + \|g\|_{4,\frac{8}{3};t_0,\infty} \int_{t_0}^{t_1} \|\theta(t)\|_2\, dt
$$

$$
\leq (t_1 - t_0)^{\frac{3}{4}} \left(\|u\|_{L^\infty(0,\infty;L^2(\Omega)^3)}^2 \int_0^\infty \|\nabla u\|_2^2\, dt \right)^{\frac{1}{4}} \tag{5.159}
$$

$$
+ (t_1 - t_0)\|g\|_{4,\frac{8}{3};t_0,\infty} \|\theta\|_{L^\infty(0,\infty;L^2(\Omega)^3)}
$$

$$
\leq \epsilon_*(t_1 - t_0).
$$

Due to the estimate above and (5.156), an application of Lemma 5.22 completes the proof. □

The next theorem translates Leray's structure theorem to the Boussinesq equations in general domains. In this general situation it will be proved that a weak solutions (u, θ) of the Boussinesq equations satisfying the strong energy inequalities is regular in almost all $t \in]0, T[$. For a formulation of Leray's structure theorem for the Navier-Stokes equations we refer to [SoWa84].

Theorem 5.24. *Let $\Omega \subseteq \mathbb{R}^3$ be a general domain, let $u_0 \in L^2_\sigma(\Omega)$, $\theta_0 \in L^2(\Omega)$. Assume $g \in L^{8/3}(0,\infty; L^4(\Omega)^3)$, and $f_1 \in L^{4/3}(0,\infty; L^2(\Omega)^3)$, $f_2 \in L^1(0,\infty; L^2(\Omega))$. Consider a weak solution (u, θ) of the Boussinesq equations (5.1) in $[0,\infty[\times\Omega$ satisfying the strong energy inequalities. Then there exist countably many disjoint open intervals $(I_j)_{j\in M}$ where $M \subseteq \mathbb{N}$, such that the union $\mathcal{J} := \bigcup_{j\in M} I_j$ has the following properties:*

(i) *The complement $]0,\infty[\backslash\mathcal{J}$ is a null set.*

(ii) *For each $j \in M$ there holds $u \in L^8_{loc}(I_j; L^4(\Omega)^3)$. Consequently (u, θ) is regular at each $t \in \mathcal{J}$.*

(iii) *There is an interval $I_{j_0} =]T_\infty, \infty[$ with $j_0 \in M$ and $T_\infty > 0$. Moreover $u \in L^8(I_j; L^4_\sigma(\Omega)^3)$.*

If additionally $u_0 \in \mathcal{D}(A^{1/4})$, then there is an interval $I_{j_0} =]0, T_0[$ with $j_0 \in M$ and $T_0 > 0$. Moreover $u \in L^8(I_{j_0}; L^4(\Omega)^3)$.

Proof. Define the set

$$
\mathcal{I} := \left\{ t_0 \in]0,\infty[;\ u(t_0) \in W^{1,2}_{0,\sigma}(\Omega) \text{ and } (u,\theta) \text{ fulfils } (5.144), (5.145) \text{ for all } t \geq t_0 \right\}. \tag{5.160}
$$

Since (u,θ) satisfy (5.5) and (u,θ) fulfil the strong energy inequalities it follows that the set $]0,\infty[\backslash\mathcal{I}$ is a null set. Consider any $t_0 \in \mathcal{I}$. By Theorem 5.20 (ii) there is $\delta_{t_0} > 0$, which will be fixed throughout the proof, such that $u \in L^8(t_0, t_0 + \delta_{t_0}; L^4(\Omega)^3)$. Define

$$
\mathcal{J} := \bigcup_{t_0 \in \mathcal{I}}]t_0, t_0 + \delta_{t_0}[. \tag{5.161}
$$

Since $\mathcal{J} \subseteq \mathbb{R}$ is an open set there exist countably many disjoint open intervals I_j, $j \in M$ where $M \subseteq \mathbb{N}$, such that $\mathcal{J} = \bigcup_{j \in M} I_j$. By construction $u \in L^8_{\mathrm{loc}}(I_j; L^4(\Omega)^3)$ for every $j \in M$.

Next it will be proved that $]0, \infty[\backslash \mathcal{J}$ is a null set. There holds

$$]0, \infty[\backslash \mathcal{J} = (]0, \infty[\backslash \mathcal{I}) \cup (\mathcal{I} \backslash \mathcal{J}). \tag{5.162}$$

We already know $]0, \infty[\backslash \mathcal{I}$ is a null set. Consider any $t \in \mathcal{I} \backslash \mathcal{J}$. There is a unique $j \in M$ with $]t, t+\delta_t[\subseteq I_j$. Since $t \notin I_j$ we see that t is the 'left endpoint' of the open interval I_j. Consequently the set $\mathcal{I} \backslash \mathcal{J}$ is countable (perhaps even empty). From (5.162) if follows that $]0, \infty[\backslash \mathcal{J}$ is a null set. By construction $u \in L^8_{\mathrm{loc}}(I_j; L^4(\Omega)^3)$ for every $j \in M$. Moreover consider $t \in]0, \infty[\backslash \mathcal{J}$. It follows that there is $j \in M$, $\delta(t) > 0$, such that $[t - \delta(t), t + \delta(t)] \subseteq I_j$ and therefore $u \in L^8(t - \delta, t + \delta; L^4(\Omega)^3)$. Altogether (i), (ii) are proven. To prove (iii) we apply Lemma 5.23 to obtain $T_\infty > 0$ with $u \in L^8(t_\infty, \infty; L^4(\Omega)^3)$.

Assume additionally $u_0 \in \mathcal{D}(A^{1/4})$. Then we use Theorem 5.20 (ii) to obtain $T_0 > 0$ such that $u \in L^8(0, T_0; L^4(\Omega)^3)$. The proof is finished. $\qquad \square$

6 Young Measures

6.1 Motivation and interpretation of Young Measures

In this section we describe the basic ideas that lead to the concept of a Young measure. Let $n, d \in \mathbb{N}$, let $\Omega \subseteq \mathbb{R}^n$ be an open set and let $(z_k)_{k \in \mathbb{N}} : \Omega \to \mathbb{R}^d$ be a bounded sequence in $L^\infty(\Omega)^d$. Let $z_k \overset{*}{\rightharpoonup} z$ as $k \to \infty$ in $L^\infty(\Omega)^d$. Let $\phi \in C_0(\mathbb{R}^d)$ be given. Then (along a not relabelled subsequence) $(\phi(z_k))_{k \in \mathbb{N}}$ is weakly* convergent in $L^\infty(\Omega)$. Since weak* convergence in $L^\infty(\Omega)$ is not continuous with respect to nonlinear functionals the weak* limit of $(\phi(z_k))_{k \in \mathbb{N}}$ need not be $\phi(z)$.

At this point we introduce a *Young measure* ν: Let $(\nu_x)_{x \in \Omega}$ be a family of probability measures ν_x, $x \in \Omega$, on $\mathbb{R}^d$ which depend measurably on x in the sense that for all $\phi \in C_0(\mathbb{R}^d)$ the function

$$\overline{\phi}(x) := \int_{\mathbb{R}^d} \phi(\lambda) \, d\nu_x(\lambda), \quad x \in \Omega,$$

is measurable. We say that $(z_k)_{k \in \mathbb{N}}$ generates $\nu = (\nu_x)_{x \in \Omega}$ if

$$\phi(z_k) \underset{k \to \infty}{\overset{*}{\rightharpoonup}} \overline{\phi} \quad \text{in } L^\infty(\Omega)$$

for all $\phi \in C_0(\mathbb{R}^n)$. For a precise definition we refer to Definition 6.10 and Definition 6.13.

In the following we present a possibility how the measures ν can be interpreted. Here we mainly follow [Ba89]. Let $\Omega \subseteq \mathbb{R}^n$ open. Define for $k \in \mathbb{N}$, for $\delta > 0$ sufficiently small and for $x \in \Omega$ the finite, signed Borel measure $\nu_{x,\delta}^k \in \mathcal{M}(\mathbb{R}^d)$ by

$$\langle \nu_{x,\delta}^k, \phi \rangle_{\mathcal{M}(\mathbb{R}^d); C_0(\mathbb{R}^d)} := \fint_{B_\delta(x)} \phi(z_k(y)) \, dy, \quad \phi \in C_0(\mathbb{R}^d).$$

Then for a measurable subset $E \subseteq \Omega$

$$\nu_{x,\delta}^k(E) = \fint_{B_\delta(x)} \chi_E(z_k(y)) \, dy = \frac{|\{ y \in B_\delta(x); z_k(y) \in E \}|}{|B_\delta(x)|}$$

We can interpret $\nu_{x,\delta}^k$ as the probability distribution of the values of $z_k(y)$ when y is taken uniformly from $B_\delta(x)$.

Lemma 6.1. *Let $n, d \in \mathbb{N}$, let $\Omega \subseteq \mathbb{R}^n$ be an open set, let $(z_k)_{k \in \mathbb{N}} : \Omega \to \mathbb{R}^d$ be a sequence of measurable functions, and assume that $\nu = (\nu_x)_{x \in \Omega}$ is the Young measure associated to $(z_k)_{k \in \mathbb{N}}$. Then*

$$\lim_{\delta \searrow 0} \lim_{k \to \infty} \nu^k_{x,\delta} = \nu_x \tag{6.1}$$

as weak* limit in $\mathcal{M}(\mathbb{R}^d)$ for almost all $x \in \Omega$.

Remark. In a more explicit form, (6.1) means that there is a null set $N \subseteq \Omega$ such that if $x \in \Omega \setminus N$ then

$$\lim_{\delta \searrow 0} \lim_{k \to \infty} \fint_{B_\delta(x)} \phi(z_k(y)) \, dy = \int_{\mathbb{R}^d} \phi(\lambda) \, d\nu_x(\lambda)$$

for all $\phi \in C_0(\mathbb{R}^d)$.

Proof. This is [Ba89, Identity (2)]. $\qquad\qquad\qquad\qquad\qquad\qquad\qquad\qquad\qquad\square$

6.2 Preliminaries

We collect several well known results taken from measure theory, which will be needed in the following subsections. Let us begin with the following definition.

Definition 6.2. Let $\mathcal{S}$ be a σ-algebra over a set X.

(i) A function $\mu : \mathcal{S} \to \mathbb{R}$ is called a *finite, signed measure* if the following two properties are fulfilled:

    (1) $\mu(\varnothing) = 0$

    (2) Whenever $A = \bigcup_{k=1}^{\infty} A_k$ with pairwise disjoint sets $A_k \in \mathcal{S}$, $k \in \mathbb{N}$, then

$$\mu(A) = \sum_{k=1}^{\infty} \mu(A_k). \tag{6.2}$$

(ii) A function $\mu : \mathcal{S} \to [0, \infty]$ is called a *positive measure* if (1), (2) are satisfied.

Remark. It can be seen that the sum in (6.2) converges absolutely.

Let X be a metric space, let $\mathcal{S}$ be a σ-algebra which contains all Borel sets of X, let $\mu : \mathcal{S} \to [0, \infty]$ be a positive measure. Define the support of μ by

$$\operatorname{supp}(\mu) := \{ x \in X ; \mu(U) > 0 \text{ for all open sets with } U \subseteq X \text{ with } x \in U \}.$$

It can be proved that $\operatorname{supp}(\mu) \subseteq X$ is a closed set.

A proof of the following classical result can be found in [FoLe07, Theorem 1.172].

Lemma 6.3. *Let μ be finite, signed measure on a σ-algebra $\mathcal{S}$ over a set X. Then there exists a disjoint decomposition $X = P \cup N$ of X with $P \in \mathcal{S}$ and $N \in \mathcal{S}$ such that $\mu(A) \geq 0$ for all $A \subseteq P$, $A \in \mathcal{S}$ and $\mu(A) \leq 0$ for all $A \subseteq N$, $A \in \mathcal{S}$. We call (X, P, N) a Hahn decomposition of μ.*

Having a Hahn decomposition at hand we proceed with the following

Definition 6.4. Let μ be a finite, signed measure on a σ-algebra $\mathcal{S}$ over a set X. Let (X, P, N) be a Hahn decomposition of X. Define for $A \in \mathcal{S}$

$$\mu^+(A) := \mu(A \cap P),$$
$$\mu^-(A) := -\mu(A \cap N),$$
$$|\mu|(A) := \mu^+(A) + \mu^-(A).$$

Remark. Let $A \in \mathcal{S}$. It can be shown that

$$\mu^+(A) = \sup\{\mu(B); B \subseteq A, B \in \mathcal{S}\},$$
$$\mu^-(A) = -\inf\{\mu(B); B \subseteq A, B \in \mathcal{S}\},$$
$$|\mu|(A) = \sup\left\{\sum_{n=1}^{\infty} |\mu(A_n)|; A = \sum_{n=1}^{\infty} A_n, A_n \in \mathcal{S} \text{ pairwise disjoint}\right\}.$$

Therefore $\mu^+, \mu^-, |\mu|$ are independent of the chosen Hahn decomposition (X, P, N) of μ and are well defined. Especially the integral $\int_X f(y) \, d\mu(y)$ of f with respect to the finite, signed measure μ is well defined.

Now we have all ingredients to define the integral of a function with respect to a finite, signed measure. Let $f : X \to \mathbb{R}$ be a measurable function.

Definition 6.5. Let μ be a finite, signed measure on a σ-algebra $\mathcal{S}$ over a set X. Let (X, P, N) be a Hahn decomposition. Let $f : X \to [-\infty, \infty]$ be a measurable function. We call f integrable with respect to μ if f is integrable with respect to the (positive) measures μ^+ and μ^-. If this is the case define

$$\int_X f(y) \, d\mu(y) = \int_X f(y) \, d\mu^+(y) - \int_X f(y) \, d\mu^-(y).$$

Definition 6.6. For $d \in \mathbb{N}$ let $\mathcal{B}(\mathbb{R}^d)$ denote the Borel σ-algebra of $\mathbb{R}^d$.

(i) Let $\mathcal{M}(\mathbb{R}^d)$ denote the set of all finite, signed measures on $(\mathbb{R}^d, \mathcal{B}(\mathbb{R}^d))$. It can be seen that $\mathcal{M}(\mathbb{R}^d)$ equipped with the norm of *total variation* $\|\mu\|_{\mathcal{M}(\mathbb{R}^d)} := |\mu|(\mathbb{R}^d)$ becomes a Banach space.

(ii) Let $\mathcal{P}(\mathbb{R}^d) \subseteq \mathcal{M}(\mathbb{R}^d)$ be the subspace of all *probability measures* on $(\mathbb{R}^d, \mathcal{B}(\mathbb{R}^d))$. This means, by definition, that $\nu \in \mathcal{P}(\mathbb{R}^d)$ is a positive measure and which fulfils additionally $\nu(\mathbb{R}^d) = 1$.

For $d \in \mathbb{N}$ we denote by

$$C_0(\mathbb{R}^d) := \big\{ \phi : \mathbb{R}^d \to \mathbb{R} \text{ continuous }; \lim_{|\lambda| \to \infty} \phi(\lambda) = 0 \big\}$$

the space of all continuous functions that vanish at infinity. By endowing this space with the supremum norm $\|\phi\|_\infty := \sup_{x \in \mathbb{R}^d} |\phi(x)|$ we obtain a separable Banach space.

The following Riesz representation theorem states that $\mathcal{M}(\mathbb{R}^d)$ can be identified as the dual space of $C_0(\mathbb{R}^d)$.

Theorem 6.7. *Let $d \in \mathbb{N}$. The map $T : \mathcal{M}(\mathbb{R}^d) \to C_0(\mathbb{R}^d)'$ defined by*

$$T(\mu)(f) := \int_{\mathbb{R}^d} f(y) \, d\mu(y), \quad f \in C_0(\mathbb{R}^d),$$

is an isometric isomorphism. This means that for every linear, continuous functional x' on $C_0(\mathbb{R}^d)$ there exists exactly one finite, signed Borel measure $\mu \in \mathcal{M}(\mathbb{R}^d)$ such that

$$x'(f) = \int_{\mathbb{R}^d} f(y) \, d\mu(y)$$

for all $f \in C_0(\mathbb{R}^d)$. Moreover $\|x'\|_{C_0(\mathbb{R}^d)'} = \|\mu\|_{\mathcal{M}(\mathbb{R}^d)}$.

Proof. See [FoLe07, Theorem 1.200]. $\qquad\qquad\square$

We need the following version of the *Vitali covering theorem*.

Theorem 6.8. *Let $E \subseteq \mathbb{R}^n, n \in \mathbb{N}$, be a measurable set, let I be a set of indices, let $\mathcal{F} = (B(x_i, r_i))_{i \in I}$ be a family of open balls with $x_i \in E, r_i > 0$ for $i \in I$, such that for each $x \in E$ there is a $\delta = \delta_x > 0$ such that the set $\{B(x,r) \in \mathcal{F}; 0 < r < \delta\}$ is uncountable. Then there exists a countable subset $M \subseteq I$ such that the balls $B(x_i, r_i), i \in M$, are pairwise disjoint and*

$$\Big| E \setminus \bigcup_{i \in M} B(x_i, r_i) \Big| = 0. \tag{6.3}$$

Proof. This follows from [FoLe07, Theorem 1.150 and Remark 1.151]. The 'classical' Vitali covering theorems deal with coverings of closed sets. The reason why we assume that the set $\{B(x,r) \in \mathcal{F}; 0 < r < \delta\}$ is uncountable is to obtain a covering of open balls fulfilling (6.3). $\qquad\qquad\square$

To proceed we need to introduce some notations. First, let $n \in \mathbb{N}, n \geq 2$, and let $\Omega \subseteq \mathbb{R}^n$, be a bounded Lipschitz domain. By Theorem 4.6, it follows that the trace $u|_{\partial\Omega} \in L^2(\partial\Omega)$ is well defined for $u \in H^1(\Omega)$. Define

$$W_0^{1,\infty}(\Omega) := \{ u \in W^{1,\infty}(\Omega); u|_{\partial\Omega} = 0 \}.$$

If $n = 1$ we say that $\Omega \subseteq \mathbb{R}$ is a bounded Lipschitz domain, if Ω is a bounded, open real interval. Assume $\Omega =]a, b[$ with $a, b \in \mathbb{R}$ and $a < b$. Then we can redefine $u \in W_0^{1,\infty}(]a, b[)$ on a null set of $]a, b[$ such that $u :]a, b[\to \mathbb{R}$ is absolutely continuous. Therefore, $u(a), u(b) \in \mathbb{R}$ are well defined. We get

$$W_0^{1,\infty}(]a, b[) := \{ u \in W^{1,\infty}(]a, b[); u(a) = 0, u(b) = 0 \}.$$

In both cases, we endow $W_0^{1,\infty}(\Omega)$ with the norm $\| \cdot \|_{W^{1,\infty}(\Omega)}$.

The following auxiliary lemma is needed in the proof of Theorem 6.12 below. It shows how weak derivatives, which are defined on pairwise disjoint domains Ω_i, can be glued together to obtain a weakly differentiable function defined on $\Omega = \bigcup_{i \in M} \Omega_i$.

Lemma 6.9. *Let $\Omega \subseteq \mathbb{R}^n, n \in \mathbb{N}$, be an open set, let M be a countable set, let $\Omega_i, i \in M$, be pairwise disjoint bounded Lipschitz domains. Assume*

$$\Omega = N \cup \bigcup_{i \in M} \Omega_i$$

where $N \subseteq \Omega$ is a null set. For every $i \in M$ consider $f_i \in W_0^{1,\infty}(\Omega_i)$. Define the function $f : \Omega \to \mathbb{R}$ by

$$f(x) := \begin{cases} f_i(x), & x \in \Omega_i, \\ 0, & x \in N. \end{cases}$$

Then $f \in W_{loc}^{1,\infty}(\Omega)$ and

$$(\nabla f)(x) = \begin{cases} \nabla f_i(x), & x \in \Omega_i, \\ 0, & x \in N \end{cases}$$

for almost all $x \in \Omega$.

Proof. Fix $i \in M$. Since Ω_i is a bounded Lipschitz domains, we get $f_i \in H^1(\Omega_i)$. Consequently, since $f_i = 0$ on $\partial \Omega_i$, we can apply [Al02, Reference A 6.10] and obtain $f_i \in H_0^1(\Omega_i)$. Thus, there exists a sequence $(g_k)_{k \in \mathbb{N}} \subseteq C_0^\infty(\Omega_i)$ with $g_k \to f_i$ as $k \to \infty$ in $H^1(\Omega_i)$. We get

$$\int_{\Omega_i} f_i \frac{\partial}{\partial x_j} \phi \, dx = \lim_{k \to \infty} \int_{\Omega_i} g_k \frac{\partial}{\partial x_j} \phi \, dx$$

$$= - \lim_{k \to \infty} \int_{\Omega_i} \frac{\partial}{\partial x_j} g_k \, \phi \, dx$$

$$= - \int_{\Omega_i} \frac{\partial}{\partial x_j} f_i \, \phi \, dx$$

for every $i \in M$, $j \in \{1, \ldots, n\}$, and $\phi \in C_0^\infty(\Omega)$. Consequently

$$
\begin{aligned}
\int_\Omega f \frac{\partial}{\partial x_j} \phi \, dx &= \sum_{i \in M} \int_{\Omega_i} f_i \frac{\partial}{\partial x_j} \phi \, dx \\
&= -\sum_{i \in M} \int_{\Omega_i} \frac{\partial}{\partial x_j} f_i \, \phi \, dx \\
&= -\sum_{i \in M} \int_\Omega 1_{\Omega_i} \frac{\partial}{\partial x_j} f_i \, \phi \, dx \\
&= -\int_\Omega \left(\sum_{i \in M} 1_{\Omega_i} \frac{\partial}{\partial x_j} f_i \right) \phi \, dx
\end{aligned}
$$

for every $j \in \{1, \ldots, n\}$ and $\phi \in C_0^\infty(\Omega)$. This means that $\frac{\partial}{\partial j} f_i$ exists for every $j \in \{1, \ldots, n\}$ and is given by

$$
\frac{\partial}{\partial j} f(x) = \begin{cases} \frac{\partial}{\partial j} f_i(x), & x \in \Omega_i, \\ 0, & x \in N. \end{cases}
$$

The claim follows. $\qquad\qquad\qquad\qquad\qquad\qquad\qquad\qquad\qquad\qquad\qquad\square$

6.3 Existence theorems

Consider a sequence $(z_k)_{k \in \mathbb{N}}$ which is bounded in $L^\infty(\Omega)^d$ where $\Omega \subseteq \mathbb{R}^n$, $n \in \mathbb{N}$, is a measurable set and $d \in \mathbb{N}$. An important tool in this thesis is the fact that there exists a subsequence $(z_{m_k})_{k \in \mathbb{N}}$ and a Young measure $\nu = (\nu_x)_{x \in \Omega}$ associated to this sequence. Moreover we will make essential use of specific properties of this Young measure ν. Therefore, we will present in this section the general definitions and results for Young measures generated by bounded sequences in $L^\infty(\Omega)^d$. Further, we will also discuss the uniqueness of Young measures. Let us start with a definition.

Definition 6.10. Let $n, d \in \mathbb{N}$, let $\Omega \subseteq \mathbb{R}^n$ be a measurable set.

1. A map $\nu = (\nu_x)_{x \in \Omega} : \Omega \to \mathcal{M}(\mathbb{R}^d)$ is called *weakly* measurable* if for every $\phi \in C_0(\mathbb{R}^d)$ the map

$$
\Omega \to \mathbb{R}, \quad x \mapsto \int_{\mathbb{R}^d} \phi(\lambda) \, d\nu_x(\lambda)
$$

   is measurable. We write ν_x as $\nu(x)$ for $x \in \Omega$.

2. Define the normed vector space

$$
L_{\omega^*}^\infty(\Omega; \mathcal{M}(\mathbb{R}^d)) := \big\{ \nu : \Omega \to \mathcal{M}(\mathbb{R}^d); \ \nu \text{ weakly}^*\text{-measurable},
$$
$$
x \mapsto \|\nu_x\|_{\mathcal{M}(\mathbb{R}^d)} \text{ measurable}, \ \|\nu\| := \inf_{N \text{ null set}} \sup_{x \in \Omega \setminus N} \|\nu_x\|_{\mathcal{M}(\mathbb{R}^d)} < \infty \big\}.
$$

3. An element $\nu = (\nu_x)_{x \in \Omega} \in L_{\omega*}^\infty(\Omega; \mathcal{M}(\mathbb{R}^d))$ is called a *Young measure* if for almost all $x \in \Omega$ there holds that ν_x is a probability measure on $\mathbb{R}^d$. Define the set

$$\mathcal{Y}(\Omega; \mathcal{P}(\mathbb{R}^d)) := \{\, \nu = (\nu_x)_{x \in \Omega} \in L_{\omega*}^\infty(\Omega; \mathcal{M}(\mathbb{R}^d)); \ \nu_x \in \mathcal{P}(\mathbb{R}^d) \text{ for a.a. } x \in \Omega \,\}$$

of all Young measures.

4. We call a Young measure $\nu = (\nu_x)_{x \in \Omega} \in \mathcal{Y}(\Omega; \mathcal{P}(\mathbb{R}^d))$ *homogeneous* if there is a probability measure $\mu \in \mathcal{P}(\mathbb{R}^d)$ such that $\nu_x = \mu$ for almost all $x \in \Omega$.

5. A *Carathéodory function* $\psi(x, \lambda) : \Omega \times \mathbb{R}^d \to \mathbb{R}$ is a function such that for almost all $x \in \Omega$ the function $\psi(x, \cdot) : \mathbb{R}^d \to \mathbb{R}$ is continuous and such that for all $\lambda \in \mathbb{R}^d$ the function $\psi(\cdot, \lambda) : \Omega \to \mathbb{R}$ is measurable.

Now we have all definitions and notation at hand to formulate the following basic existence theorem.

Theorem 6.11. *Let $n, d \in \mathbb{N}$, let $\Omega \subseteq \mathbb{R}^n$ be a measurable set, let $(z_k)_{k \in \mathbb{N}} : \Omega \to \mathbb{R}^d$ be a sequence of measurable functions. Then there exists a subsequence $(z_{m_k})_{k \in \mathbb{N}}$ and $\nu = (\nu_x)_{x \in \Omega} \in L_{\omega*}^\infty(\Omega; \mathcal{M}(\mathbb{R}^d))$ with the following properties:*

(i) For almost all $x \in \Omega$ there holds that ν_x is a positive measure and $\nu_x(\mathbb{R}^d) \leq 1$.

(ii) Fix $\phi \in C_0(\mathbb{R}^d)$. Define

$$\overline{\phi}(x) := \int_{\mathbb{R}^d} \phi(\lambda)\, d\nu_x(\lambda) \quad \text{for a.a. } x \in \Omega. \tag{6.4}$$

Then $\overline{\phi} \in L^\infty(\Omega)$ and

$$\phi(z_{m_k}) \underset{k \to \infty}{\overset{*}{\rightharpoonup}} \overline{\phi} \quad \text{in } L^\infty(\Omega). \tag{6.5}$$

Proof. This is proven in [Ba89, Section 2]. $\qquad\qquad\square$

The uniqueness statements in this section are all based on the Lemma below.

Lemma 6.12. *Let $n, d \in \mathbb{N}$, let $\Omega \subseteq \mathbb{R}^n$ be a measurable set. Let $\mathcal{D}_1$ be dense in $L^1(\Omega)$ and let $\mathcal{D}_2$ be dense in $C_0(\mathbb{R}^d)$. Assume that $(\nu_x^1)_{x \in \Omega}, (\nu_x^2)_{x \in \Omega} \in L_{\omega*}^\infty(\Omega; \mathcal{M}(\mathbb{R}^d))$ satisfy*

$$\int_\Omega h(x) \left(\int_{\mathbb{R}^d} \phi(\lambda)\, d\nu_x^1(\lambda) \right) dx = \int_\Omega h(x) \left(\int_{\mathbb{R}^d} \phi(\lambda)\, d\nu_x^2(\lambda) \right) dx \tag{6.6}$$

for all $h \in \mathcal{D}_1$ and all $\phi \in \mathcal{D}_2$. Then $\nu_x^1 = \nu_x^2$ for almost all $x \in \Omega$.

Proof. By density of $\mathcal{D}_1$ in $L^1(\Omega)$ it can be assumed that $\mathcal{D}_1 = L^1(\Omega)$. We use Lebesgue's dominated convergence theorem to conclude that (6.6) also holds with $\mathcal{D}_2 = C_0(\mathbb{R}^d)$. For all $\phi \in C_0(\mathbb{R}^d)$ there exists by (6.6) a null set $N(\phi) \subseteq \Omega$ such that

$$\int_{\mathbb{R}^d} \phi(\lambda)\, d\nu_x^1(\lambda) = \int_{\mathbb{R}^d} \phi(\lambda)\, d\nu_x^2(\lambda) \tag{6.7}$$

for all $x \in \Omega \setminus N(\phi)$. Since $C_0(\mathbb{R}^d)$ is separable we find a null set $N \subseteq \Omega$ such that (6.7) hold for all $x \in \Omega \setminus N$ and all $\phi \in C_0(\mathbb{R}^d)$. Theorem 6.7 implies that every finite, signed Borel measure $\mu \in \mathcal{M}(\mathbb{R}^d)$ is uniquely determined by the integrals

$$\int_{\mathbb{R}^d} f(y) \, d\mu(y), f \in C_0(\mathbb{R}^d).$$ Therefore for all $x \in \Omega \setminus N$ it follows $\nu_x^1 = \nu_x^2$. $\qquad\square$

Definition 6.13. Let $n, d \in \mathbb{N}$, let $\Omega \subseteq \mathbb{R}^n$ be a measurable set, let $(z_k)_{k \in \mathbb{N}} : \Omega \to \mathbb{R}^d$ be a sequence of measurable functions. Let $\nu = (\nu_x)_{x \in \Omega} \in L_{\omega*}^\infty(\Omega; \mathcal{P}(\mathbb{R}^d))$ be a Young measure. Assume that (6.5) is satisfied for all $\phi \in C_0(\mathbb{R}^d)$ with $\bar{\phi}$ as defined in (6.4) (with $m_k = k$). Then we say that $(z_k)_{k \in \mathbb{N}}$ generates ν or that ν is the Young measure associated to $(z_k)_{k \in \mathbb{N}}$.

The following Lemma shows that if a sequence $(z_k)_{k \in \mathbb{N}}$ generates a Young measure $\nu = (\nu_x)_{x \in \Omega}$, then ν is uniquely determined as element of the space $L_{\omega*}^\infty(\Omega; \mathcal{M}(\mathbb{R}^d))$.

Lemma 6.14. *Let $n, d \in \mathbb{N}$, let $\Omega \subseteq \mathbb{R}^n$ be a measurable set, let $(z_k)_{k \in \mathbb{N}} : \Omega \to \mathbb{R}^d$ be a sequence of measurable functions, let $\nu^1 = (\nu_x^1)_{x \in \Omega}, \nu^2 = (\nu_x^2)_{x \in \Omega} \in L_{\omega*}^\infty(\Omega; \mathcal{P}(\mathbb{R}^d))$ be Young measures. Assume that $(z_k)_{k \in \mathbb{N}}$ generates ν^1 as well as ν^2. Then $\nu_x^1 = \nu_x^2$ for a.a. $x \in \Omega$.*

Proof. By uniqueness of the weak* limit in $L^\infty(\Omega)$ it follows that (6.6) is fulfilled with $\mathcal{D}_1 = L^1(\Omega)$ and $\mathcal{D}_2(\Omega) = C_0(\mathbb{R}^d)$. By Lemma 6.12 the conclusion follows. $\qquad\square$

The following Lemma is needed for the identification of Young measures generated by sequences which are close to each other in the sense of (6.8) below.

Lemma 6.15. *Let $n, d \in \mathbb{N}$, let $\Omega \subseteq \mathbb{R}^n$ be a measurable set, let $(z_k)_{k \in \mathbb{N}} : \Omega \to \mathbb{R}^d$ and $(w_k)_{k \in \mathbb{N}} : \Omega \to \mathbb{R}^d$ be sequences of measurable functions such that*

$$\lim_{k \to \infty} \left| \{ x \in \Omega; z_k(x) \neq w_k(x) \} \right| = 0. \tag{6.8}$$

Assume that $(z_k)_{k \in \mathbb{N}}$ generates the Young measure $\nu = (\nu_x)_{x \in \Omega} \in L_{\omega*}^\infty(\Omega; \mathcal{P}(\mathbb{R}^d))$. Then $(w_k)_{k \in \mathbb{N}}$ also generates ν.

Proof. Fix $h \in L^1(\Omega)$ and $\phi \in C_0(\mathbb{R}^d)$. Then

$$\left| \int_\Omega h(x) \big(\phi(z_k(x)) - \phi(w_k(x)) \big) \, dx \right| \leq 2\|\phi\|_\infty \int_{\{x \in \Omega; z_k(x) \neq w_k(x)\}} |h(x)| \, dx. \tag{6.9}$$

By Lebesgue's dominated convergence theorem and (6.5), (6.9) it follows

$$\lim_{k \to \infty} \int_\Omega h(x) \phi(w_k(x)) \, dx = \int_\Omega h(x) \left(\int_{\mathbb{R}^d} \phi(\lambda) \, d\nu_x(\lambda) \right) dx \tag{6.10}$$

for all $h \in L^1(\Omega), \phi \in C_0(\mathbb{R}^d)$. Thus $(w_k)_{k \in \mathbb{N}}$ generates ν as well. $\qquad\square$

The following theorem will be one of the main tools in our Young measure approach to describe the boundary behaviour of weak solutions of the Boussinesq equations. Its content is the existence of Young measures generated by bounded sequences in $L^\infty(\Omega)^d$.

Theorem 6.16. *Let $n, d \in \mathbb{N}$, let $\Omega \subseteq \mathbb{R}^n$ be a measurable set, let $(z_k)_{k \in \mathbb{N}} : \Omega \to \mathbb{R}^d$ be a bounded sequence in $L^\infty(\Omega)^d$. Then there exist a subsequence $(z_{m_k})_{k \in \mathbb{N}}$ and a Young measure $\nu = (\nu_x)_{x \in \Omega} \in \mathcal{Y}(\Omega; \mathcal{P}(\mathbb{R}^d))$ such that the following properties are satisfied:*

  (i) *The sequence $(z_{m_k})_{k \in \mathbb{N}}$ generates ν.*

  (ii) *There is a compact subset $K \subseteq \mathbb{R}^d$ such that $\mathrm{supp}(\nu_x) \subseteq K$ for a.a. $x \in \Omega$.*

  (iii) *Let $A \subseteq \Omega$ be a measurable set, let $\psi = \psi(x, \lambda) : A \times \mathbb{R}^d \to \mathbb{R}$ be a Carathéodory function with the property that $\big(\psi(\cdot, z_{m_k}(\cdot))\big)_{k \in \mathbb{N}}$ is a weakly convergent sequence in $L^1(A)$. Then $\psi(x, \cdot)$ is integrable with respect to ν_x for a.a. $x \in A$. Define*

$$\overline{\psi}(x) := \int_{\mathbb{R}^d} \psi(x, \lambda) \, d\nu_x(\lambda) \quad \text{for a.a. } x \in A. \tag{6.11}$$

Then $\overline{\psi} \in L^1(A)$ and

$$\psi(\cdot, z_{m_k}(\cdot)) \underset{k \to \infty}{\rightharpoonup} \overline{\psi} \quad \text{in } L^1(A). \tag{6.12}$$

Proof. This follows from [Ba89, Section 2]. Further, let us remark that the expression $\psi(\cdot, z_{m_k}(\cdot))$ is an abbreviation for the function $x \mapsto \psi(x, z_{m_k}(x))$, $x \in A$. $\square$

The following technical corollary states that every Young measure generated by a sequence $(z_k)_{k \in \mathbb{N}}$ which is bounded in $L^\infty(\Omega)$ automatically fulfils properties (ii), (iii) of the theorem above.

Corollary 6.17. *Let $n, d \in \mathbb{N}$, let $\Omega \subseteq \mathbb{R}^n$ be a measurable set, let $(z_k)_{k \in \mathbb{N}} : \Omega \to \mathbb{R}^d$ be a bounded sequence in $L^\infty(\Omega)^d$. Let $\nu = (\nu_x)_{x \in \Omega} \in \mathcal{Y}(\Omega; \mathcal{P}(\mathbb{R}^d))$ be a Young measure and assume that $(z_k)_{k \in \mathbb{N}}$ generates ν. Then properties (ii), (iii) of Theorem 6.16 are satisfied with $m_k = k$, $k \in \mathbb{N}$.*

Proof. By Theorem 6.16 we obtain the existence of a subsequence $(z_{m_k})_{k \in \mathbb{N}}$ and of a Young measure $\mu = (\mu_x)_{x \in \Omega} \in L^\infty_{\omega*}(\Omega; \mathcal{P}(\mathbb{R}^d))$ such that $(z_{m_k})_{k \in \mathbb{N}}$ generates μ and properties (ii), (iii) stated in Theorem 6.16 are fulfilled. By Lemma 6.14 we obtain $\nu_x = \mu_x$ for a.a. $x \in \Omega$. Furthermore, it follows that the choice $m_k = k$, $k \in \mathbb{N}$, is possible. $\square$

6.4 Generating some special gradient Young measures

This section is dedicated to the proof of Theorem 6.20 below. This result will be one of the main tools in the construction of the special domains with rough boundaries motivated by the introduction of this thesis. In the first step we consider the 'homogeneous case' of Theorem 6.20.

Lemma 6.18. *Let $n \in \mathbb{N}$, let $\Omega \subseteq \mathbb{R}^n$ be a bounded Lipschitz domain, let $z \in \mathbb{R}^n$ be a vector. Define the homogeneous Young measure $\nu = (\nu_x)_{x \in \Omega} \in \mathcal{Y}(\Omega; \mathcal{P}(\mathbb{R}^n))$ by*

$$\nu_x := \frac{1}{2}(\delta_z + \delta_{-z}), \quad x \in \Omega. \tag{6.13}$$

Then there exist non-negative functions $u_k : \Omega \to \mathbb{R}_0^+$, $k \in \mathbb{N}$, fulfilling the following properties:

(i) The sequence $(u_k)_{k \in \mathbb{N}}$ is bounded in $W_0^{1,\infty}(\Omega)$ and $(\nabla u_k)_{k \in \mathbb{N}}$ generates ν.

(ii) There holds $u_k \to 0$ strongly in $L^\infty(\Omega)$ as $k \to \infty$ and $\|\nabla u_k\|_\infty \leq \frac{3}{2}|z|$ for all $k \in \mathbb{N}$.

Proof. Step 1. In the first step we construct non-negative functions $w_k : \Omega \to \mathbb{R}_0^+$, $k \in \mathbb{N}$, such that the sequence $(w_k)_{k \in \mathbb{N}}$ is bounded in $W^{1,\infty}(\Omega)$, there holds that $(\nabla w_k)_{k \in \mathbb{N}}$ generates ν and the properties formuated in (ii) above are fulfilled. (This means that $w_k = 0$ on $\partial\Omega$ is not necessarily fulfilled).

Define for $x \in [0, 1[$

$$\chi(x) := \begin{cases} 1, & x \in [0, \frac{1}{2}], \\ -1, & x \in]\frac{1}{2}, 1[, \end{cases}$$

and extend χ by periodicity to all of $\mathbb{R}$. Introduce the function $h : \mathbb{R}^n \to \mathbb{R}$ by

$$h(x) := 1 + \int_0^{x \cdot z} \chi(s)\,ds, \quad x \in \mathbb{R}^n.$$

Then $\nabla h(x) = \chi(x \cdot z)z$ for almost all $x \in \mathbb{R}^n$. Define the function

$$w_k(x) := \frac{1}{k}h(kx) = \frac{1}{k} + \frac{1}{k}\int_0^{kx \cdot z} \chi(s)\,ds, \quad x \in \mathbb{R}^n. \tag{6.14}$$

Consequently

$$\nabla w_k(x) = (\nabla h)(kx) = \chi(kx \cdot z)z,$$

for almost all $x \in \mathbb{R}^n$. There holds $\|w_k\|_\infty \leq \frac{2}{k}$, $k \in \mathbb{N}$, and therefore $w_k \to 0$ strongly in $L^\infty(\Omega)$ as $k \to \infty$. Moreover $w_k \in W^{1,\infty}(\Omega)$ and $\|\nabla w_k\|_\infty = |z|$. It remains to prove the following

Assertion. *The sequence $(\nabla w_k)_{k \in \mathbb{N}}$ generates ν as defined in (6.13).*

Proof of the assertion. For $\phi \in C_0(\mathbb{R}^n)$ we introduce the function

$$f : \mathbb{R}^n \to \mathbb{R}, \quad f(x) := \phi(\chi(x \cdot z)z).$$

Further, for $i \in \{1, \ldots, n\}$ define

$$l_i := \begin{cases} \frac{1}{|z_i|}, & \text{if } z_i \neq 0, \\ 1, & \text{if } z_i = 0, \end{cases}$$

and $Y :=]0, l_1[\times \ldots \times]0, l_n[$. Then $f \in L^\infty(\mathbb{R}^n)$ is Y-periodic in the sense of [CiDo99, Definition 2.1]. Define $f_k : \mathbb{R}^n \to \mathbb{R}, f_k(x) := f(kx)$ for every $k \in \mathbb{N}$. By the Riemann-Lebesgue Lemma, see [CiDo99, Theorem 2.6], we get

$$f_k \underset{k \to \infty}{\overset{*}{\rightharpoonup}} \overline{f} \quad \text{in } L^\infty(\Omega) \tag{6.15}$$

where $\overline{f} : \Omega \to \mathbb{R}$ denotes the constant function with $\overline{f}(x) := \frac{1}{|Y|} \int_Y f(y) \, dy$ for every $x \in \Omega$. Relation (6.15) means

$$\phi(\nabla w_k) \underset{k \to \infty}{\overset{*}{\rightharpoonup}} \left(x \mapsto \frac{1}{|Y|} \int_Y \phi(\chi(y \cdot z) z) \, dy \right) \quad \text{in } L^\infty(\Omega). \tag{6.16}$$

By construction

$$|\{y \in Y; \chi(y \cdot z) = 1\}| = |\{y \in Y; \chi(y \cdot z) = -1\}| = \frac{|Y|}{2}.$$

Thus

$$\frac{1}{|Y|} \int_Y \phi(\chi(y \cdot z) z) \, dy = \frac{1}{2}(\phi(z) + \phi(-z)). \tag{6.17}$$

The definition of ν implies

$$\int_{\mathbb{R}^n} \phi(\lambda) \, d\nu_x(\lambda) = \frac{1}{2}(\phi(z) + \phi(-z)) \tag{6.18}$$

for all $x \in \Omega$. Combining (6.16), (6.17), (6.18) yields

$$\phi(\nabla w_k) \underset{k \to \infty}{\overset{*}{\rightharpoonup}} \left(x \mapsto \int_{\mathbb{R}^n} \phi(\lambda) \, d\nu_x(\lambda) \right) \quad \text{in } L^\infty(\Omega)$$

for a.a. $x \in \Omega$. Altogether, (see Definition 6.13), the sequence $(\nabla w_k)_{k \in \mathbb{N}}$ generates ν. □

Step 2. This step is based on the ideas of [Pe97, Lemma 8.3]. Let $\eta_k \in C^1(\overline{\Omega})$ be a sequence of cut-off functions with $0 \leq \eta_k \leq 1$ such that the following properties are fulfilled:

1. $\eta_k(x) = 0$ if $x \in \partial\Omega$,

2. $\eta_k(x) = 1$ if $x \in \Omega$ with $\text{dist}(x, \partial\Omega) > \frac{1}{k}$,

3. There exists a constant $c = c(\Omega) > 0$ such that $|\nabla \eta_k(x)| \leq ck$ for all $k \in \mathbb{N}$ and $x \in \Omega$.

For $j, k \in \mathbb{N}$ define

$$w_{j,k}(x) := \eta_k(x) w_j(x) \quad \text{for a.a. } x \in \Omega, \tag{6.19}$$

so that $w_{j,k} = 0$ on $\partial\Omega$ and $\nabla w_{j,k}(x) = \eta_k(x) \nabla w_j(x) + w_j(x) \nabla \eta_k(x)$ for a.a. $x \in \Omega$. Choose a strictly increasing sequence $(j_k)_{k \in \mathbb{N}}$ of natural numbers $j_k \in \mathbb{N}, k \in \mathbb{N}$, such that

$$\|w_{j_k,k}\|_\infty \leq \frac{1}{k^3}, \quad \text{for all } k \in \mathbb{N} \tag{6.20}$$

and define $u_k := w_{j_k,k}$ for $k \in \mathbb{N}$. Then

$$\|\nabla u_k\|_\infty \leq \|\nabla w_{j_k}\|_\infty + \|w_{j_k}\|_\infty \|\nabla \eta_k\|_\infty \leq \|\nabla w_{j_k}\|_\infty + c\frac{1}{k^2}. \tag{6.21}$$

By (6.19), (6.21) we obtain that $(u_k)_{k \in \mathbb{N}}$ is bounded in $W_0^{1,\infty}(\Omega)$ and $u_k \to 0$ strongly in $L^\infty(\Omega)$ as $k \to \infty$. From

$$\left| \{ x \in \Omega; \nabla u_k(x) \neq \nabla w_{j_k}(x) \} \right| \leq \left| \{ x \in \Omega; \operatorname{dist}(x, \partial\Omega) \leq \frac{1}{k} \} \right| \underset{k \to \infty}{\to} 0$$

it follows (see Lemma 6.15) that $(\nabla u_k)_{k \in \mathbb{N}}$ generates ν. There holds $\|\nabla w_k\|_\infty \leq \|z\|$ for all $k \in \mathbb{N}$. From (6.21) we see that we can choose a subsequence $(l_k)_{k \in \mathbb{N}}$ of $(j_k)_{k \in \mathbb{N}}$ with $\|\nabla u_{l_k}\|_\infty \leq \frac{3}{2}\|z\|$ for all $k \in \mathbb{N}$. The sequence $(u_{l_k})_{k \in \mathbb{N}}$ satisfies the properties (i), (ii), (iii). $\qquad \square$

Furthermore, we need the Lemma below which is based on Lebesgue's differentiation theorem and Vitali's covering theorem.

Lemma 6.19. *Let $n \in \mathbb{N}$, let $\Omega \subseteq \mathbb{R}^n$ be an open set and let $f_j : \Omega \to \mathbb{R}, j \in \mathbb{N}$, be integrable functions. Then there exist countably many (empty or nonempty) open balls $B(x_{k,l}, r_{k,l})_{k,l \in \mathbb{N}} \subseteq \Omega$ such that the following properties are satisfied:*

(i) For every $k, l \in \mathbb{N}$ we have that $x_{k,l}$ is a Lebesgue point of all functions $f_j, j \in \mathbb{N}$.

(ii) Fix $k \in \mathbb{N}$. Then the balls $B(x_{k,l}, r_{k,l}), l \in \mathbb{N}$, are pairwise disjoint and there is a null set N_k such that

$$\Omega = \bigcup_{l \in \mathbb{N}} B(x_{k,l}, r_{k,l}) \cup N_k. \tag{6.22}$$

(iii) We have

$$\int_\Omega h(x) f_j(x) \, dx = \lim_{k \to \infty} \sum_{l \in \mathbb{N}} f_j(x_{k,l}) \int_{B(x_{k,l}, r_{k,l})} h(x) \, dx \tag{6.23}$$

for all $h \in C_0^\infty(\Omega)$ and $j \in \mathbb{N}$.

Remark. To simplify notation we allow $r_{k,l} = 0$, i.e. that some of the open balls $B(x_{k,l}, r_{k,l})$ may be empty.

Proof. Define the set

$$A := \bigcap_{j \in \mathbb{N}} \left\{ x \in \Omega; \lim_{r \searrow 0} \frac{1}{|B(x,r)|} \int_{B(x,r)} |f_j(y) - f_j(x)|\, dy = 0 \right\}.$$

It is easily verified that $A \subseteq \Omega$ is measurable. By Lebesgue's differentiation theorem (see (2.31)) we have $|\Omega \setminus A| = 0$. For $k \in \mathbb{N}$ define

$$\mathcal{O}_k := \left\{ B(x,r) \subseteq \Omega \text{ such that } x \in A, 0 < r < \frac{1}{k} \text{ satisfy} \right.$$
$$\left. \frac{1}{|B(x,r)|} \int_{B(x,r)} |f_j(y) - f_j(x)|\, dy \le \frac{1}{k} \text{ for all } j = 1, \ldots, k \right\}. \tag{6.24}$$

We see that for fixed $k \in \mathbb{N}$ the family of sets $\mathcal{O}_k$ covers A in the sense of Theorem 6.8. Therefore, there exist pairwise disjoint sets $B(x_{k,l}, r_{k,l}) \in \mathcal{O}_k$, $l \in \mathbb{N}$, such that

$$\left| A \setminus \bigcup_{l \in \mathbb{N}} B(x_{k,l}, r_{k,l}) \right| = 0.$$

Consequently there is a null set N_k such that (6.22) is satisfied.

Fix $j \in \mathbb{N}$ and $h \in C_0^\infty(\Omega)$. With no loss of generality assume $\mathrm{supp}(h) \ne \varnothing$. Define $\delta := \mathrm{dist}(\mathrm{supp}(h), \mathbb{R}^n \setminus \Omega)$ if $\Omega \ne \mathbb{R}^n$ and $\delta := 1$ if $\Omega = \mathbb{R}^n$. Introduce the sets

$$D := \left\{ x \in \Omega; \mathrm{dist}(\mathrm{supp}(h), x) < \frac{\delta}{2} \right\},$$
$$M_k := \left\{ l \in \mathbb{N}; \mathrm{supp}(h) \cap B(x_{k,l}, r_{k,l}) \ne \varnothing \right\}, \quad k \in \mathbb{N}.$$

For $l \in M_k$ and $x \in B(x_{k,l}, r_{k,l})$ there holds $\mathrm{dist}(\mathrm{supp}(h), x) < 2r_{k,l} < \frac{2}{k}$. Consequently, for all $k \in \mathbb{N}, k > \frac{4}{\delta}$, and for all $l \in M_k$ we get $B(x_{k,l}, r_{k,l}) \subseteq D$. Thus, for all $k \in \mathbb{N}$ with $k > \max\{\frac{4}{\delta}, j\}$ it follows with (6.24)

$$\left| \int_\Omega h(x) f_j(x)\, dx - \sum_{l \in \mathbb{N}} f_j(x_{k,l}) \int_{B(x_{k,l}, r_{k,l})} h(x)\, dx \right|$$
$$= \left| \sum_{l \in \mathbb{N}} \int_{B(x_{k,l}, r_{k,l})} h(x)(f_j(x) - f_j(x_{k,l}))\, dx \right|$$
$$\le \|h\|_\infty \sum_{l \in M_k} \int_{B(x_{k,l}, r_{k,l})} |f_j(x) - f_j(x_{k,l})|\, dx$$
$$\le \|h\|_\infty \sum_{l \in M_k} \frac{1}{k} |B(x_{k,l}, r_{k,l})|$$
$$\le \frac{1}{k} \|h\|_\infty |D|.$$

Passing to the limit in the inequality above yields (6.23). □

Now we are in the position to formulate the main result of this section.

Theorem 6.20. *Let $n \in \mathbb{N}$, let $\Omega \subseteq \mathbb{R}^n$ be an arbitrary domain (which is not necessarily bounded), let $z : \Omega \to \mathbb{R}^n$ be a bounded, measurable function. Consider the Young measure $\nu = (\nu_x)_{x \in \Omega} \in \mathcal{Y}(\Omega; \mathcal{P}(\mathbb{R}^n))$ defined by*

$$\nu_x := \frac{1}{2}\big(\delta_{z(x)} + \delta_{-z(x)}\big), \quad x \in \Omega. \tag{6.25}$$

Then there exist non-negative functions $u_k : \Omega \to \mathbb{R}_0^+$, $k \in \mathbb{N}$, such that the following properties are satisfied:

(i) *The sequence $(u_k)_{k \in \mathbb{N}}$ is bounded in $W^{1,\infty}(\Omega)$ and $(\nabla u_k)_{k \in \mathbb{N}}$ generates ν.*

(ii) *There holds $u_k \to 0$ strongly in $L^\infty(\Omega)$ as $k \to \infty$.*

Proof. Step 1. Due to the assumptions on z there holds $|z(x)| \leq \|z\|_\infty$ for all $x \in \Omega$. Choose a sequence $(h_i)_{i \in \mathbb{N}} \subseteq C_0^\infty(\Omega)$ which is dense in $L^1(\Omega)$ and a sequence $(\phi_j)_{j \in \mathbb{N}} \subseteq C_0(\mathbb{R}^n)$ which is dense in $C_0(\mathbb{R}^n)$. Define for $j \in \mathbb{N}$

$$\overline{\phi_j}(x) := \int_{\mathbb{R}^n} \phi_j(\lambda)\, d\nu_x(\lambda) = \frac{1}{2}\big(\phi_j(z(x)) + \phi_j(-z(x))\big) \quad \text{for a.a. } x \in \Omega. \tag{6.26}$$

By Lemma 6.19 we obtain for every $k \in \mathbb{N}$ countably many pairwise disjoint open balls $B(x_{k,l}, r_{k,l}) \subseteq \Omega$, $l \in \mathbb{N}$, satisfying the following properties: For $k, l \in \mathbb{N}$ there holds that $x_{k,l}$ is a Lebesgue point of all functions $\overline{\phi_j}$, $j \in \mathbb{N}$, moreover there is a null set N_k such that

$$\Omega = \bigcup_{l \in \mathbb{N}} B(x_{k,l}, r_{k,l}) \cup N_k, \quad k \in \mathbb{N}. \tag{6.27}$$

We have that the identity

$$\int_\Omega h(x)\overline{\phi_j}(x)\, dx = \lim_{k \to \infty} \sum_{l \in \mathbb{N}} \overline{\phi_j}(x_{k,l}) \int_{B(x_{k,l}, r_{k,l})} h(x)\, dx \tag{6.28}$$

is satisfied for all $h \in L^\infty(\Omega)$ and all $j \in \mathbb{N}$. For $k, l \in \mathbb{N}$ define the homogeneous Young measure $\nu^{k,l} \in \mathcal{Y}(B(x_{k,l}, r_{k,l}); \mathcal{P}(\mathbb{R}^n))$ by

$$\nu^{k,l}(x) := \nu_{x_{k,l}} \quad \text{for all } x \in B(x_{k,l}, r_{k,l}).$$

Step 2. In this step we will define non-negative functions $u_k : \Omega \to \mathbb{R}_0^+$, $k \in \mathbb{N}$, with $u_k \to 0$ strongly in $L^\infty(\Omega)$ as $k \to \infty$ and $\|\nabla u_k\|_\infty \leq \frac{3}{2}\|z\|_\infty$, $k \in \mathbb{N}$, in such a way that

$$\lim_{k \to \infty} \sum_{l \in \mathbb{N}} \int_{B(x_{k,l}, r_{k,l})} h_i(x)\phi_j(\nabla u_k(x))\, dx = \lim_{k \to \infty} \sum_{l \in \mathbb{N}} \overline{\phi_j}(x_{k,l}) \int_{B(x_{k,l}, r_{k,l})} h_i(x)\, dx \tag{6.29}$$

for all $i, j \in \mathbb{N}$.

Fix $k, l \in \mathbb{N}$. In the following u_k will be defined on $B(x_{k,l}, r_{k,l})$. Use Lemma 6.18 to obtain non-negative functions $z_m^{k,l} : B(x_{k,l}, r_{k,l}) \to \mathbb{R}$, $m \in \mathbb{N}$, such that the following properties are fulfilled:

(i) The sequence $(z_m^{k,l})_{m \in \mathbb{N}}$ is bounded in $W_0^{1,\infty}(B(x_{k,l}, r_{k,l}))$ and $(z_m^{k,l})_{m \in \mathbb{N}} \to 0$ strongly in $L^\infty(\Omega)$ as $m \to \infty$.

(ii) The sequence $(\nabla z_m^{k,l})_{m \in \mathbb{N}}$ generates $\nu^{k,l} = (\nu_{x_{k,l}})_{x \in B(x_{k,l}, r_{k,l})}$.

(iii) We have $\|\nabla z_m^{k,l}\|_\infty \leq \frac{3}{2}\|z\|_\infty$ for all $m \in \mathbb{N}$.

Thus

$$\lim_{m \to \infty} \int_\Omega h(x)\phi(\nabla z_m^{k,l}(x))\, dx = \overline{\phi}(x_{k,l}) \int_\Omega h(x)\, dx \tag{6.30}$$

for all $h \in L^1(\Omega)$ and all $\phi \in C_0(\mathbb{R}^n)$. Choose $m = m(k,l) \in \mathbb{N}$ (see (6.30)) such that $\|z_{m(k,l)}^{k,l}\|_\infty \leq \frac{1}{k}$ and that

$$\left| \int_{B(x_{k,l}, r_{k,l})} h_i(x)\phi_j(\nabla z_{m(k,l)}^{k,l}(x))\, dx - \overline{\phi}_j(x_{k,l}) \int_{B(x_{k,l}, r_{k,l})} h_i(x)\, dx \right| \leq \frac{1}{2^l k} \tag{6.31}$$

is satisfied for all $i, j = 1, \ldots, k$. Now we define

$$u_k(x) := \begin{cases} z_{m(k,l)}^{k,l}(x), & \text{if } x \in B(x_{k,l}, r_{k,l}), \\ 0, & \text{if } x \in N_k. \end{cases} \tag{6.32}$$

Fix $k \in \mathbb{N}$. Since $z_{m(k,l)}^{k,l} \in W_0^{1,\infty}(B(x_{k,l}, r_{k,l}))$ for all $l \in \mathbb{N}$ we obtain from Lemma 6.9 applied to (6.32) that the weak gradient ∇u_k exists as function $\nabla u_k \in L_{\text{loc}}^\infty(\Omega)^n$. For fixed $i, j \in \mathbb{N}$ we sum over $l \in \mathbb{N}$ in (6.31), let $k \to \infty$ in this sum and use (6.32) to prove (6.29). Moreover by construction $\|u_k\|_\infty \leq \frac{1}{k}$ and $\|\nabla u_k\|_\infty \leq \frac{3}{2}\|z\|_\infty$ for $k \in \mathbb{N}$.

Step 3. Fix $i, j \in \mathbb{N}$. Then

$$\begin{aligned}
\lim_{k \to \infty} \int_\Omega h_i(x)\phi_j(\nabla u_k(x))\, dx &= \lim_{k \to \infty} \sum_{l \in \mathbb{N}} \int_{B(x_{k,l}, r_{k,l})} h_i(x)\phi_j(\nabla u_k(x))\, dx \\
&= \lim_{k \to \infty} \sum_{l \in \mathbb{N}} \overline{\phi}_j(x_{k,l}) \int_{B(x_{k,l}, r_{k,l})} h_i(x)\, dx \\
&= \int_\Omega h_i(x)\overline{\phi}_j(x)\, dx \\
&= \int_\Omega h_i(x) \left(\int_{\mathbb{R}^n} \phi_j(\lambda)\, d\nu_x(\lambda) \right) dx.
\end{aligned} \tag{6.33}$$

To obtain (6.33) we have used (6.27), (6.29) and (6.28). By Lemma 6.12 we obtain from (6.33) that $(\nabla u_k)_{k \in \mathbb{N}}$ generates ν. $\qquad \square$

7 Influence of boundary rugosity to weak solutions of the Boussinesq equations

We consider the Boussinesq equations

$$u_t - \nu \Delta u + (u \cdot \nabla)u + \frac{1}{\rho_0}\nabla p = \beta\theta g + f_1 \quad \text{in }]0,T[\times\Omega,$$

$$\text{div } u = 0 \qquad \text{in }]0,T[\times\Omega,$$

$$\theta_t - \kappa\Delta\theta + (u \cdot \nabla)\theta = f_2 \qquad \text{in }]0,T[\times\Omega, \tag{7.1}$$

$$u(0) = u_0 \qquad \text{in } \Omega,$$

$$\theta(0) = \theta_0 \qquad \text{in } \Omega$$

for a viscous, incompressible fluid with velocity u, pressure p, and temperature θ on a domain $\Omega \subseteq \mathbb{R}^n$, $n \in \{2,3\}$, and a finite time interval $[0,T[$.

In this chapter we consider a sequence $(\Omega_k)_{k\in\mathbb{N}}$ of domains with 'rough boundaries' converging to a 'physical' domain $\Omega \subseteq \mathbb{R}^n$. For detailed information about the domains $\Omega, (\Omega_k)_{k\in\mathbb{N}}$ and the considered data $f_1, f_2, g, h_0, u_0, \theta_0$ we refer to Definition 7.27. Let $(u_k, \theta_k)_{k\in\mathbb{N}}$ be a sequence of solutions of the Boussinesq equations (7.1) in $[0,T[\times\Omega_k$ satisfying the energy inequalities. Since in our model we assume that the boundaries $\partial\Omega_k$ are impermeable, the normal component of u_k satisfies

$$u_k \cdot N = 0 \quad \text{on }]0,T[\times\partial\Omega_k. \tag{7.2}$$

We suppose that the tangential component

$$[u_k]_{\text{tan}} = u_k - (u_k \cdot N)N \quad \text{on }]0,T[\times\partial\Omega_k$$

of u_k satisfies the *Navier slip boundary condition*

$$[\delta u_k + 2\nu D(u_k) \cdot N]_{\text{tan}} = 0 \quad \text{on }]0,T[\times\partial\Omega_k. \tag{7.3}$$

where $D(u_k) := \frac{1}{2}\big(\nabla u_k + (\nabla u_k)^T\big)$ denotes the *symmetric part of the gradient* and $\delta \in [0,\infty[$ denotes a friction coefficient. The combination of the conditions (7.2), (7.3) is called the *Navier slip boundary condition*. For further information about the Navier slip boundary we refer to Section 7.3. Further, we assume that θ_k satisfies the Robin boundary condition

$$\frac{\partial\theta_k}{\partial N} + \alpha(\theta_k - h_0) = 0 \quad \text{on }]0,T[\times\partial\Omega_k. \tag{7.4}$$

In Theorem 7.13 we will investigate the limit equations and boundary conditions fulfilled by a weak limit (u, θ) of $(u_k, \theta_k)_{k \in \mathbb{N}}$ in $[0, T[\times \Omega$. In the following we give an overview about the organization of this chapter.

Let $\Omega \subseteq \mathbb{R}^n, n \in \{2, 3\}$, be a domain satisfying the uniform Lipschitz condition, let $(u_k, \theta_k)_{k \in \mathbb{N}}$ be a sequence of solutions of (7.1) in $[0, T[\times \Omega$ satisfying the energy inequalities. Assume that (7.2), (7.4) are fulfilled. In Section 7.1 we 'ignore' the concrete boundary behaviour of (u_k, θ_k) on $[0, T[\times \partial\Omega$. Theorem 7.9 states that every limit (u, θ) of $(u_k, \theta_k)_{k \in \mathbb{N}}$ in $[0, T[\times \Omega$ (with respect to the associated energy norms (7.40), (7.41)) is a solution of (7.1) in $[0, T[\times \Omega$. This property can be seen as *sequential compactness* of the system of equations (7.1). Furthermore, we will show that (7.43), (7.44) are fulfilled. These relations will play a crucial role in this chapter. The proof of this Theorem is based on the construction of a local pressure introduced by J. Wolf in [Wo07] and the classical Aubin-Lions compactness result. For local sequential compactness of the Navier-Stokes system we refer to [BFNW08].

Let (u, θ) be a weak limit of $(u_k, \theta_k)_{k \in \mathbb{N}}$ in $[0, T[\times \Omega$. In Section 7.2 we investigate the boundary behaviour of (u, θ) on $]0, T[\times \partial\Omega$. We remark that we not need any assumptions on the tangential component $[u_k]_{\tan}$ of u_k on $]0, T[\times \partial\Omega_k$ throughout this section. The main tool of our approach is the theory of Young measures as developed in Chapter 6. The idea to use Young measures to analyse the slip behaviour of the velocity field u was introduced in the paper [BFNW08]. Due to the results of Subsection 7.2.2 we will observe that the temperature θ satisfies the boundary condition

$$\frac{\partial \theta}{\partial N} + \alpha \Lambda (\theta - h_0) = 0 \quad \text{on } \partial\Omega \tag{7.5}$$

with an 'additional weight function' $\Lambda : \partial\Omega \to [1, \infty]$ which can be computed using Young measures.

Finally Section 7.3 combines the results obtained in Section 7.1 and 7.2 to prove the main results of this chapter which are Theorem 7.28 and Theorem 7.30.

7.1 Sequential stability of the Boussinesq equations on a fixed domain

In the first part of this sections we deal with a decomposition of $L^2(G)$ where $G \subseteq \mathbb{R}^n$ is an arbitrary domain. Our approach is based on Hilbert space theory. These results are used to obtain existence and decomposition of a pressure and the estimates as stated in Theorem 7.9.

If G is bounded we denote by $\delta(G)$ the diameter of G. We begin with an important estimate.

Lemma 7.1. *Let $G \subseteq \mathbb{R}^n, n \in \mathbb{N}$, be a bounded domain. Then*

$$\|\phi\|_{W^{2,2}(G)} \leq c \|\Delta\phi\|_2 \tag{7.6}$$

for all $\phi \in W_0^{2,2}(G)$ with a constant $c = c(n, \delta(G)) > 0$.

Proof. Due to our definition in the preliminaries, the space $W_0^{2,2}(G)$ denotes the closure of $C_0^\infty(G)$ with respect to the norm $\|\cdot\|_{W^{2,2}(G)}$. Thus, by the Poincaré inequality

$$\|\phi\|_{W^{2,2}(G)} \leq c\|\nabla^2\phi\|_2$$

for all $\phi \in W_0^{2,2}(G)$ with a constant $c > 0$ depending only on the diameter $\delta(G)$ of G. Further, by a well known inequality, there is a constant $c = c(n) > 0$ such that

$$\|\nabla^2\phi\|_2 \leq c\|\Delta\phi\|_2$$

for all $\phi \in W_0^{2,2}(G)$. Altogether, (7.6) holds with a constant $c = c(n, \delta(\Omega)) > 0$. $\square$

Looking at (7.6) it follows immediately that $\|\Delta\phi\|_2$ and $\|\phi\|_{W^{2,2}(G)}$ are equivalent norms for $\phi \in W_0^{2,2}(G)$. Before proceeding, we need the following definitions:

$$\Delta W_0^{2,2}(G) := \{\,\Delta p; p \in W_0^{2,2}(G)\,\},$$

$$L_0^2(G) := \{\,p \in L^2(G); \int_G p\,dx = 0\,\}.$$

It follows immediately that $\Delta W_0^{2,2}(G) \subseteq L_0^2(G)$ holds.

Consider $p \in L_{\text{loc}}^1(G)$ which satisfies $\Delta p = 0$ in the sense of distributions, i.e.

$$\int_G p\Delta\phi\,dx = 0 \quad \forall\phi \in C_0^\infty(G).$$

Due to Weyl's Lemma (see e.g. [Jo98, Korollar 2.1]) such a p can be redefined on a null set of G such that $p \in C^\infty(G)$ holds. The next lemma of this section presents a useful estimate for harmonic functions.

Lemma 7.2. *Let $G_1, G_2 \subseteq \mathbb{R}^n$, $n \in \mathbb{N}$, be bounded domains, assume $\overline{G_1} \subseteq G_2$. Then there is a constant $c = c(n, G_1, G_2) > 0$ with the following property: For all harmonic functions $p \in C^2(G_2)$ with $p \in L^1(G_2)$ there holds*

$$\|p\|_{C^2(G_1)} \leq \|p\|_{L^1(G_2)}. \tag{7.7}$$

Proof. Define $\mu := \text{dist}(G_1, \mathbb{R}^n \setminus G_2)$. By assumptions on G_1, G_2 it follows $\mu > 0$. Fix any $0 < r < \mu$. By a well known estimate for harmonic function (see [Ev98, Theorem 2.2.7]) we get for each $k \in \mathbb{N}$ a constant $c(n, k) > 0$ such that

$$|D^\alpha p(x_0)| \leq \frac{c(n,k)}{r^{n+k}}\|p\|_{L^1(B(x_0,r))} \tag{7.8}$$

for all $x_0 \in G_1$, all multi-indices $\alpha \in \mathbb{N}^n$ with $|\alpha| = k$, and all harmonic functions $p \in C^2(G_2)$. Define $c := \max\left\{\dfrac{c(n,0)}{r^n}, \dfrac{c(n,1)}{r^{n+1}}, \dfrac{c(n,2)}{r^{n+2}}\right\}$. Then

$$|D^\alpha p(x_0)| \leq c\|p\|_{L^1(G_2)} \tag{7.9}$$

for all $x_0 \in G_1$, all multi-indices $\alpha \in \mathbb{N}^n$ with $|\alpha| = k$, $k \in \{0, 1, 2\}$, and all harmonic functions $p \in C^2(G_2)$. Thus, (7.7) can be fulfilled with $c = c(n, G_1, G_2) > 0$. $\square$

Lemma 7.3. *Let $G \subseteq \mathbb{R}^n, n \in \mathbb{N}$, be a bounded domain. The following statements are satisfied:*

(i) For every functional $F \in W_0^{2,2}(G)'$ there is exactly one $\psi \in W_0^{2,2}(G)$ such that

$$\int_G \Delta\psi\Delta\phi\, dx = [F, \phi]_{W_0^{2,2}(G)', W^{2,2}(G)}$$

for all $\phi \in W_0^{2,2}(G)$.

(ii) We have

$$\|\psi\|_2 = \sup_{\phi \in C_0^\infty(G), \phi \neq 0} \frac{\int_G \psi\Delta\phi\, dx}{\|\Delta\phi\|_2} \tag{7.10}$$

for all $\psi \in \Delta W_0^{2,2}(G)$.

Proof. (i) Consider $F \in W_0^{2,2}(G)'$. By definition $F : W_0^{2,2}(G) \to \mathbb{R}$ is a continuous, linear functional with respect to the norm $\|\cdot\|_{W^{2,2}(G)}$. Define the Hilbert space $H := W_0^{2,2}(G)$ with scalar product $\langle\phi, \psi\rangle_H := \langle\Delta\phi, \Delta\psi\rangle_G$. Due to the equivalence of the norms $\|\phi\|_{W^{2,2}(G)}$ and $\|\Delta\phi\|_2$ for $\phi \in W_0^{2,2}(G)$ it follows $F \in H'$. The Riesz representation theorem applied to the Hilbert space H yields (i).

(ii) Consider $\chi \in W_0^{2,2}(G)$ with $\psi = \Delta\chi$. Choose $\phi_k \in C_0^\infty(G), k \in \mathbb{N}$, with $\phi_k \to \chi$ as $k \to \infty$ in $W^{2,2}(G)$. Thus

$$\|\Delta\chi\|_2 = \lim_{k \to \infty} \frac{\langle\Delta\chi, \Delta\phi_k\rangle_G}{\|\Delta\phi_k\|_2} \leq \sup_{\phi \in C_0^\infty(G), \phi \neq 0} \frac{\langle\Delta\chi, \Delta\phi\rangle_G}{\|\Delta\phi\|_2}.$$

It follows immediately that even

$$\|\Delta\chi\|_2 = \sup_{\phi \in C_0^\infty(G), \phi \neq 0} \frac{\langle\Delta\chi, \Delta\phi\rangle_G}{\|\Delta\phi\|_2}$$

holds. Since $\psi = \Delta\chi$ we get (7.10). $\qquad\square$

We obtain the following decomposition of $L^2(G)$:

Lemma 7.4. *Let $G \subseteq \mathbb{R}^n, n \in \mathbb{N}$, be a bounded domain. For every $p \in L^2(G)$ there exist unique $p_0 \in \Delta W_0^{2,2}(G)$ and $p_h \in L^2(G), \Delta p_h = 0$, such that*

$$p = p_0 + p_h.$$

Furthermore, $p = p_0 + p_h$ is an orthogonal decomposition in $L^2(G)$, i.e. $\langle p_0, p_h\rangle_G = 0$, and we have

$$\|p_0\|_2 + \|p_h\|_2 \leq 3\|p\|_2. \tag{7.11}$$

Remark. In Lemma [Wo07, Corollary 2.5] an analogous decomposition to (7.11) for functions $p \in L^r(\Omega)$ where $1 < r < \infty$ is proved. Since the Hilbert space approach is much simpler (but restricted to $r = 2$), we are able to consider arbitrary bounded domains $G \subseteq \mathbb{R}^n$ and obtain that the estimate (7.11) holds with a constant independent of the domain.

Proof. Fix $p \in L^2(G)$. By Lemma 7.3 there exists $\psi \in W_0^{2,2}(G)$ such that

$$\int_G \Delta\psi\Delta\phi \, dx = \int_G p\Delta\phi \, dx$$

for all $\phi \in W_0^{2,2}(G)$. A consequence of Weyl's Lemma is $\Delta(p - \Delta\psi) = 0$. Therefore $p = \Delta\psi + (p - \Delta\psi)$. Define $p_0 := \Delta\psi$ and $p_h := p - \Delta\psi$. Since p_h is harmonic there holds

$$\int_G p_h\Delta\phi \, dx = 0 \tag{7.12}$$

for all $\phi \in C_0^\infty(G)$. From $p_0 \in \Delta W_0^{2,2}(G)$ we obtain from (7.12) by density that $\langle p_0, p_h\rangle_G = 0$.

In the next step (7.11) will be proved. Assume that $p = p_0 + p_h$ with $p_0 \in \Delta W_0^{2,2}(G)$ and $p_h \in L^2(G)$, $\Delta p_h = 0$. There holds

$$\int_G p\Delta\phi \, dx = \int_G p_0\Delta\phi \, dx$$

for all $\phi \in C_0^\infty(G)$. It follows with (7.10)

$$\|p_0\|_2 = \sup_{\phi \in C_0^\infty(G), \phi \neq 0} \frac{\langle p_0, \Delta\phi\rangle_G}{\|\Delta\phi\|_2}$$

$$= \sup_{\phi \in C_0^\infty(G), \phi \neq 0} \frac{\langle p, \Delta\phi\rangle_G}{\|\Delta\phi\|_2}$$

$$\leq \|p\|_2.$$

Since $p_h = p - p_0$ the proof of (7.11) is complete.

It remains to prove the uniqueness. Consider

$$p = p_0^{(1)} + p_h^{(1)} = p_0^{(2)} + p_h^{(2)}$$

with $p_0^{(1)}, p_0^{(2)} \in \Delta W_0^{2,2}(G)$ and $p_h^{(1)}, p_h^{(2)} \in L^2(G)$ with $\Delta p_h^1 = 0$, $\Delta p_h^2 = 0$. It follows

$$(p_0^{(1)} - p_0^{(2)}) + (p_h^{(1)} - p_h^{(2)}) = 0$$

and consequently with (7.11)

$$\|p_0^{(1)} - p_0^{(2)}\|_2 + \|p_h^{(1)} - p_h^{(2)}\|_2 \leq 3\|0\|_2 = 0.$$

$\square$

In the proof of Theorem 7.9 we make use of the following lemma.

Lemma 7.5. *Let $G \subseteq \mathbb{R}^n, n \in \mathbb{N}$, be a bounded domain. Let $p \in \Delta W_0^{2,2}(G)$. Assume that $Q_1 \in L^2(G)^{n \times n}$ and $Q_2 \in L^2(G)^n$ satisfy*

$$\int_G p \Delta \phi \, dx = \int_G Q_1 : \nabla^2 \phi \, dx + \int_G Q_2 \cdot \nabla \phi \, dx \qquad (7.13)$$

for all $\phi \in C_0^\infty(G)$. Then there is a constant $c = c(n, \delta(G))$ (independent of p, Q_1, Q_2) with

$$\|p\|_2 \leq c(\|Q_1\|_2 + \|Q_2\|_2). \qquad (7.14)$$

Proof. It follows from (7.6), (7.10) and (7.13)

$$
\begin{aligned}
\|p\|_2 &= \sup_{\phi \in C_0^\infty(G), \phi \neq 0} \frac{\langle p, \Delta \phi \rangle_G}{\|\Delta \phi\|_2} \\
&= \sup_{\phi \in C_0^\infty(G), \phi \neq 0} \frac{\langle Q_1, \nabla^2 \phi \rangle_G + \langle Q_2, \nabla \phi \rangle_G}{\|\Delta \phi\|_2} \\
&\leq \sup_{\phi \in C_0^\infty(G), \phi \neq 0} \frac{\|Q_1\|_2 \|\nabla^2 \phi\|_2 + \|Q_2\|_2 \|\nabla \phi\|_2}{\|\Delta \phi\|_2} \\
&\leq c(\|Q_1\|_2 + \|Q_2\|_2)
\end{aligned}
$$

with a constant $c = c(n, \delta(G)) > 0$. □

Moreover the following Lemma is needed for the proof of Theorem 7.8.

Lemma 7.6. *Let X be a reflexive Banach space, let $0 < T < \infty, 1 < s < \infty$, and let $f \in L^s(0, T; X)$. Assume that*

$$\int_0^{T'} \left\| \frac{f(t + \tau) - f(t)}{\tau} \right\|_X^s dt \leq c$$

for all $0 < T' < T$ and $0 < \tau < T - T'$ with a constant $c > 0$ independent of T', τ. Then $f \in W^{1,s}(0, T; X)$ and

$$\left\| \frac{d}{dt} f \right\|_{L^s(0,T;X)} \leq c. \qquad (7.15)$$

Proof. For a real or Banach space-valued function $t \mapsto h(t), t \in]0, T[$, and $0 < \tau < T$ we will denote by δ_τ the function

$$\delta_\tau h(t) := \frac{h(t + \tau) - h(t)}{\tau}, \quad t \in]0, T - \tau[.$$

Fix $0 < T' < T$. A consequence of the assumptions is $\|\delta_{\frac{1}{k}} f\|_{L^s(0,T';X)} \leq c$ for all sufficiently large $k \in \mathbb{N}$. Since the set $\{f \in L^s(0, T'; X); \|f\|_{L^s(0,T';X)} \leq c\}$ is weakly sequentially compact there exists $g \in L^s(0, T'; X)$ and a subsequence $(\tau_k)_{k \in \mathbb{N}}$ of $(\frac{1}{k})_{k \in \mathbb{N}}$ such that

$$\delta_{\tau_k} f \underset{k \to \infty}{\rightharpoonup} g \quad \text{in } L^s(0, T'; X).$$

Furthermore, by weak lower semicontinuity of the norm we get

$$\|g\|_{L^s(0,T';X)} \le c.$$

Fix $w \in X'$. Define

$$\widetilde{f}(t) := [w, f(t)]_{X';X}, \quad \widetilde{g}(t) := [w, g(t)]_{X';X} \quad \text{for } t \in]0, T'[.$$

We obtain

$$\int_0^{T'} \widetilde{f}(t)\, \eta'(t)\, dt = \lim_{k \to \infty} \int_0^{T'} \widetilde{f}(t)\, (\delta_{-\tau_k}\eta)(t)\, dt$$

$$= -\lim_{k \to \infty} \int_0^{T'} (\delta_{\tau_k}\widetilde{f})(t)\, \eta(t)\, dt$$

$$= -\int_0^{T'} \widetilde{g}(t)\, \eta(t)\, dt$$

for every $\eta \in C_0^\infty(0, T')$. Thus

$$\frac{d}{dt}[w, f(t)]_{X';X} = [w, g(t)]_{X';X} \quad \text{on }]0, T'[$$

for all $w \in X'$. By Lemma 2.16 we obtain $f \in W^{1,s}(0, T'; X)$ and $\frac{d}{dt}f = g$ in $W^{1,s}(0, T'; X)$. Due to the arbitrary choice of $0 < T' < T$ we obtain (7.15). $\qquad \square$

The lemma below shows that the divergence problem can be solved in bounded Lipschitz domains.

Lemma 7.7. *Let $G \subseteq \mathbb{R}^n, n \in \mathbb{N}$, be a bounded Lipschitz domain. Then there exists a constant $c_1 = c_1(G) > 0$ with the following property: For all $\phi \in L_0^2(G)$ the divergence problem*

$$\begin{aligned} \operatorname{div} w &= \phi \quad &\text{on } G, \\ w &= 0 \quad &\text{on } \partial G, \end{aligned} \tag{7.16}$$

has a solution $w \in H_0^1(G)^n$ with

$$\|w\|_{H^1(G)^n} \le c_1 \|\phi\|_2. \tag{7.17}$$

Proof. This is a well established result, for a proof we refer to [Ga98, Chapter III, Theorem 3.2]. $\qquad \square$

Let $c_1 = c_1(G)$ denote the constant in (7.17).

We proceed with the following theorem which is a variant of [Wo07, Theorem 2.6].

Theorem 7.8. *Let $G \subseteq \mathbb{R}^n, n \in \mathbb{N}$, be a bounded Lipschitz domain, let $0 < T < \infty$, $1 < \gamma < \infty$, let $u_0 \in L^2(G)^n$, $Q_1 \in L^\gamma(0, T; L^2(G)^{n \times n})$, let $Q_2 \in L^\gamma(0, T; L^2(G)^n)$. Consider $u \in L^\infty(0, T; L^2(G)^n)$ with $\mathrm{div}\, u(t) = 0$ for almost all $t \in]0, T[$, assume that*

$$- \int_0^T \langle u, \partial_t w \rangle_G \, dt + \int_0^T \langle Q_1, \nabla w \rangle_G \, dt + \int_0^T \langle Q_2, w \rangle_G \, dt - \langle u_0, w(0) \rangle_G = 0 \quad (7.18)$$

for all $w \in C_0^\infty([0, T[; C_{0,\sigma}^\infty(G))$. Let $c_1 = c_1(G) > 0$ denote a constant fulfilling (7.17). Then there exist unique functions $p_r \in L^\gamma(0, T; L_0^2(G))$, $p_h \in L^\infty(0, T; L_0^2(G))$ with $p_r(t) \in \Delta W_0^{2,2}(G)$, $\Delta_x p_h(t) = 0$ for a.a. $t \in]0, T[$, such that

$$- \int_0^T \langle u + \nabla_x p_h, \partial_t w \rangle_G \, dt + \int_0^T \langle Q_1, \nabla w \rangle_G \, dt + \int_0^T \langle Q_2, w \rangle_G \, dt$$

$$= \langle u_0, w(0) \rangle_G + \int_0^T \langle p_r, \mathrm{div}\, w \rangle_G \, dt \quad (7.19)$$

for all $w \in C_0^\infty([0, T[\times G)^n$. Further there is a constant $c = c(n, \gamma, T, c_1, \delta(G)) > 0$ such that the following estimates are satisfied:

$$\|p_r\|_{2,\gamma;G;T} \leq c(\|Q_1\|_{2,\gamma;G;T} + \|Q_2\|_{2,\gamma;G;T}), \quad (7.20)$$

$$\|p_h\|_{2,\infty;G;T} \leq c(\|u\|_{2,\infty;G;T} + \|Q_1\|_{2,\gamma;G;T} + \|Q_2\|_{2,\gamma;G;T}). \quad (7.21)$$

Especially (7.20), (7.21) can be fulfilled with a constant $c = c(\gamma, T, G) > 0$.

Proof. Step 1. The goal of this step is to prove the following:
Assertion. *There exists $p \in L^\infty(0, T; L_0^2(G))$ such that*

$$- \int_0^T \langle u, \partial_t w \rangle_G \, dt + \int_0^T \langle Q_1, \nabla w \rangle_G \, dt + \int_0^T \langle Q_2, w \rangle_G \, dt - \langle u_0, w(0) \rangle_G$$

$$= \int_0^T \langle p(t), \partial_t \mathrm{div}_x w \rangle_G \, dt \quad (7.22)$$

for all $w \in C_0^\infty([0, T[\times G)^n$. The estimate

$$\|p\|_{2,\infty;G;T} \leq c(\|u\|_{2,\infty;G;T} + \|Q_1\|_{2,\gamma;G;T} + \|Q_2\|_{2,\gamma;G;T}) \quad (7.23)$$

is satisfied with a constant $c = c(n, \gamma, T, c_1) > 0$.
Proof of the assertion. By assumption (7.18)

$$- \int_0^T \langle u(t), w \rangle_G \, \eta'(t) \, dt - \langle u_0, w \rangle_G \, \eta(0)$$

$$= \int_0^T \left(-\langle Q_1(t), \nabla w \rangle_G - \langle Q_2(t), w \rangle_G \right) \eta(t) \, dt \quad (7.24)$$

for all $w \in C_{0,\sigma}^\infty(G)$, $\eta \in C_0^\infty([0, T[)$. Looking at (7.24) and Lemma 2.16 we see that for fixed $w \in C_{0,\sigma}^\infty(G)$ the function $t \mapsto \langle u(t), w \rangle_G$ belongs to the space $W^{1,\gamma}(0, T)$ and has the following derivative and initial value:

$$\frac{d}{dt} \langle u, w \rangle_G = -\langle Q_1, \nabla w \rangle_G - \langle Q_2, w \rangle_G \quad \text{on }]0, T[,$$

$$\langle u(t), w \rangle_G \big|_{t=0} = \langle u_0, w \rangle. \tag{7.25}$$

To simplify the notation define $u(0) := u_0$. Identity (7.25) and Lemma 2.16 imply that there exists a Lebesgue null set $N = N(w) \subseteq]0, T[$ such that

$$\langle u(t), w \rangle_G - \langle u_0, w \rangle_G = -\int_0^t \left(\langle Q_1(s), \nabla w \rangle_G + \langle Q_2(s), w \rangle_G \right) ds \tag{7.26}$$

for all $t \in [0, T[\backslash N$. Using $u \in L^\infty(0, T; L^2(G)^n)$ and a separability argument u can be redefined on a Lebesgue null set of $[0, T[$ such that (7.26) holds for all $t \in [0, T[$ and all $w \in C_{0,\sigma}^\infty(G)$. For $t \in [0, T[$ define $\widetilde{Q_1}(t) := \int_0^t Q_1(s) \, ds$ and $\widetilde{Q_2}(t) := \int_0^t Q_2(s) \, ds$. We employ Fubini's Theorem and [So01, Lemma II.2.2.2] to get for each $t \in [0, T[$ a unique $p(t) \in L_0^2(G)$ with

$$\langle u(t) - u_0, w \rangle_G + \langle \widetilde{Q_1}(t), \nabla w \rangle_G + \langle \widetilde{Q_2}(t), w \rangle_G = \langle p(t), \operatorname{div} w \rangle_G \tag{7.27}$$

for all $w \in W_0^{1,2}(G)^n$ and $t \in [0, T[$.

In the following we show that the function $]0, T[\to L^2(G), t \mapsto p(t)$, is Bochner measurable. By Pettis' Theorem (see Theorem 2.11) and the fact that $L^2(G)$ is separable, this is equivalent to prove that this function is weakly measurable. For this reason fix $\phi \in L^2(G)$. Choose $w \in H_0^1(G)^n$ with $\operatorname{div} w = \phi - \frac{1}{|G|} \int_G \phi(y) \, dy$. It follows (since $\int_G p(t) \, dx = 0$)

$$\langle p(t), \phi \rangle_G = \langle p(t), \operatorname{div} w \rangle_G = \langle u(t) - u_0, w \rangle_G + \langle \widetilde{Q_1}(t), \nabla w \rangle_G + \langle \widetilde{Q_2}(t), w \rangle_G \tag{7.28}$$

for all $t \in [0, T[$. From (7.28) it follows that $t \mapsto p(t)$ is weakly measurable.

By Lemma 7.7 we can choose for each $t \in [0, T[$ an element $w = w(t) \in H_0^1(G)^n$ with $\operatorname{div} w(t) = p(t)$ and $\|w(t)\|_{H^1(G)^n} \leq c_1 \|p(t)\|_2$. Insert this special w in (7.27) to obtain the estimate

$$\|p(t)\|_2 \leq c_1 \left(\|u(t) - u_0\|_2 + \|\widetilde{Q_1}(t)\|_2 + \|\widetilde{Q_2}(t)\|_2 \right) \tag{7.29}$$

for all $t \in [0, T[$. We obtain from (7.29) with Hölder's inequality

$$\|p\|_{2,\infty;G;T} \leq c \left(\|u\|_{2,\infty;G;T} + \|Q_1\|_{2,\gamma;G;T} + \|Q_2\|_{2,\gamma;G;T} \right) \tag{7.30}$$

with a constant $c = c(\gamma, T, c_1) > 0$. Altogether $p \in L^\infty(0, T; L_0^2(G))$ holds and the proof of the assertion is complete. $\qquad \square$

Step 2. Lemma 7.4 yields the existence of unique $\widetilde{p_r}(t) \in \Delta W_0^{2,2}(G)$, $p_h(t) \in L^2(G)$ with $\Delta_x p_h(t) = 0$, $t \in [0, T[$, such that

$$p(t) = \widetilde{p_r}(t) + p_h(t), \quad \|\widetilde{p_r}(t)\|_2 + \|p_h(t)\|_2 \leq 3\|p(t)\|_2 \tag{7.31}$$

for all $t \in [0, T[$. Using that $p :]0, T[\rightarrow L^2(G)$ is measurable, the estimate (7.30) and $p(t) \in L_0^2(G)$, $t \in [0, T[$, we conclude $\widetilde{p}_r \in L^\infty(0, T; L_0^2(G))$ and $p_h \in L^\infty(0, T; L_0^2(G))$.

From (7.29), (7.31) it follows $\widetilde{p}_r(0) = 0$. For a fixed $w \in C_0^\infty([0, T[\times G)^n$ we insert $w(t)$ in (7.27) and integrate with respect to $t \in [0, T[$. Afterwards, we integrate by parts in space to get

$$
\int_0^T \langle u(t) - u_0, w(t)\rangle_G \, dt + \int_0^T \langle \widetilde{Q}_1(t), \nabla w(t)\rangle_G \, dt + \int_0^T \langle \widetilde{Q}_2(t), w(t)\rangle_G \, dt
$$
$$
= \int_0^T \langle p(t), \operatorname{div} w(t)\rangle_G \, dt \tag{7.32}
$$
$$
= \int_0^T \langle \widetilde{p}_r(t), \operatorname{div} w(t)\rangle_G \, dt - \int_0^T \langle \nabla_x p_h(t), w(t)\rangle_G \, dt.
$$

Step 3. Fix $\tau \in]0, T[$. Given $\phi \in C_0^\infty(G)$ we insert $w := \nabla\phi$ in (7.27). Consider $t \in]0, T - \tau[$ such that $\operatorname{div} u(t) = 0$ and $\operatorname{div} u(t + \tau) = 0$ holds. We make use of $p_h(t)$, $p_h(t + \tau) \in L_0^2(G)$ to obtain

$$
\int_G \big(\widetilde{p}_r(t + \tau) - \widetilde{p}_r(t)\big)\Delta\phi \, dx
$$
$$
= \int_G \big(\widetilde{Q}_1(t + \tau) - \widetilde{Q}_1(t)\big) : \nabla^2\phi \, dx + \int_G \big(\widetilde{Q}_2(t + \tau) - \widetilde{Q}_2(t)\big) \cdot \nabla\phi \, dx. \tag{7.33}
$$

By Lemma 7.5

$$
\big\|\widetilde{p}_r(t + \tau) - \widetilde{p}_r(t)\big\|_2^\gamma \le c \sum_{i=1}^2 \big\|\widetilde{Q}_i(t + \tau) - \widetilde{Q}_i(t)\big\|_2^\gamma
$$
$$
= c \sum_{i=1}^2 \Big\|\int_t^{t+\tau} Q_i(s) \, ds\Big\|_2^\gamma \tag{7.34}
$$

with a constant $c = c(n, \gamma, T, c_1, \delta(G))$ independent of t, τ. Especially, estimate (7.34) is satisfied for almost all $t \in]0, T - \tau[$. Fix $0 < T' < T$. We obtain with Minkowski's inequality for integrals, Hölder's inequality and Fubini's Theorem

$$
\int_0^{T'} \big\|\widetilde{p}_r(t + \tau) - \widetilde{p}_r(t)\big\|_2^\gamma \, dt \le c \sum_{i=1}^2 \int_0^{T'} \Big(\int_t^{t+\tau} \|Q_i(s)\|_2 \, ds\Big)^\gamma dt
$$
$$
\le c \sum_{i=1}^2 \int_0^{T'} \int_t^{t+\tau} \|Q_i(s)\|_2^\gamma \, ds \, \tau^{\gamma/\gamma'} \, dt
$$
$$
= c\tau^{\gamma/\gamma'} \sum_{i=1}^2 \int_0^T \int_0^{T'} 1_{[t,t+\tau]}(s)\|Q_i(s)\|_2^\gamma \, dt \, ds \tag{7.35}
$$
$$
\le c\tau^\gamma \sum_{i=1}^2 \int_0^T \|Q_i(s)\|_2^\gamma \, ds
$$

for all $\tau \in]0, T - T'[$ with a constant $c = c(n, \gamma, T, c_1, \delta(G))$ independent of T', τ.

Estimate (7.35) in combination with Lemma 7.6 yields $\widetilde{p}_r \in W^{1,\gamma}(0, T; L^2(G))$ and

$$\|\partial_t \widetilde{p}_r\|_{2,\gamma;G;T} \leq c\left(\|Q_1\|_{2,\gamma;G;T} + \|Q_2\|_{2,\gamma;G;T}\right) \tag{7.36}$$

with $c = c(n, \gamma, T, c_1, \delta(G))$.

Step 4. Let $p_r := \partial_t \widetilde{p}_r \in L^\gamma(0, T; L^2(G))$. For arbitrary $w \in C_0^\infty([0, T[\times G)$ consider $\partial_t w$ instead of w in (7.32) and integrate by parts to get (7.19). In this argument $\operatorname{div} u = 0$, $w(T) = 0$, and $\widetilde{p}_r(0) = 0$ were used. The uniqueness follows from (7.20), (7.21). □

Now we have all ingredients at hand to prove the main theorem of this section. This result will be one of the main tools to prove the main results of this chapter in Section 7.3.

Theorem 7.9. *Let* $\Omega \subseteq \mathbb{R}^n$, $n \in \{2, 3\}$, *be a domain satisfying the uniform Lipschitz condition, let* $0 < T < \infty$, *let* $f_1 \in L^2(0, T; L^2(\Omega)^n)$, $f_2 \in L^2(0, T; L^2(\Omega))$, *and let* $u_0 \in L^2(\Omega)^n$. *Further assume* $\theta_0 \in L^2(\Omega)$, $g \in L^\infty(]0, T[\times\Omega)^n$, *and assume that* β, ν, κ *are positive constants. Let* $(u_k)_{k\in\mathbb{N}}$ *and let* $(\theta_k)_{k\in\mathbb{N}}$ *be sequences such that*

$$
\begin{aligned}
(u_k)_{k\in\mathbb{N}} \ &\textit{is bounded in } L^\infty(0, T; L^2(\Omega)^n) \cap L^2(0, T; H^1(\Omega)^n), \\
(\theta_k)_{k\in\mathbb{N}} \ &\textit{is bounded in } L^\infty(0, T; L^2(\Omega)) \cap L^2(0, T; H^1(\Omega)),
\end{aligned}
\tag{7.37}
$$

and $\operatorname{div} u_k(t) = 0$ *for almost all* $t \in]0, T[$ *and all* $k \in \mathbb{N}$. *Assume that*

$$
\begin{aligned}
-\int_0^T &\langle u_k, w_t \rangle_\Omega \, dt + \nu \int_0^T \langle \nabla u_k, \nabla w \rangle_\Omega \, dt + \int_0^T \langle u_k \cdot \nabla u_k, w \rangle_\Omega \, dt \\
&= \beta \int_0^T \langle \theta_k g, w \rangle_\Omega \, dt + \int_0^T \langle f_1, w \rangle_\Omega \, dt + \langle u_0, w(0) \rangle_\Omega
\end{aligned}
\tag{7.38}
$$

for all $k \in \mathbb{N}$ *and all* $w \in C_0^\infty([0, T[; C_{0,\sigma}^\infty(\Omega))$. *Further assume*

$$
\begin{aligned}
-\int_0^T &\langle \theta_k, \phi_t \rangle_\Omega \, dt + \kappa \int_0^T \langle \nabla \theta_k, \nabla \phi \rangle_\Omega \, dt + \int_0^T \langle u_k \cdot \nabla \theta_k, \phi \rangle_\Omega \, dt \\
&= \int_0^T \langle f_2, \phi \rangle_\Omega \, dt + \langle \theta_0, \phi(0) \rangle_\Omega
\end{aligned}
\tag{7.39}
$$

for all $k \in \mathbb{N}$ *and all* $\phi \in C_0^\infty([0, T[\times\Omega)$.

Consider

$$
\begin{aligned}
u &\in L^\infty(0, T; L^2(\Omega)^n) \cap L^2(0, T; H^1(\Omega)^n), \\
\theta &\in L^\infty(0, T; L^2(\Omega)) \cap L^2(0, T; H^1(\Omega))
\end{aligned}
$$

such that

$$
u_k \underset{k\to\infty}{\overset{*}{\rightharpoonup}} u \textit{ in } L^\infty(0, T; L^2(\Omega)^n), \quad u_k \underset{k\to\infty}{\rightharpoonup} u \textit{ in } L^2(0, T; H^1(\Omega)^n), \tag{7.40}
$$

$$
\theta_k \underset{k\to\infty}{\overset{*}{\rightharpoonup}} \theta \textit{ in } L^\infty(0, T; L^2(\Omega)), \quad \theta_k \underset{k\to\infty}{\rightharpoonup} \theta \textit{ in } L^2(0, T; H^1(\Omega)). \tag{7.41}
$$

Then the following statements are satisfied:

(i) There holds

$$-\int_0^T \langle u, w_t \rangle_\Omega \, dt + \nu \int_0^T \langle \nabla u, \nabla w \rangle_\Omega \, dt + \int_0^T \langle u \cdot \nabla u, w \rangle_\Omega \, dt$$
$$= \beta \int_0^T \langle \theta g, w \rangle_\Omega \, dt + \int_0^T \langle f_1, w \rangle_\Omega \, dt + \langle u_0, w(0) \rangle_\Omega \tag{7.42}$$

for all $w \in C_0^\infty([0,T[; C_{0,\sigma}^\infty(\Omega))$.

(ii) Fix $0 \leq s < 1$. Then for all bounded Lipschitz domains G with $G \subseteq \Omega$ there holds

$$\theta_k \underset{k \to \infty}{\to} \theta \quad \text{strongly in } L^2(0,T; W^{s,2}(G)). \tag{7.43}$$

Furthermore

$$\theta_k \underset{k \to \infty}{\to} \theta \quad \text{strongly in } L^2(0,T; L^2(\partial G)). \tag{7.44}$$

Remark. In the application of the theorem above it is important that in statement (ii) the intersection $\overline{G} \cap \partial \Omega$ is allowed to be nonempty.

Proof of statement (i) of Theorem 7.9.

Step 1. We collect some preliminaries needed for the proof of (7.42).

1. Integration by parts yields

$$\int_0^T \langle v \otimes v, \nabla w \rangle_\Omega \, dt = -\int_0^T \langle v \cdot \nabla v, w \rangle_\Omega \, dt \tag{7.45}$$

for all $v \in L^2(0,T; H^1(\Omega)^n)$ satisfying $\operatorname{div} v(t) = 0$, a.a. $t \in]0,T[$, and for all $w \in C_0^\infty([0,T[; C_{0,\sigma}^\infty(\Omega))$.

2. Consider an arbitrary domain $G \subseteq \Omega$ and $\psi \in L^2(0,T; H^1(G))$ with $\Delta \psi(t) = 0$ for almost all $t \in]0,T[$. By Weyl's lemma we can redefine $\psi(t)$ on a null set of G such that $\psi(t) \in C^\infty(G)$ holds for a.a. $t \in]0,T[$. By Lemma 7.2 we obtain

$\psi \in L^2(0, T; H^2(\Omega'))$ for all bounded domains $\Omega' \subseteq\subseteq G$ and get that

$$\int_0^T \langle \nabla_x \psi \otimes \nabla_x \psi, \nabla w \rangle_G \, dt$$

$$= \sum_{i,j=1}^n \int_0^T \langle \partial_i \psi \partial_j \psi, \partial_i w_j \rangle_G \, dt$$

$$= - \sum_{i,j=1}^n \int_0^T \left(\langle \partial_i \partial_i \psi \, \partial_j \psi, w_j \rangle_G + \langle \partial_i \psi \, \partial_i \partial_j \psi, w_j \rangle_G \right) dt \tag{7.46}$$

$$= - \sum_{i,j=1}^n \int_0^T \frac{1}{2} \langle \partial_j ((\partial_i \psi)^2), w_j \rangle_G \, dt$$

$$= - \sum_{i,j=1}^n \int_0^T \frac{1}{2} \langle (\partial_i \psi)^2, \partial_j w_j \rangle_G \, dt$$

$$= 0$$

for all $w \in C_0^\infty([0, T[; C_{0,\sigma}^\infty(G))$.

3. We make use of $n \in \{2, 3\}$ and Sobolev's imbedding theorem to get

$$\|v\|_4 \le c \|v\|_2^{1/4} \|v\|_{H^1(\Omega)^n}^{3/4} \tag{7.47}$$

for all $v \in H^1(\Omega)^n$ with a constant $c = (\Omega, n) > 0$, c.f. Lemma 4.12 (ii).

Step 2. By (7.37), (7.47) we get

$$\int_0^T \|u_k \otimes u_k\|_{2,\Omega}^{4/3} \, dt \le c \int_0^T \|u_k\|_{2,\Omega}^{2/3} \|u_k\|_{H^1(\Omega)^n}^2 \, dt \le c \tag{7.48}$$

with a constant $c > 0$ independent of $k \in \mathbb{N}$. Therefore we find a matrix field in $L^{4/3}(0, T; L^2(\Omega)^{n\times n})$, denoted by $\overline{u \otimes u}$, such that (along a not relabelled subsequence)

$$u_k \otimes u_k \underset{k\to\infty}{\rightharpoonup} \overline{u \otimes u} \quad \text{in } L^{4/3}(0, T; L^2(\Omega)^{n\times n}). \tag{7.49}$$

The crucial point in the proof of statement (i) of Theorem 7.9 is the proof of the following
Assertion. We have

$$\lim_{k\to\infty} \int_0^T \langle u_k \cdot \nabla u_k, w \rangle_\Omega \, dt = \int_0^T \langle u \cdot \nabla u, w \rangle_\Omega \, dt \tag{7.50}$$

for all $w \in C_0^\infty([0, T[; C_{0,\sigma}^\infty(\Omega))$.
Proof of the assertion. Fix bounded Lipschitz domains G_1, G_2 with $\overline{G_1} \subseteq G_2$ and $\overline{G_2} \subseteq \Omega$. In the following we prove that (7.50) holds for all $w \in C_0^\infty([0, T[; C_{0,\sigma}^\infty(G_1))$.

Fix $k \in \mathbb{N}$. To apply Theorem 7.8 define

$$Q_1 := \nu \nabla u_k - u_k \otimes u_k \,,$$
$$Q_2 := -\beta \theta_k g - f_1 \,.$$

Since (7.38) holds for all $w \in C_0^\infty([0,T[; C_{0,\sigma}^\infty(G_2))$ and $Q_1 \in L^{4/3}(0,T;L^2(G_2)^{n \times n})$, $Q_2 \in L^2(0,T;L^2(G_2)^n)$ we see with (7.45) that (7.18) is fulfilled. Therefore Theorem 7.8 yields the existence of unique

$$p_{r,k} \in L^{4/3}(0,T;L_0^2(G_2)) \,, \quad p_{r,k}(t) \in \Delta W_0^{2,2}(G_2) \text{ for a.a. } t \in]0,T[\,, \tag{7.51}$$
$$p_{h,k} \in L^\infty(0,T;L_0^2(G_2)) \,, \quad \Delta_x p_{h,k}(t) = 0 \text{ in } G_2 \text{ for a.a. } t \in]0,T[\tag{7.52}$$

such that

$$-\int_0^T \langle u_k + \nabla_x p_{h,k}, w \rangle_{G_2} \eta'(t)\, dt = \int_0^T \Big(\langle u_k \otimes u_k, \nabla w \rangle_{G_2} - \nu \langle \nabla u_k, \nabla w \rangle_{G_2}$$
$$+ \beta \langle \theta_k g, w \rangle_{G_2} + \langle f_1, w \rangle_{G_2} + \langle p_{r,k}, \mathrm{div} w \rangle_{G_2} \Big) \eta(t)\, dt \tag{7.53}$$

for all $w \in C_0^\infty(G_2)^n$, $\eta \in C_0^\infty(]0,T[)$.

Since the constant c appearing in the estimates (7.20) and (7.21) is independent of $k \in \mathbb{N}$ it follows that $(p_{h,k})_{n \in \mathbb{N}}$ is a bounded sequence in $L^\infty(0,T;L^2(G_2))$ and that $(p_{r,k})_{k \in \mathbb{N}}$ is a bounded sequence in $L^{4/3}(0,T;L^2(G_2))$. Hence (along a not relabelled subsequence)

$$p_{h,k} \underset{k \to \infty}{\overset{*}{\rightharpoonup}} p_h \text{ in } L^\infty(0,T;L^2(G_2)) \,, \tag{7.54}$$
$$p_{r,k} \underset{k \to \infty}{\rightharpoonup} p_r \text{ in } L^{4/3}(0,T;L^2(G_2)) \,. \tag{7.55}$$

Using (7.54) and $\Delta_x p_{h,k}(t) = 0$ for a.a. $t \in]0,T[$ and all $k \in \mathbb{N}$, we obtain that $\Delta_x p_h(t) = 0$ holds for a.a. $t \in]0,T[$. Therefore, by Weyl's Lemma, we can redefine $p_h(t)$ on a null set of G_2 such that $p_h(t) \in C^\infty(G_2)$ is satisfied for almost all $t \in]0,T[$. Consequently (7.7) and (7.54) imply that $(p_{h,k})_{k \in \mathbb{N}}$ is bounded in $L^\infty(0,T;H^2(G_1))$. Therefore

$$p_{h,k} \underset{k \to \infty}{\rightharpoonup} p_h \text{ in } L^2(0,T;H^2(G_1)). \tag{7.56}$$

Fix $k \in \mathbb{N}$. We use the continuous imbedding $L^2(G_1)^n \hookrightarrow H^{-1}(G_1)^n$ to identify $u_k + \nabla_x p_{h,k}$ with the functional $u_k + \nabla_x p_{h,k} \in H^{-1}(G_1)^n$ which is defined by

$$[u_k + \nabla_x p_{h,k}, w]_{H^{-1};H_0^1} := \langle u_k + \nabla_x p_{h,k}, w \rangle_{G_1} \,, \quad w \in H_0^1(G_1)^n.$$

For a.a. $t \in]0,T[$ define $\zeta_k(t) \in H^{-1}(G_1)^n$ by

$$[\zeta_k(t), w]_{H^{-1};H_0^1} := \langle u_k(t) \otimes u_k(t), \nabla w \rangle_{G_1} - \nu \langle \nabla u_k(t), \nabla w \rangle_{G_1} + \langle \beta \theta_k(t) g(t), w \rangle_{G_1}$$
$$+ \langle f_1(t), w \rangle_{G_1} + \langle p_{r,k}(t), \mathrm{div} w \rangle_{G_1} \,, \quad w \in H_0^1(G_1)^n. \tag{7.57}$$

Employing these definitions and Lemma 2.16, identity (7.53) can be rewritten as

$$\frac{d}{dt}(u_k + \nabla_x p_{h,k}) = \zeta_k \quad \text{in } L^{4/3}(0,T;H^{-1}(G_1)^n). \tag{7.58}$$

We get the estimate

$$\int_0^T \left\| \frac{d}{dt}(u_k + \nabla_x p_{h,k}) \right\|_{H^{-1}(G_1)^n}^{4/3} dt$$

$$\leq c \int_0^T \left(\|u_k \otimes u_k\|_2^{4/3} + \nu\|\nabla u_k\|_2^{4/3} + \beta\|\theta_k g\|_2^{4/3} + \|f_1\|_2^{4/3} + \|p_{r,k}\|_2^{4/3} \right) dt$$

with a constant $c > 0$ independent of $k \in \mathbb{N}$. Thus, by (7.37), (7.49), (7.55), we see that $\left(\frac{d}{dt}(u_k + \nabla_x p_{h,k}) \right)_{k\in\mathbb{N}}$ is a bounded sequence in $L^{4/3}(0,T;H^{-1}(G_1)^n)$. Further by (7.37) and (7.56) the sequence $(u_k + \nabla_x p_{h,k})_{k\in\mathbb{N}}$ is bounded in $L^2(0,T;H^1(G_1)^n)$. Consider the imbedding scheme

$$H^1(G_1)^n \underset{\text{compact}}{\hookrightarrow} L^2(G_1)^n \underset{\text{continuous}}{\hookrightarrow} H^{-1}(G_1)^n. \tag{7.59}$$

We get with (7.59) and the Aubin-Lions compactness theorem (see Theorem 2.25), that there exists a subsequence $(u_{m_k} + \nabla_x p_{h,m_k})_{k\in\mathbb{N}}$ which is strongly convergent in $L^2(0,T;L^2(G_1)^n)$. Since $(u_k + \nabla_x p_{h,k})_{k\in\mathbb{N}}$ is also weakly convergent to $u + \nabla_x p_h$ as $k \to \infty$ in $L^2(0,T;L^2(G_1)^n)$ (see (7.40), (7.56)) it is possible to choose $m_k = k$, $k \in \mathbb{N}$. Altogether

$$u_k + \nabla_x p_{h,k} \underset{k\to\infty}{\to} u + \nabla_x p_h \quad \text{strongly in } L^2(0,T;L^2(G_1)^n). \tag{7.60}$$

We employ (7.40), (7.46), (7.49), (7.56) and (7.60) to obtain

$$\int_0^T \langle \overline{u \otimes u}, \nabla w \rangle_{G_1} dt$$

$$= \lim_{k\to\infty} \int_0^T \langle u_k \otimes u_k, \nabla w \rangle_{G_1} dt$$

$$= \lim_{k\to\infty} \int_0^T \langle (u_k + \nabla_x p_{h,k}) \otimes u_k, \nabla w \rangle_{G_1} dt$$

$$- \lim_{k\to\infty} \int_0^T \langle \nabla_x p_{h,k} \otimes (u_k + \nabla_x p_{h,k}), \nabla w \rangle_{G_1} dt$$

$$= \int_0^T \langle (u + \nabla_x p_h) \otimes u, \nabla w \rangle_{G_1} - \int_0^T \langle \nabla_x p_h \otimes (u + \nabla_x p_h), \nabla w \rangle_{G_1} dt$$

$$= \int_0^T \langle u \otimes u, \nabla w \rangle_{G_1} dt.$$

From (7.45), (7.49) and the computation above it follows that (7.50) is satisfied for all $w \in C_0^\infty([0,T[; C_{0,\sigma}^\infty(G_1))$.

For $w \in C_0^\infty([0, T[; C_{0,\sigma}^\infty(\Omega))$ we choose bounded Lipschitz domains G_1, G_2 with $\overline{G_1} \subseteq G_2, \overline{G_2} \subseteq \Omega$ such that $\operatorname{supp}(w(t, \cdot)) \subseteq G_1$ for all $t \in [0, T[$. Consequently the assertion holds. $\qquad\square$

Step 3. Now the proof of statement (i) of Theorem 7.9 can be finished. Consider $w \in C_0^\infty([0, T[; C_{0,\sigma}^\infty(\Omega))$. Letting $k \to \infty$ in (7.38) and using (7.40), (7.41), (7.50) we get that (u, θ) satisfies (7.42). $\qquad\square$

Proof of statement (ii) of Theorem 7.9.

Fix $\frac{1}{2} < s < 1$ and a bounded Lipschitz domain G with $G \subseteq \Omega$. From (7.39) we get

$$
\begin{aligned}
&- \int_0^T \langle \theta_k, \phi \rangle_G \, \eta'(t) \, dt \\
&= \int_0^T \big(\langle \theta_k u_k, \nabla\phi \rangle_G - \kappa \langle \nabla\theta_k, \nabla\phi \rangle_G + \langle f_2, \phi \rangle_G \big) \eta(t) \, dt
\end{aligned} \tag{7.61}
$$

for all $\phi \in C_0^\infty(G)$ and $\eta \in C_0^\infty(]0, T[)$. Fix $k \in \mathbb{N}$. For a.a. $t \in]0, T[$ define $\zeta_k(t) \in H^{-1}(G)$ by

$$
[\zeta_k(t), \phi]_{H^{-1};H^1} := \langle \theta_k(t) u_k(t), \nabla\phi \rangle_G - \kappa \langle \nabla\theta_k(t), \nabla\phi \rangle_G + \langle f_2(t), \phi \rangle_G, \quad \phi \in H_0^1(G). \tag{7.62}
$$

Then (7.61) in combination with the continuous imbedding $L^2(G) \hookrightarrow H^{-1}(G)$ and Lemma 2.16 mean

$$
\frac{d}{dt}\theta_k = \zeta_k \quad \text{in } L^{4/3}(0, T; H^{-1}(G)). \tag{7.63}
$$

We get the estimate

$$
\int_0^T \Big\| \frac{d}{dt}\theta_k \Big\|_{H^{-1}(G)}^{4/3} \le c \int_0^T \Big(\|\theta_k u_k\|_2^{4/3} + \kappa \|\nabla\theta_k\|_2^{4/3} + \|f_2\|_2^{4/3} \Big) \, dt \tag{7.64}
$$

for all $k \in \mathbb{N}$ with a constant $c > 0$ independent of k. By (7.37) we see that the sequence $(\theta_k)_{k \in \mathbb{N}}$ is bounded in $L^2(0, T; H^1(G))$. From (7.64) combined with (7.37) it follows that $(\frac{d}{dt}\theta_k)_{k \in \mathbb{N}}$ is bounded in $L^{4/3}(0, T; H^{-1}(G))$. Consider the imbedding scheme

$$
H^1(G) \underset{\text{compact}}{\hookrightarrow} W^{s,2}(G) \underset{\text{continuous}}{\hookrightarrow} H^{-1}(G). \tag{7.65}
$$

The compactness of the first imbedding follows from Theorem 2.5. We get from (7.65) and Theorem 2.25 the existence of a subsequence $(\theta_{m_k})_{k \in \mathbb{N}}$ which is strongly convergent in $L^2(0, T; W^{s,2}(G))$. Since $\theta_k \rightharpoonup \theta$ as $k \to \infty$ in $L^2(0, T; L^2(G))$ it is possible to choose $m_k = k$, $k \in \mathbb{N}$. Therefore (7.43) is satisfied. Looking at the continuous imbedding in Theorem 2.6 we see that (7.44) holds. The proof is complete. $\qquad\square$

7.2 The Young measure approach to determine the boundary behaviour

The goal of this section is to present the results needed to analyse the influence of surface roughness to weak solutions of the Boussinesq equations. First, we introduce sequences of domains $(\Omega_k)_{k\in\mathbb{N}}$ considered in the sequel.

Definition 7.10. Let $n \in \mathbb{N}, n \geq 2$, let $h_k : \mathbb{R}^{n-1} \to \mathbb{R}_0^+, k \in \mathbb{N}$, be non-negative functions such that the sequence $(h_k)_{k\in\mathbb{N}}$ is equi-Lipschitz continuous and $h_k \to 0$ uniformly on all compact subsets of $\mathbb{R}^{n-1}$. Further let $h : \mathbb{R}^{n-1} \to \mathbb{R}$ be a Lipschitz continuous function. Define the domains

$$\Omega := \left\{ (x', x_n) \in \mathbb{R}^n ; x_n > h(x') , x' \in \mathbb{R}^{n-1} \right\}, \tag{7.66}$$
$$\Omega_k := \left\{ (x', x_n) \in \mathbb{R}^n ; x_n > h(x') - h_k(x') , x' \in \mathbb{R}^{n-1} \right\}, \quad k \in \mathbb{N}. \tag{7.67}$$

Due to the equi-Lipschitz continuity of $(h_k)_{k\in\mathbb{N}}$, it follows that the sequence $(\nabla h_k)_{k\in\mathbb{N}}$ is bounded in $L^\infty(\mathbb{R}^{n-1})^{n-1}$. Therefore by Theorem 6.16, there exists a subsequence $(\nabla h_{m_k})_{k\in\mathbb{N}}$ and a Young measure $\nu = (\nu_{x'})_{x'\in\mathbb{R}^{n-1}} \in \mathcal{Y}(\mathbb{R}^{n-1}; \mathcal{P}(\mathbb{R}^{n-1}))$ such that $(\nabla h_{m_k})_{k\in\mathbb{N}}$ generates ν. Furthermore (see Corollary 6.17), there exists a compact set $\mathcal{K}_\nu \subseteq \mathbb{R}^{n-1}$ with $\operatorname{supp}(\nu_{x'}) \subseteq \mathcal{K}_\nu$ for a.a. $x' \in \mathbb{R}^{n-1}$. To simplify the notation we will assume that $m_k = k, k \in \mathbb{N}$, holds.

This section is organized as follows: After presenting some preliminary results, we will formulate in Subsection 7.2.2 the results needed to analyse the influence of boundary rugosity to the Robin boundary condition. In Subsection 7.2.3 we consider a sequence $(u_k)_{k\in\mathbb{N}}$ which is bounded in $H^1(\Omega_k)_{k\in\mathbb{N}}$ and satisfies $u_k \cdot N = 0$ on $\partial\Omega_k$, and we consider $u \in H^1(\Omega)$ such that $u_k \rightharpoonup u$ as $k \to \infty$ in $H^1(\Omega)$. We use the support $\operatorname{supp}(\nu_{x'}), x' \in \mathbb{R}^{n-1}$, of the Young measure ν generated by the sequence $(\nabla h_k)_{k\in\mathbb{N}}$ to analyse the behaviour of the tangential component $[u]_{\mathrm{tan}}$ of u on $\partial\Omega$.

7.2.1 Preliminary results

We collect several lemmata which will be needed in Section 7.2.2 and Section 7.2.3. Let us begin with the following useful Lemma.

Lemma 7.11. Let $\Omega \subseteq \mathbb{R}^n, n \in \mathbb{N}$, be a measurable set, let $1 < p \leq \infty$, let $(f_k)_{k\in\mathbb{N}}$ be a bounded sequence in $L^p(\Omega)$. Let $f \in L^1_{loc}(\Omega)$ such that

$$f_k \underset{k\to\infty}{\rightarrow} f \quad in \ L^1(K) \tag{7.68}$$

for all compact sets $K \subseteq \Omega$. Then $f \in L^p(\Omega)$ and

$$\begin{cases} f_k \underset{k\to\infty}{\rightharpoonup} f & in \ L^p(\Omega) & if \ 1 \leq p < \infty, \\ f_k \underset{k\to\infty}{\overset{*}{\rightharpoonup}} f & in \ L^\infty(\Omega) & if \ p = \infty. \end{cases}$$

Proof. Assume $p = \infty$. The proof is the same if $1 < p \leq \infty$. Fix a subsequence $(f_{m_k})_{k \in \mathbb{N}}$. Due to a well known argument on weak* compactness, there exists a subsequence $(f_{m_l})_{l \in \mathbb{N}}$ of $(f_{m_k})_{k \in \mathbb{N}}$ and $g \in L^\infty(\Omega)$ with $f_{m_l} \rightharpoonup^* g$ as $l \to \infty$ in $L^\infty(\Omega)$. Since $|f - g| : \Omega \to \mathbb{R}$ is a non-negative measurable function we can define a positive measure $\mu : \mathcal{L}(\Omega) \to [0, \infty]$ by

$$\mu(A) := \int_A |f - g| \, dx, \quad A \in \mathcal{L}(\Omega).$$

By uniqueness of weak* limits it follows from (7.68) that $\mu(A \cap K) = 0$ for all compact subsets $K \subseteq \Omega$. Let $(K_m)_{m \in \mathbb{N}}$ be a sequence of compact subsets of $\mathbb{R}^n$ with $\mathbb{R}^n = \bigcup_{m \in \mathbb{N}} K_m$. Then

$$\mu(\Omega) = \lim_{m \to \infty} \mu(\Omega \cap K_m) = 0.$$

Consequently $f(x) = g(x)$ for a.a. $x \in \Omega$. Due to the arbitrary choice of $(m_k)_{k \in \mathbb{N}}$ we obtain $f_k \underset{k \to \infty}{\rightharpoonup^*} f$ as $k \to \infty$ in $L^\infty(\Omega)$. $\qquad \square$

Lemma 7.12. *Let $n \in \mathbb{N}, n \geq 2$, let Ω be as in Definition 7.10. Then*

$$L^2_\sigma(\Omega) = \{\, u \in L^2(\Omega)^n; \langle u, \nabla \phi \rangle_\Omega = 0 \text{ for all } \phi \in C^\infty_0(\overline{\Omega}) \,\}. \tag{7.69}$$

Proof. Combine [Ga98, Lemma III.2.1] with [Fa93, Lemma 2.1 (i)]. $\qquad \square$

The Lemma below is needed to prove convergence of surface integrals in Theorem 7.19 and Lemma 7.24.

Lemma 7.13. *Let $n, \Omega, (\Omega_k)_{k \in \mathbb{N}}$ be as in Definition 7.10, let $d \in \mathbb{N}$ and $K \subseteq \mathbb{R}^{n-1}$ be a compact set.*

(i) Let $(v_k)_{k \in \mathbb{N}}$ be a bounded sequence in $H^1(\Omega_k)^d$. Then

$$\lim_{k \to \infty} \int_K |v_k(x', h(x') - h_k(x')) - v_k(x', h(x'))| \, dx' = 0. \tag{7.70}$$

(ii) Let $(u_k)_{k \in \mathbb{N}}$ be a bounded sequence in $L^2(0, T; H^1(\Omega_k)^d)$ where $0 < T < \infty$. Then

$$\lim_{k \to \infty} \int_{[0,T[\times K} |u_k(t, x', h(x') - h_k(x')) - u_k(t, x', h(x'))| \, d(x', t) = 0. \tag{7.71}$$

Remark. Since $v_k \in H^1(\Omega_k)^d$ the trace $v_k = v_k|_{\partial\Omega_k} \in L^2(\partial\Omega_k)^d$ is well defined. Therefore the integral in (7.70) is well defined. From $u_k = u_k|_{\partial\Omega_k} \in L^2(0, T; L^2(\partial\Omega_k)^d)$ the same holds for (7.71).

Proof. To simplify the notation assume $d = 1$. First we prove (7.70). There holds

$$\int_K v(x', h(x') - h_k(x')) - v(x', h(x')) \, dx' = \int_K \int_{h(x')}^{h(x') - h_k(x')} \frac{\partial v}{\partial x_n}(x', \tau) \, d\tau \, dx' \tag{7.72}$$

for all $v \in C_0^\infty(\overline{\Omega_k})$. By density, the identity (7.72) still holds true for $v_k \in H^1(\Omega_k)$. We deduce with (7.72)

$$
\begin{aligned}
&\int_K |v_k(x', h(x') - h_k(x')) - v_k(x', h(x'))| \, dx' \\
&\leq \int_K \int_{h(x')-h_k(x')}^{h(x')} \left| \frac{\partial v_k}{\partial x_n}(x', \tau) \right| d\tau \, dx' \\
&\leq \left(\int_K \int_{h(x')-h_k(x')}^{h(x')} \left| \frac{\partial v_k}{\partial x_n}(x', \tau) \right|^2 d\tau \, dx' \right)^{1/2} \left(\int_K \int_{h(x')-h_k(x')}^{h(x')} 1 \, d\tau \, dx' \right)^{1/2} \\
&\leq |K|^{1/2} \left(\sup_{x' \in K} h_k(x') \right)^{1/2} \|v_k\|_{H^1(\Omega_k)}
\end{aligned}
\tag{7.73}
$$

for all $k \in \mathbb{N}$. Using $h_k(x') \to 0$ as $k \to \infty$ uniformly for $x' \in K$, we get (7.70).

Let $(u_k)_{k\in\mathbb{N}}$ as in (ii). We use $u_k(t, \cdot) \in H^1(\Omega_k)$ for almost all $t \in]0, T[$ and integrate (7.73) with respect to $t \in]0, T[$ to obtain

$$
\begin{aligned}
&\int_0^T \int_K |u_k(t, x', h(x') - h_k(x')) - u_k(t, x', h(x'))| \, dx' \, dt \\
&\leq |K|^{1/2} \left(\sup_{x' \in K} h_k(x') \right)^{1/2} \int_0^T \|u_k(t)\|_{H^1(\Omega_k)} \, dt
\end{aligned}
$$

for all $k \in \mathbb{N}$. Passing to the limit in the estimate above yields (7.71). $\qquad \square$

Let $(u_k)_{k\in\mathbb{N}}$ be a bounded sequence in $L^{4/3}(0, T; L^2(\Omega_k)^d)$, let $u \in L^{4/3}(0, T; L^2(\Omega)^d)$ with $u_k \rightharpoonup u$ as $k \to \infty$ in $L^{4/3}(0, T; L^2(\Omega)^d)$. The following Lemma will be frequently used in this chapter.

Lemma 7.14. *Let $n, \Omega, (\Omega_k)_{k\in\mathbb{N}}$ be as in Definition 7.10, let $d \in \mathbb{N}, 0 < T < \infty$.*

(i) *Let $(v_k)_{k\in\mathbb{N}}$ be a bounded sequence in $L^2(\Omega_k)^d$, let $v \in L^2(\Omega)^d$ with $v_k \rightharpoonup v$ as $k \to \infty$ in $L^2(\Omega)^d$. Then*

$$
\lim_{k\to\infty} \langle v_k, w \rangle_{\Omega_k} = \langle v, w \rangle_\Omega
\tag{7.74}
$$

for all $w \in C_0^0(\mathbb{R}^n)^d$.

(ii) *Consider a sequence $(u_k)_{k\in\mathbb{N}}$ which is bounded in $L^{4/3}(0, T; L^2(\Omega_k)^d)$, consider $u \in L^{4/3}(0, T; L^2(\Omega)^d)$ with $u_k \rightharpoonup u$ as $k \to \infty$ in $L^{4/3}(0, T; L^2(\Omega)^d)$. Then*

$$
\lim_{k\to\infty} \int_0^T \langle u_k, w \rangle_{\Omega_k} \, dt = \int_0^T \langle u, w \rangle_\Omega \, dt
\tag{7.75}
$$

for all $w \in C_0^0([0, T[\times \mathbb{R}^n)$.

Proof. Choose a ball $B \subseteq \mathbb{R}^n$ such that $\mathrm{supp}(w) \subseteq [0, T[\times B$. Define

$$B_k := (\Omega_k \setminus \Omega) \cap B, \quad k \in \mathbb{N}.$$

Then

$$\int_0^T \langle u_k, w \rangle_{\Omega_k} \, dt = \int_0^T \langle u_k, w \rangle_\Omega \, dt + \int_0^T \langle u_k, w \rangle_{B_k} \, dt. \tag{7.76}$$

We obtain with Hölder's inequality

$$\int_0^T |\langle u_k, w \rangle_{B_k}| \, dt \leq \|u_k\|_{L^{4/3}(0,T;L^2(B_k)^d)} \|w\|_{L^4(0,T;L^2(B_k)^d)}$$

$$\leq |B_k|^{1/2} \, T^{1/4} \, \|u_k\|_{L^{4/3}(0,T;L^2(B_k))} \sup_{(t,x) \in [0,T[\times \mathbb{R}^n} |w(t,x)|$$

for all $k \in \mathbb{N}$. Since $|B_k| \to 0$ as $k \to \infty$, we get

$$\lim_{k \to \infty} \int_0^T \langle u_k, w \rangle_{B_k} \, dt = 0. \tag{7.77}$$

Consequently from (7.76), (7.77) we conclude

$$\lim_{k \to \infty} \int_0^T \langle u_k, w \rangle_{\Omega_k} \, dt = \int_0^T \langle u, w \rangle_\Omega \, dt.$$

Thus (7.75) holds. The proof of (7.74) is based on a similar argumentation. $\qquad\square$

The first application of the Lemma above is to prove the next

Lemma 7.15. *Let $n, \Omega, (\Omega_k)_{k \in \mathbb{N}}$ be as in Definition 7.10.*

(i) *Let $(v_k)_{k \in \mathbb{N}}$ be a bounded sequence in $L^2_\sigma(\Omega_k)$, let $v \in L^2(\Omega)^n$ with $v_k \rightharpoonup v$ as $k \to \infty$ in $L^2(\Omega)^n$. Then $v \in L^2_\sigma(\Omega)$.*

(ii) *Let $(u_k)_{k \in \mathbb{N}}$ be a bounded sequence in $L^2(0, T; L^2_\sigma(\Omega_k))$, let $u \in L^2(0, T; L^2(\Omega)^n)$ with $u_k \rightharpoonup u$ as $k \to \infty$ in $L^2(0, T; L^2(\Omega)^n)$. Then $u(t) \in L^2_\sigma(\Omega)$ for a.a. $t \in]0, T[$.*

Proof. We will only prove the second statement. Let $\phi \in C_0^\infty([0, T[\times \mathbb{R}^n)$. We use (7.69), (7.75) to get

$$\int_0^T \langle u, \nabla \phi \rangle_\Omega \, dt = \lim_{k \to \infty} \int_0^T \langle u_k, \nabla \phi \rangle_{\Omega_k} \, dt = 0. \tag{7.78}$$

Fix $\psi \in C_0^\infty(\overline{\Omega})$. By (7.78) we get

$$\int_0^T \langle u, \nabla \psi \rangle_\Omega \, \eta \, dt = 0 \tag{7.79}$$

for all $\eta \in C_0^\infty(]0,T[)$ and $\psi \in C_0^\infty(\overline{\Omega})$. The fundamental theorem of calculus of variations applied to (7.79) yields that there is a null set $N = N(\psi) \subseteq]0,T[$ such that $\langle u(\tau), \nabla\psi \rangle_\Omega = 0$ for all $\tau \in]0,T[\backslash N$. By a separability argument there exists a null set $N \subseteq]0,T[$ (independent of ψ) such that

$$\langle u(t), \nabla\psi \rangle_\Omega = 0 \qquad (7.80)$$

for all $t \in]0,T[\backslash N$ and all $\psi \in C_0^\infty(\overline{\Omega})$. Thus, from (7.69), (7.80) we get $u(t) \in L_\sigma^2(\Omega)$ for all $t \in]0,T[\backslash N$. $\qquad\square$

In the following we sketch some general informations about mollifiers. Let X be a Banach space, let $0 < T < \infty$, let $1 \leq s < \infty$, and let $u \in L^s(0,T;X)$. Define

$$\tilde{u}(t) := \begin{cases} u(t), & t \in]0,T[, \\ 0, & t \in \mathbb{R}\backslash]0,T[. \end{cases}$$

Fix $\rho \in C^\infty(\mathbb{R})$ with $0 \leq \rho \leq 1$, $\mathrm{supp}(\rho) \subseteq B_1(0)$ and $\int_0^1 \rho(x)\,dx = 1$. Define for $t \in \mathbb{R}$ and $0 < \delta < \infty$

$$\rho_\delta(t) := \frac{1}{\delta}\rho(\frac{t}{\delta}),$$
$$u^\delta(t) := (\tilde{u} * \rho_\delta)(t) = \int_\mathbb{R} \rho_\delta(\tau)\tilde{u}(t-\tau)\,d\tau. \qquad (7.81)$$

It is well known that $u^\delta \in C^\infty(\mathbb{R};X)$ and

$$\lim_{\delta \searrow 0} \|u - u^\delta\|_{L^s(0,T;X)} = 0. \qquad (7.82)$$

Let $\delta > 0$ with $\delta < T - \delta$. By an elementary calculation there holds

$$u^\delta(t) = \int_{t-\delta}^{t+\delta} \rho_\delta(t-\tau)u(\tau)\,d\tau = \int_{-\delta}^{\delta} \rho_\delta(\tau)u(t-\tau)\,d\tau$$

for all $t \in [\delta, T-\delta]$.

Our first Lemma concerning the mollification procedure reads as follows:

Lemma 7.16. *Let $n, \Omega, (\Omega_k)_{k\in\mathbb{N}}$ be as in Definition 7.10, let $(u_k)_{k\in\mathbb{N}}$ be a bounded sequence in $L^2(0,T;H^1(\Omega_k)^n)$, let $u \in L^2(0,T;H^1(\Omega)^n)$ such that $u_k \rightharpoonup u$ as $k \to \infty$ in $L^2(0,T;H^1(\Omega)^n)$. Fix $\delta > 0$ with $\delta < T - \delta$ and $t \in [\delta, T-\delta]$. The following statements are fulfilled:*

(i) We have

$$u_k^\delta(t) \underset{k\to\infty}{\rightharpoonup} u^\delta(t) \quad in \ H^1(\Omega)^n \qquad (7.83)$$

for all $t \in [\delta, T-\delta]$.

(ii) There exists a constant $c = c(\delta) > 0$ such that

$$\|u_k^\delta(t)\|_{H^1(\Omega_k)^n} \le c \tag{7.84}$$

for all $t \in [\delta, T - \delta]$ and all $k \in \mathbb{N}$.

Proof.   (i) Fix $\psi \in H^1(\Omega)^n$. Using the properties of the Bochner integral we obtain

$$\langle u_k^\delta(t), \psi \rangle_{H^1(\Omega)^n} = \int_{t-\delta}^{t+\delta} \langle \rho_\delta(t - \tau) u_k(\tau), \psi \rangle_{H^1(\Omega)^n} \, d\tau$$

$$= \int_0^T \langle u_k(\tau), \rho_\delta(t - \tau) \psi \rangle_{H^1(\Omega)^n} \, d\tau$$

for all $t \in [\delta, T - \delta]$ and $k \in \mathbb{N}$. We get

$$\lim_{k \to \infty} \langle u_k^\delta(t), \psi \rangle_{H^1(\Omega)^n} = \int_0^T \langle u(\tau), \rho_\delta(t - \tau) \psi \rangle_{H^1(\Omega)^n} \, d\tau = \langle u^\delta(t), \psi \rangle_{H^1(\Omega)^n}.$$

(ii) We get with Minkowski's inequality for integrals

$$\|u_k^\delta(t)\|_{H^1(\Omega_k)^n} \le \int_{t-\delta}^{t+\delta} \|\rho_\delta(t - \tau) u_k(\tau)\|_{H^1(\Omega_k)^n} \, d\tau$$

$$\le \frac{1}{\delta} \int_{t-\delta}^{t+\delta} \|u_k(\tau)\|_{H^1(\Omega_k)^n} \, d\tau$$

$$\le \frac{1}{\delta} (2\delta)^{\frac{1}{2}} \left(\int_{t-\delta}^{t+\delta} \|u_k(\tau)\|_{H^1(\Omega_k)^n}^2 \, d\tau \right)^{\frac{1}{2}}$$

$$\le c(\delta) \|u_k\|_{L^2(0,T;H^1(\Omega_k)^n)}$$

for all $t \in [\delta, T - \delta]$ and all $k \in \mathbb{N}$.   □

We proceed with the following

Lemma 7.17. *Let n, Ω, $(\Omega_k)_{k \in \mathbb{N}}$ be as in Definition 7.10, let $u \in L^2(0, T; L^2(\Omega)^n)$, and let u_δ, $\delta > 0$, be defined as (7.81). Fix $\delta > 0$ with $\delta < T - \delta$.*

(i) Assume $u(t) \in L_\sigma^2(\Omega)$ for a.a. $t \in]0, T[$. Then $u^\delta(t) \in L_\sigma^2(\Omega)$ for all $t \in [\delta, T - \delta]$.

(ii) Assume $u(t) \in H_0^1(\Omega)^n$ for a.a. $t \in]0, T[$. Then $u^\delta(t) \in H_0^1(\Omega)^n$ for every $t \in [\delta, T - \delta]$.

Proof.   (i) Let $\phi \in C_0^\infty(\overline{\Omega})$ be given. Then

$$\langle u^\delta(t), \nabla \phi \rangle_\Omega = \int_{t-\delta}^{t+\delta} \langle \rho_\delta(t - \tau) u(\tau), \nabla \phi \rangle_\Omega \, d\tau$$

$$= \int_{t-\delta}^{t+\delta} \rho_\delta(t - \tau) \underbrace{\langle u(\tau), \nabla \phi \rangle_\Omega}_{=0 \text{ for a.a. } \tau \in [t-\delta, t+\delta]} \, d\tau$$

$$= 0.$$

Lemma 7.12 yields $u^\delta(t) \in L_\sigma^2(\Omega)$ for all $t \in [\delta, T - \delta]$.

(ii) For $w \in L^2(\partial\Omega)^n$ there holds

$$
\begin{aligned}
\langle u^\delta(t), w \rangle_{\partial\Omega} &= \int_{t-\delta}^{t+\delta} \langle \rho_\delta(t-\tau) u(\tau), w \rangle_{\partial\Omega} \, d\tau \\
&= \int_{t-\delta}^{t+\delta} \rho_\delta(t-\tau) \underbrace{\langle u(\tau), w \rangle_{\partial\Omega}}_{=0 \text{ for a.a. } \tau \in [t-\delta, t+\delta]} d\tau \\
&= 0.
\end{aligned}
$$

Therefore $u_\delta(t)|_{\partial\Omega} = 0$ for all $t \in [\delta, T - \delta]$. This implies (see Lemma 2.15) that $u_\delta(t) \in H_0^1(\Omega)^n$ holds for all $t \in [\delta, T - \delta]$. $\qquad \square$

7.2.2 The influence to the Robin boundary condition

We need the following technical lemma.

Lemma 7.18. *Let $n \in \mathbb{N}, n \geq 2$, let $h : \mathbb{R}^{n-1} \to \mathbb{R}$ be a Lipschitz continuous function, let $h_k : \mathbb{R}^{n-1} \to \mathbb{R}, k \in \mathbb{N}$, be functions such that the sequence $(h_k)_{k \in \mathbb{N}}$ is equi-Lipschitz continuous. Assume that $(\nabla h_k)_{k \in \mathbb{N}}$ generates the Young measure $\nu = (\nu_{x'})_{x' \in \mathbb{R}^{n-1}} \in \mathcal{Y}(\mathbb{R}^{n-1}; \mathcal{P}(\mathbb{R}^{n-1}))$.*

(i) Define

$$
g_k(x') := \sqrt{1 + |\nabla h(x') - \nabla h_k(x')|^2}, \quad \text{for a.a. } x' \in \mathbb{R}^{n-1} \text{ and all } k \in \mathbb{N},
$$

(7.85)

$$
\widetilde{\Lambda}(x') := \int_{\mathbb{R}^{n-1}} \sqrt{1 + |\nabla h(x') - \lambda|^2} \, d\nu_{x'}(\lambda), \quad \text{for a.a. } x' \in \mathbb{R}^{n-1}.
$$

(7.86)

Then $\widetilde{\Lambda} \in L^\infty(\mathbb{R}^{n-1})$ and

$$
g_k \xrightarrow[k \to \infty]{*} \widetilde{\Lambda} \quad \text{in } L^\infty(\mathbb{R}^{n-1}).
$$

(7.87)

Moreover, for all $1 \leq p < \infty$ and all compact subsets $K \subseteq \mathbb{R}^{n-1}$ we have

$$
g_k \xrightarrow[k \to \infty]{} \widetilde{\Lambda} \quad \text{in } L^p(K).
$$

(7.88)

(ii) Almost all $x' \in \mathbb{R}^{n-1}$ are Lebesgue points of the function $\widetilde{\Lambda}$. Choose any Lebesgue point $x_0' \in \mathbb{R}^{n-1}$ of $\overline{\Lambda}$. Then

$$
\widetilde{\Lambda}(x_0') = \lim_{R \to 0} \lim_{k \to \infty} \frac{1}{|B_R(x_0')|} \mathcal{H}^{n-1}\big(\{ (x', h(x') - h_k(x')); \, x' \in B_R(x_0') \}\big).
$$

(7.89)

Proof. (i) The equi-Lipschitz continuity of $(h_k)_{k\in\mathbb{N}}$ implies that

$$(g_k)_{k\in\mathbb{N}} \text{ is bounded in } L^\infty(\mathbb{R}^{n-1}). \tag{7.90}$$

For a compact set $A \subseteq \mathbb{R}^{n-1}$ define the Carathéodory function

$$\psi(x,\lambda) := \sqrt{1+|\nabla h(x') - \lambda|^2}, \quad \text{for a.a. } x' \in A \text{ and all } \lambda \in \mathbb{R}.$$

Fix any sequence $(m_k)_{k\in\mathbb{N}}$ of natural numbers $m_k \in \mathbb{N}, k \in \mathbb{N}$. By (7.90) there is a subsequence $(m_l)_{l\in\mathbb{N}}$ of $(m_k)_{k\in\mathbb{N}}$ such that $(g_{m_l})_{l\in\mathbb{N}} = \left(\psi(\cdot, \nabla h_{m_l}(\cdot))\right)_{l\in\mathbb{N}}$ is a weakly convergent sequence in $L^1(A)$. Using the theory of Young measures (see Theorem 6.16) we get

$$g_{m_l} \underset{l\to\infty}{\rightharpoonup} \widetilde{\Lambda} \quad \text{in } L^1(A). \tag{7.91}$$

Due to the arbitrary choice of $(m_k)_{k\in\mathbb{N}}$ we obtain from (7.91)

$$g_k \underset{k\to\infty}{\rightharpoonup} \widetilde{\Lambda} \quad \text{in } L^1(A). \tag{7.92}$$

Combining (7.90), (7.92) and Lemma 7.11 it follows that (7.88) is fulfilled.

(ii) Lebesgue's Differentiation Theorem (see Theorem 2.21) applied to the locally integrable function $\widetilde{\Lambda}$ yields that almost all $x' \in \mathbb{R}^{n-1}$ are Lebesgue points of $\widetilde{\Lambda}$. Fix any Lebesgue point $x_0' \in \mathbb{R}^{n-1}$ of $\widetilde{\Lambda}$. With the help of (4.5) and (7.88) we conclude

$$\begin{aligned}
&\lim_{R\searrow 0}\lim_{k\to\infty} \frac{1}{|B_R(x_0')|}\mathcal{H}^{n-1}\left(\{\,(x', h(x') - h_k(x')); \, x \in B_R(x_0')\,\}\right)\\
&= \lim_{R\searrow 0}\lim_{k\to\infty} \frac{1}{|B_R(x_0')|}\int_{B_R(x_0')} \sqrt{1+|\nabla h(x') - \nabla h_k(x')|^2}\,dx'\\
&= \lim_{R\searrow 0} \frac{1}{|B_R(x_0')|}\int_{B_R(x_0')} \widetilde{\Lambda}(x')\,dx'\\
&= \widetilde{\Lambda}(x_0').
\end{aligned} \tag{7.93}$$

$\square$

The following theorem will be the main tool to analyse the influence of surface roughness to the Robin boundary condition.

Theorem 7.19. *Let* $n, \Omega, (\Omega_k)_{k\in\mathbb{N}}$ *be as in Definition 7.10, assume that* $(\nabla h_k)_{k\in\mathbb{N}}$ *generates the Young measure* $\nu = (\nu_{x'})_{x'\in\mathbb{R}^{n-1}} \in \mathcal{Y}(\mathbb{R}^{n-1}; \mathcal{P}(\mathbb{R}^{n-1}))$. *We define the function* $\Lambda : \partial\Omega \to \mathbb{R}$ *by*

$$\Lambda(x', h(x')) := \frac{1}{\sqrt{1+|\nabla h(x')|^2}}\int_{\mathbb{R}^{n-1}} \sqrt{1+|\nabla h(x') - \lambda|^2}\,d\nu_{x'}(\lambda) \tag{7.94}$$

for a.a. $x = (x', h(x')) \in \partial\Omega$ with $x' \in \mathbb{R}^{n-1}$. Let $(\theta_k)_{k \in \mathbb{N}}$ be a bounded sequence in $L^2(0, T; H^1(\Omega_k))$, let $\theta \in L^2(0, T; H^1(\Omega))$ be given. Assume that

$$\theta_k \underset{k \to \infty}{\to} \theta \quad \text{strongly in } L^2(0, T; L^2(\partial G)) \tag{7.95}$$

for all bounded Lipschitz domains G with $G \subseteq \Omega$. Then

$$\lim_{k \to \infty} \int_0^T \langle \theta_k, \phi \rangle_{\partial\Omega_k} \, dt = \int_0^T \langle \Lambda\theta, \phi \rangle_{\partial\Omega} \, dt \tag{7.96}$$

for all $\phi \in C_0^\infty([0, T[\times\mathbb{R}^n)$. Especially

$$\lim_{k \to \infty} \int_0^T \langle h_0, \phi \rangle_{\partial\Omega_k} \, dt = \int_0^T \langle \Lambda h_0, \phi \rangle_{\partial\Omega} \, dt \tag{7.97}$$

for all $h_0 \in L^2(0, T; H^1(\mathbb{R}^n))$ and all $\phi \in C_0^\infty([0, T[\times\mathbb{R}^n)$.

Proof. Fix $\phi \in C_0^\infty([0, T[\times\mathbb{R}^n)$. Due to the compact support $\text{supp}(\phi) \subseteq [0, T[\times\mathbb{R}^n$ there is an open ball $B := B_r(0) \subseteq \mathbb{R}^{n-1}$ with $0 < r < \infty$ such that

$$\text{supp}(\phi) \subseteq \{ (t, x', x_n) \in [0, T[\times\mathbb{R}^{n-1} \times \mathbb{R}; \ x' \in B \}.$$

Define

$$s_k(x') := \sqrt{1 + |\nabla h(x') - \nabla h_k(x')|^2} \quad \text{for a.a. } x' \in B \text{ and all } k \in \mathbb{N},$$

and $Q := [0, T[\times B$. By assumptions on $h, (h_k)_{k \in \mathbb{N}}$ it follows that

$$(s_k)_{k \in \mathbb{N}} \text{ is bounded in } L^\infty(B). \tag{7.98}$$

We get

$$\left| \int_0^T \langle \theta_k, \phi \rangle_{\partial\Omega_k} \, dt - \int_0^T \langle \Lambda\theta, \phi \rangle_{\partial\Omega} \, dt \right|$$

$$\leq \int_Q |(\theta_k\phi)(t, x', h(x') - h_k(x')) - (\theta_k\phi)(t, x', h(x'))| \, s_k(x') \, d(x', t)$$

$$+ \int_Q |(\theta_k\phi)(t, x', h(x')) - (\theta\phi)(t, x', h(x'))| \, s_k(x') \, d(x', t)$$

$$+ \left| \int_Q (\theta\phi)(t, x', h(x')) \left(s_k(x') - \int_{\mathbb{R}^{n-1}} \sqrt{1 + |\nabla h(x') - \lambda|^2} \, d\nu_{x'}(\lambda) \right) d(x', t) \right|$$

$$=: I_1^k + I_2^k + I_3^k$$
$$\tag{7.99}$$

for all $k \in \mathbb{N}$.

Proof of $\lim_{k\to\infty} I_1^k = 0$. We use the uniform continuity of $\phi : [0, T[\times\mathbb{R}^n \to \mathbb{R}$ and $h_k \to 0$ as $k \to \infty$ uniformly on the bounded set B to obtain

$$\lim_{k\to\infty} \sup_{(t,x')\in Q} |\phi(t, x', h(x') - h_k(x')) - \phi(t, x', h(x'))| = 0. \tag{7.100}$$

We make use of (7.98) and get

$$
\begin{aligned}
I_1^k &\leq c \int_Q |\theta_k(t, x', h(x') - h_k(x')) - \theta_k(t, x', h(x'))| \, |\phi(t, x', h(x') - h_k(x'))| \, d(x', t) \\
&\quad + c \int_Q |\theta_k(t, x', h(x'))| \, |\phi(t, x', h(x') - h_k(x')) - \phi(t, x', h(x'))| \, d(x', t) \\
&\leq c \int_Q |\theta_k(t, x', h(x') - h_k(x')) - \theta_k(t, x', h(x'))| \, d(x', t) \\
&\quad + c \sup_{(t,x')\in Q} |\phi(t, x', h(x') - h_k(x')) - \phi(t, x', h(x'))| \int_Q |\theta_k(t, x', h(x'))| \, d(x', t)
\end{aligned}
$$

for all $k \in \mathbb{N}$ with a constant $c > 0$ independent of k. We use (7.71), (7.100) to obtain $\lim_{k\to\infty} I_1^k = 0$.

Proof of $\lim_{k\to\infty} I_2^k = 0$. Define the bounded Lipschitz domain G by

$$G := \{(x', x_n) \in \mathbb{R}^n; h(x') < x_n < h(x') + 1, x' \in B\}. \tag{7.101}$$

By (7.95) there holds

$$\lim_{k\to\infty} \int_Q |\theta_k(t, x', h(x')) - \theta(t, x', h(x'))|^2 \sqrt{1 + |\nabla h(x')|^2} \, d(x', t) = 0. \tag{7.102}$$

We get

$$
\begin{aligned}
I_2^k &\leq c \left(\int_Q |(\theta_k - \theta)(t, x', h(x'))|^2 \, d(x', t) \right)^{1/2} \left(\int_Q |\phi(t, x', h(x'))|^2 \, d(x', t) \right)^{1/2} \\
&\leq c \left(\int_Q |(\theta_k - \theta)(t, x', h(x'))|^2 \sqrt{1 + |\nabla h(x')|^2} \, d(x', t) \right)^{1/2}
\end{aligned}
\tag{7.103}
$$

with a constant $c > 0$ independent of $k \in \mathbb{N}$. By (7.102) it follows $\lim_{k\to\infty} I_2^k = 0$.

Proof of $\lim_{k\to\infty} I_3^k = 0$. Define, cf. (7.86),

$$\tilde{\Lambda}(x') := \int_{\mathbb{R}^{n-1}} \sqrt{1 + |\nabla h(x') - \lambda|^2} \, d\nu_{x'}(\lambda) \quad \text{for a.a. } x' \in \mathbb{R}^{n-1}.$$

Using (7.88) it follows that

$$s_k = \sqrt{1 + |\nabla h - \nabla h_k|^2} \underset{k\to\infty}{\rightharpoonup} \tilde{\Lambda} \quad \text{in } L^2(B).$$

Especially we get for a.a. $t \in [0, T[$

$$\lim_{k \to \infty} \int_B (\theta \phi)(t, x', h(x')) s_k(x') \, dx'$$
$$= \int_B (\theta \phi)(t, x', h(x')) \int_{\mathbb{R}^{n-1}} \sqrt{1 + |\nabla h(x') - \lambda|^2} \, d\nu_{x'}(\lambda) \, dx'. \tag{7.104}$$

By (7.98), (7.104) Lebesgue's dominated convergence theorem implies $\lim_{k \to \infty} I_3^k = 0$.

Altogether we obtain from (7.99) that (7.96) is fulfilled. It follows immediately that (7.97) holds. $\qquad \square$

7.2.3 The influence to the velocity field

The Lemma below is needed for the proof of Theorem 7.25.

Lemma 7.20. *Let $n \in \mathbb{N}$, let μ be a positive, finite measure on the σ-algebra $\mathcal{L}(\mathbb{R}^n)$ with compact support $\mathrm{supp}(\mu)$. Define*

$$\mathcal{U} := \left\{ \int_{\mathbb{R}^n} f(\lambda) \lambda \, d\mu(\lambda) \, ; \, f \in C_0(\mathbb{R}^n) \right\}. \tag{7.105}$$

Then

(i) $\mathcal{U}$ is a linear subspace of $\mathbb{R}^n$.

(ii) There holds $\mathrm{supp}(\mu) \subseteq \mathcal{U}$, i.e. if $v \in \mathrm{supp}(\mu)$ then there is $f \in C_0(\mathbb{R}^n)$ satisfying
$$v = \int_{\mathbb{R}^n} f(\lambda) \lambda \, d\mu(\lambda).$$

Remark. The integral in (7.105) can be evaluated in the following sense:

$$\int_{\mathbb{R}^n} f(\lambda) \lambda \, d\mu(\lambda) = \begin{pmatrix} \int_{\mathbb{R}^n} f(\lambda) \lambda_1 \, d\mu(\lambda) \\ \cdots \\ \int_{\mathbb{R}^n} f(\lambda) \lambda_n \, d\mu(\lambda) \end{pmatrix}. \tag{7.106}$$

Proof. (i) This follows immediately by linearity of the integral.

(ii) Fix $v \in \mathrm{supp}(\mu)$. First we prove the following assertion:

Assertion. *For all $\epsilon > 0$ there is $u = u(\epsilon) \in \mathcal{U}$ such that $u \in W_\epsilon(v)$, where $W_\epsilon(v) = \{y \in \mathbb{R}^n; \|y - v\|_\infty \le \epsilon\}$ denotes the closed cube with side length ϵ and center v.*

Proof of the assertion. Fix $\epsilon > 0$. We choose a non-negative function $f \in C_0(\mathbb{R}^n)$ with $f(v) > 0$ such that $f(\lambda) = 0$ for all $\lambda \notin W_\epsilon(v)$. Replacing f by the function $x \mapsto \dfrac{1}{\int_{W_\epsilon(v)} f(\lambda) \, d\mu(\lambda)} f(x) \, , x \in \mathbb{R}^n$, we can additionally assume

$$\int_{W_\epsilon(v)} f(\lambda) \, d\mu(\lambda) = 1.$$

We obtain

$$\left\| \int_{W_\epsilon(v)} f(\lambda)\lambda \, d\mu(\lambda) - v \right\|_\infty = \left\| \int_{W_\epsilon(v)} f(\lambda)(\lambda - v) \, d\mu(\lambda) \right\|_\infty$$

$$\leq \int_{W_\epsilon(v)} f(\lambda)\|\lambda - v\|_\infty \, d\mu(\lambda)$$

$$\leq \epsilon \int_{W_\epsilon(v)} f(\lambda) \, d\mu(\lambda)$$

$$= \epsilon.$$

Define $u := \int_{\mathbb{R}^n} f(\lambda)\lambda \, d\mu(\lambda)$. By construction $u \in \mathcal{U}$ and $u \in W_\epsilon(v)$. $\qquad\square$

Consequently, there exists a sequence $(v_k)_{k \in \mathbb{N}}, v_k \in \mathcal{U}$, such that $v_k \to v$ as $k \to \infty$. Since $\mathcal{U}$ is a finite dimensional subspace and therefore closed, we get $v \in \mathcal{U}$. $\qquad\square$

As already motivated in the introduction we need the *non-degeneracy* condition. This condition was introduced in [BFNW08] in the case $n = 3$.

Definition 7.21. Let $\Omega \subseteq \mathbb{R}^n, n \in \mathbb{N}$, be a measurable set. A Young measure $\nu = (\nu_x)_{x \in \Omega} \in \mathcal{Y}(\Omega; \mathcal{P}(\mathbb{R}^n))$ is called *non-degenerate* if for a.a. $x \in \mathbb{R}^n$ there is a basis $w^i = w^i_x, i = 1 \ldots, n$, of $\mathbb{R}^n$ such that $w^i \in \mathrm{supp}(\nu_x)$ for all $i \in \{1, \ldots, n\}$.

For sufficient criteria when the non-degeneracy condition is satisfied in the three-dimensional case we refer to [BFNW08, Section 3 and Section 4]. In the following we present two lemmata characterizing in the case $n = 2$ the non-degeneracy of the Young measure ν generated by $(\nabla h_{m_k})_{k \in \mathbb{N}}$. The first of these lemmata states a relation between the non-degeneracy of ν and the function $\tilde{\Lambda}$ defined in (7.86).

Lemma 7.22. *Let $n = 2$, let $(h_k)_{k \in \mathbb{N}}$ be as in Definition 7.10, and assume that $(\nabla h_k)_{k \in \mathbb{N}}$ generates the Young measure $\nu = (\nu_{x'})_{x' \in \mathbb{R}} \in \mathcal{Y}(\mathbb{R}; \mathcal{P}(\mathbb{R}))$. Then ν is non-degenerate if and only if*

$$\int_{\mathbb{R}} \sqrt{1 + \lambda^2} \, d\nu_{x'}(\lambda) > 1 \quad \text{for a.a. } x' \in \mathbb{R}. \tag{7.107}$$

Proof. Fix any probability measure $\mu \in \mathcal{P}(\mathbb{R})$. Then

$$\int_{\mathbb{R}} \sqrt{1 + \lambda^2} \, d\mu(\lambda) = 1 \quad \Longleftrightarrow \quad \int_{\mathbb{R}} \left(\sqrt{1 + \lambda^2} - 1 \right) d\mu(\lambda) = 0. \tag{7.108}$$

Consequently (7.108) is satisfied if and only if

$$\sqrt{1 + \lambda^2} = 1 \quad \text{for almost all } \lambda \in \mathbb{R} \text{ (with respect to } \mu\text{)}.$$

The condition above is equivalent to the condition that the Borel set $\{\lambda \in \mathbb{R}; \lambda \neq 0\}$ is a null set. Consequently

$$\int_{\mathbb{R}} \sqrt{1 + \lambda^2}\, d\mu(\lambda) = 1 \iff \mu(\mathbb{R} \setminus \{0\}) = 0 \iff \mu = \delta_0. \tag{7.109}$$

Altogether, (7.107) is satisfied if and only if $\nu_{x'} \neq \delta_0$ for almost all $x' \in \mathbb{R}$. Since $n = 2$ we see that ν is non-degenerate if and only if (7.107) is fulfilled. $\qquad\square$

We proceed with the following lemma.

Lemma 7.23. *Let* $n = 2$, *let* $(h_k)_{k \in \mathbb{N}}$ *be as in Definition 7.10, and assume that* $(\nabla h_k)_{k \in \mathbb{N}}$ *generates the Young measure* $\nu = (\nu_{x'})_{x' \in \mathbb{R}} \in \mathcal{Y}(\mathbb{R}; \mathcal{P}(\mathbb{R}))$. *Then* ν *is non-degenerate if and only if the following condition is satisfied: For all* $A \in \mathcal{L}(\mathbb{R}^{n-1})$ *with* $0 < |A| < \infty$ *we have that* $(\nabla h_k)_{k \in \mathbb{N}}$ *does not converge to 0 strongly in* $L^1(A)$ *as* $k \to \infty$.

Proof. The proof is based on the following

Assertion. Fix a Lebesgue measurable set $A \subseteq \mathbb{R}$ with $0 < |A| < \infty$. Then there holds

$$\nu_{x'} = \delta_0 \text{ for a.a. } x' \in A \quad \Longleftrightarrow \quad \nabla h_k \underset{k \to \infty}{\to} 0 \text{ strongly in } L^1(A).$$

Proof of the assertion. First, let us assume $\nu_{x'} = \delta_0$ for a.a. $x' \in A$. Define the Carathéodory function

$$\psi : A \times \mathbb{R} \to \mathbb{R}, \quad \psi(x', \lambda) := |\lambda|.$$

Due to $|A| < \infty$ and the boundedness of the sequence $(\nabla h_k)_{k \in \mathbb{N}}$ in $L^\infty(\mathbb{R})$ it follows that there is a subsequence $(\nabla h_{m_k})_{k \in \mathbb{N}}$ such that $(\psi(\cdot, \nabla h_{m_k}))_{k \in \mathbb{N}} = (|\nabla h_{m_k}|)_{k \in \mathbb{N}}$ is weakly convergent in $L^1(A)$. Thus, since $\nu_{x'} = \delta_0$ for a.a. $x' \in A$ we get from Theorem 6.16

$$\lim_{k \to \infty} \int_A |\nabla h_{m_k}|\, dx' = \int_A \left(\int_{\mathbb{R}} |\lambda|\, d\delta_0(\lambda) \right) dx' = 0.$$

Let us conversely assume that $\nabla h_k \to 0$ as $k \to \infty$ strongly in $L^1(A)$. Fix $\phi \in C_0(\mathbb{R})$. It follows that the sequence $(\phi(\nabla h_k))_{k \in \mathbb{N}}$ converges strongly as $k \to \infty$ in $L^1(A)$ to the function $x' \mapsto \phi(0)$, $x' \in A$. Thus

$$\lim_{k \to \infty} \int_A g(x')\phi(\nabla h_k(x'))\, dx' = \int_A g(x')\phi(0)\, dx'$$

for all $g \in L^\infty(A)$. We can rewrite the identity in the form above as

$$\lim_{k \to \infty} \int_A g(x')\phi(\nabla h_k(x'))\, dx' = \int_A g(x') \left(\int_{\mathbb{R}} \phi(\lambda)\, d\delta_0(\lambda) \right) dx'$$

for all $g \in L^\infty(\Omega)$. Therefore, (see Definition 6.13) we obtain $\nu_{x'} = \delta_0$ for a.a. $x' \in A$. $\qquad\square$

Since $n = 2$ it follows immediately from the definition that ν is non-degenerate if and only if for all measurable sets $A \subseteq \mathbb{R}^{n-1}$ with $0 < |A| < \infty$ there holds $\nu_{x'} \neq \delta_0$ for all $x' \in A$. Thus, looking at the assertion above, we see that the conclusions of Lemma 7.23 are fulfilled. $\qquad\square$

The next lemma will play a crucial role in our goal to analyse the influence of surface roughness to the velocity field. The proof of this result is based on the approach presented in [BFNW08].

Lemma 7.24. *Let $n, \Omega, (\Omega_k)_{k \in \mathbb{N}}$ be as in Definition 7.10, assume that $(\nabla h_k)_{k \in \mathbb{N}}$ generates the Young measure $\nu = (\nu_{x'})_{x' \in \mathbb{R}^{n-1}} \in \mathcal{Y}(\mathbb{R}^{n-1}; \mathcal{P}(\mathbb{R}^{n-1}))$. Let $(u_k)_{k \in \mathbb{N}}$ be a bounded sequence in $H^1(\Omega_k)^n$, assume $u_k \in L^2_\sigma(\Omega_k)$ for all $k \in \mathbb{N}$, and let $u \in H^1(\Omega)^n$ such that $u_k \rightharpoonup u$ as $k \to \infty$ in $H^1(\Omega)^n$. Then $u \in L^2_\sigma(\Omega)$. Furthermore, there exists a compact set $\mathcal{K}_\nu \subseteq \mathbb{R}^{n-1}$ such that the following properties are satisfied:*

(i) *There holds $\operatorname{supp}(\nu_{x'}) \subseteq \mathcal{K}_\nu$ for almost all $x' \in \mathbb{R}^{n-1}$.*

(ii) *For almost all $x' \in \mathbb{R}^{n-1}$ there holds*

$$\big(u^1, \ldots, u^{n-1} \big)|_{\partial\Omega}\big(x', h(x')\big) \cdot \int_{\mathcal{K}_\nu} f(\lambda)\lambda \, d\nu_{x'}(\lambda) = 0 \quad \forall f \in C_0(\mathbb{R}^n). \tag{7.110}$$

Proof. We write $u = u|_{\partial\Omega}$ for the fixed class representative of the trace of u. By Corollary 6.17 there exists a compact set $\mathcal{K}_\nu \subseteq \mathbb{R}^{n-1}$ and a null set $M \subseteq \mathbb{R}^{n-1}$ such that $\operatorname{supp}(\nu_{x'}) \subseteq \mathcal{K}_\nu$ for all $x' \in \mathbb{R}^{n-1} \backslash M$. We know from Lemma 7.15 that $u \in L^2_\sigma(\Omega)$. Consequently

$$\big(u^1, \ldots, u^n \big)\big(x', h(x')\big) \cdot \big(\nabla h(x'), -1\big) = 0 \tag{7.111}$$

for almost all $x' \in \mathbb{R}^{n-1}$ and all $k \in \mathbb{N}$. Define

$$N_k(x') := \big(\nabla h(x') - \nabla h_k(x'), -1\big) \quad \text{for a.a. } x' \in \mathbb{R}^{n-1} \text{ and all } k \in \mathbb{N}.$$

Due to the assumptions on $h, (h_k)_{k \in \mathbb{N}}$ the sequence $(N_k)_{k \in \mathbb{N}}$ is bounded in $L^\infty(\mathbb{R}^{n-1})^n$. Fix $g \in C_0^\infty(\mathbb{R}^{n-1})$ and $f \in C_0(\mathbb{R}^{n-1})$. Choose an open ball $B := B_r(0) \subseteq \mathbb{R}^{n-1}$ with $0 < r < \infty$ such that $\operatorname{supp}(g) \subseteq B$. From $u_k \in L^2_\sigma(\Omega_k)$ we get

$$0 = \int_B g(x')f(\nabla h_k(x'))\big(u_k^1, \ldots, u_k^n\big)\big(x', h(x') - h_k(x')\big) \cdot N_k(x') \, dx' \tag{7.112}$$

for all $k \in \mathbb{N}$. Define the bounded Lipschitz domain G by

$$G := \{(x', x_n) \in \mathbb{R}^n; h(x') < x_n < h(x') + 1, x' \in B\}.$$

From $u_k \rightharpoonup u$ as $k \to \infty$ in $H^1(G)^n$ and the compact imbedding (2.12) we get

$$u_k \underset{k \to \infty}{\to} u \quad \text{strongly in } L^2(\partial G)^n. \tag{7.113}$$

Passing to the limit in (7.112), using the boundedness of the sequence $(f(\nabla h_k)N_k)_{k\in\mathbb{N}}$ in $L^\infty(B)^n$ combined with the strong convergence properties in (7.70), (7.113) and the definition of N_k yield

$$
\begin{aligned}
0 &= \lim_{k\to\infty} \int_B g(x')f(\nabla h_k(x'))\big(u_k^1,\dots,u_k^n\big)\big(x',h(x')-h_k(x')\big)\cdot N_k(x')\,dx' \\
&= \lim_{k\to\infty} \int_B g(x')f(\nabla h_k(x'))\big(u_k^1,\dots,u_k^n\big)\big(x',h(x')\big)\cdot N_k(x')\,dx' \\
&= \lim_{k\to\infty} \int_B g(x')f(\nabla h_k(x'))\big(u^1,\dots,u^n\big)\big(x',h(x')\big)\cdot N_k(x')\,dx' \\
&= \underbrace{\lim_{k\to\infty} \int_B g(x')f(\nabla h_k(x'))\big(u^1,\dots,u^n\big)\big(x',h(x')\big)\cdot\big(\nabla h(x'),-1\big)\,dx'}_{=:I_1} \\
&\quad - \underbrace{\lim_{k\to\infty} \int_B g(x')\big(u^1,\dots,u^{n-1}\big)\big(x',h(x')\big)\cdot f(\nabla h_k(x'))\nabla h_k(x')\,dx'}_{=:I_2}.
\end{aligned}
\tag{7.114}
$$

Computation of I_1. From (7.111) it follows

$$
\int_B g(x')f(\nabla h_k(x'))\big(u^1,\dots,u^n\big)\big(x',h(x')\big)\cdot\big(\nabla h(x'),-1\big)\,dx' = 0
$$

for all $k\in\mathbb{N}$ and consequently $I_1 = 0$.

Computation of I_2. For $j\in\{1,\dots,n-1\}$ we introduce the Carathéodory function

$$
\psi_j : B\times\mathbb{R}^{n-1}\to\mathbb{R}, \quad \psi_j(x',\lambda) := f(\lambda)\lambda_j.
$$

Consider an arbitrary sequence $(m_k)_{k\in\mathbb{N}}$ of natural numbers $m_k\in\mathbb{N}$, $k\in\mathbb{N}$. A weak compactness argument yields a weakly convergent subsequence $\big(\psi_j(\cdot,\nabla h_{m_l}(\cdot))\big)_{l\in\mathbb{N}}$ in $L^2(B)$ of $\big(\psi_j(\cdot,\nabla h_{m_k}(\cdot))\big)_{k\in\mathbb{N}}$. Since $(\nabla h_k)_{k\in\mathbb{N}}$ generates the Young measure ν and $|B|<\infty$ it follows from Theorem 6.16 that

$$
f(\nabla h_{m_l})\frac{\partial}{\partial_j}h_{m_l} \xrightarrow[l\to\infty]{} \left(x'\mapsto\int_{\mathcal{K}_\nu} f(\lambda)\lambda_j\,d\nu_{x'}(\lambda)\right) \quad\text{in } L^2(B).
$$

Due to the arbitrary choice of the subsequence $\big(\psi_j(\cdot,\nabla h_{m_k}(\cdot))\big)_{k\in\mathbb{N}}$ there even holds

$$
f(\nabla h_k)\frac{\partial}{\partial_j}h_k \xrightarrow[k\to\infty]{} \left(x'\mapsto\int_{\mathcal{K}_\nu} f(\lambda)\lambda_j\,d\nu_{x'}(\lambda)\right) \quad\text{in } L^2(B).
$$

Thus (see 7.106))

$$
f(\nabla h_k)\nabla h_k \xrightarrow[k\to\infty]{} \left(x'\mapsto\int_{\mathcal{K}_\nu} f(\lambda)\lambda\,d\nu_{x'}(\lambda)\right) \quad\text{in } L^2(B).
\tag{7.115}
$$

Therefore it follows from (7.115)

$$
\begin{aligned}
I_2 &= -\lim_{k\to\infty} \int_B g(x')\big(u^1,\dots,u^{n-1}\big)\big(x',h(x')\big)\cdot f(\nabla h_k(x'))\nabla h_k(x')\,dx' \\
&= -\int_B g(x')\big(u^1,\dots,u^{n-1}\big)\big(x',h(x')\big)\cdot \int_{\mathcal{K}_\nu} f(\lambda)\lambda\,d\nu_{x'}(\lambda)\,dx'.
\end{aligned}
\tag{7.116}
$$

Altogether (7.114) implies

$$
\int_{\mathbb{R}^{n-1}} g(x')\big(u^1,\dots,u^{n-1}\big)\big(x',h(x')\big)\cdot \int_{\mathcal{K}_\nu} f(\lambda)\lambda\,d\nu_{x'}(\lambda)\,dx' = 0
$$

for all $g \in C_0^\infty(\mathbb{R}^{n-1})$ and $f \in C_0(\mathbb{R}^{n-1})$. By the fundamental theorem of calculus of variations we get for each $f \in C_0(\mathbb{R}^{n-1})$ a null set $N = N(f) \subseteq \mathbb{R}^{n-1}$ such that

$$
\big(u^1,\dots,u^{n-1}\big)\big(x',h(x')\big)\cdot \int_{\mathcal{K}_\nu} f(\lambda)\lambda\,d\nu_{x'}(\lambda)\,dx' = 0
\tag{7.117}
$$

for all $x' \in \mathbb{R}^{n-1}\setminus N$. Let $f_k \in C_0(\mathbb{R}^{n-1})$, $k \in \mathbb{N}$, be chosen such that $A := \{f_k; k \in \mathbb{N}\}$ is a dense set in $C_0(\mathbb{R}^{n-1})$ with respect to the sup-norm $\|\cdot\|_\infty$. Define the null set $N := \bigcup_{k\in\mathbb{N}} N(f_k)$. Then

$$
\big(u^1,\dots,u^{n-1}\big)\big(x',h(x')\big)\cdot \int_{\mathcal{K}_\nu} f_k(\lambda)\lambda\,d\nu_{x'}(\lambda) = 0
$$

for all $x' \in \mathbb{R}^{n-1} \setminus N$ and all $k \in \mathbb{N}$. Fix any $f \in C_0(\mathbb{R}^n)$ and choose a sequence $(f_k)_{k\in\mathbb{N}}$ of functions $f_k \in A$, $k \in \mathbb{N}$, with $\|f_k - f\|_\infty \to 0$ as $k \to \infty$. Consequently, since $\mathcal{K}_\nu$ is compact, there exists a constant $c > 0$ such that

$$
|f_k(\lambda)\lambda| \le c
$$

for all $\lambda \in \mathcal{K}_\nu$ and $k \in \mathbb{N}$. Lebesgue's dominated convergence theorem yields

$$
\lim_{k\to\infty} \int_{\mathcal{K}_\nu} f_k(\lambda)\lambda\,d\nu_{x'}(\lambda) = \int_{\mathcal{K}_\nu} f(\lambda)\lambda\,d\nu_{x'}(\lambda).
\tag{7.118}
$$

Therefore (7.117) holds for all $x' \in \mathbb{R}^{n-1} \setminus N$ and all $f \in C_0(\mathbb{R}^{n-1})$ where the null set N is independent of f. Altogether, for $x' \in \mathbb{R}^{n-1} \setminus (N \cup M)$ the properties (i), (ii) are satisfied. $\qquad\square$

The next theorem gives a sufficient criterion when u vanishes in a direction $(w(x'), 0)$ with a vector field $w = w(x')$. As a consequence it gives a criterion when u satisfies the no slip condition $u = 0$ on $\partial\Omega$. The proof of Theorem 7.25 is based on Theorem 7.24 and Lemma 7.20.

Theorem 7.25. *Let $n, \Omega, (\Omega_k)_{k\in\mathbb{N}}$ be as in Definition 7.10, assume that $(\nabla h_k)_{k\in\mathbb{N}}$ generates the Young measure $\nu = (\nu_{x'})_{x'\in\mathbb{R}^{n-1}} \in \mathcal{Y}(\mathbb{R}^{n-1}; \mathcal{P}(\mathbb{R}^{n-1}))$. Let $(u_k)_{k\in\mathbb{N}}$ be a bounded sequence in $H^1(\Omega_k)^n$, assume $u_k \in L_\sigma^2(\Omega_k)$ for all $k \in \mathbb{N}$, and let $u \in H^1(\Omega)^n$ such that $u_k \rightharpoonup u$ as $k \to \infty$ in $H^1(\Omega)^n$. Then the following statements are satisfied:*

(i) Let $w : \mathbb{R}^{n-1} \to \mathbb{R}^{n-1}$ be a vector field such that $w(x') \in \text{supp}(\nu_{x'})$ for a.a. $x' \in \mathbb{R}^{n-1}$. Then

$$u|_{\partial\Omega}(x', h(x')) \cdot \big(w(x'), 0\big) = 0 \tag{7.119}$$

for a.a. $x' \in \mathbb{R}^{n-1}$.

(ii) Assume additionally that ν is non-degenerate. Then $u \in H_0^1(\Omega)^n$.

Proof. We write $u = u|_{\partial\Omega}$ for the fixed class representative of the trace of u. By Lemma 7.24 there is a null set $N \subseteq \mathbb{R}^{n-1}$ and a compact set $\mathcal{K}_\nu \subseteq \mathbb{R}^{n-1}$ with $\text{supp}(\nu_{x'}) \subseteq \mathcal{K}_\nu$ for all $x' \in \mathbb{R}^{n-1} \setminus N$, such that

$$\big(u^1, \ldots, u^{n-1}\big)\big(x', h(x')\big) \cdot \int_{\mathcal{K}_\nu} f(\lambda)\lambda \, d\nu_{x'}(\lambda) = 0 \quad \forall f \in C_0(\mathbb{R}^n) \tag{7.120}$$

holds for all $x' \in \mathbb{R}^{n-1} \setminus N$. Furthermore $u \in L_\sigma^2(\Omega)$, and consequently (after a possible enlarging of the null set N) we have

$$\big(u^1, \ldots, u^n\big)\big(x', h(x')\big) \cdot \big(\nabla h(x'), -1\big) = 0 \tag{7.121}$$

for all $x' \in \mathbb{R}^{n-1} \setminus N$. Now statements (i), (ii) can be proven.

(i) Combining (7.120) and Lemma 7.20 (ii) yields that (7.119) is valid.

(ii) Fix $x' \in \mathbb{R}^{n-1} \setminus N$ such that $\text{supp}(\nu_{x'})$ contains a basis of $\mathbb{R}^{n-1}$. Due to Lemma 7.20 (ii) there exists functions $f_i \in C_0(\mathbb{R}^{n-1}), i = 1, \ldots, n-1$, such that the vectors

$$\int_{\mathcal{K}_\nu} f_i(\lambda)\lambda \, d\nu_{x'}(\lambda), i = 1, \ldots, n-1,$$

form a basis of $\mathbb{R}^{n-1}$. Consequently, with (7.120) we get $u^i(x', h(x')) = 0$ for all $i = 1, \ldots, n-1$. Looking at (7.121) we get $u(x', h(x')) = 0$. Finally, the non-degeneracy of ν implies $u(x', h(x')) = 0$ for almost all $x' \in \mathbb{R}^{n-1}$. $\qquad\square$

Making use of the mollification procedure introduced in Section 7.2.1 we obtain the following 'time dependent' version of Theorem 7.25. This result will play a crucial role in Section 7.3.

Theorem 7.26. *Let $n \in \mathbb{N}, n \geq 2$, let $\Omega, (\Omega_k)_{k\in\mathbb{N}}$ be as in Definition 7.10, assume that $(\nabla h_k)_{k\in\mathbb{N}}$ generates the non-degenerate Young measure $\nu = (\nu_{x'})_{x'\in\mathbb{R}^{n-1}}$. Let $(u_k)_{k\in\mathbb{N}}$ be a bounded sequence in $L^2(0, T; H^1(\Omega_k)^n)$, let $u \in L^2(0, T; H^1(\Omega)^n)$ with $u_k \rightharpoonup u$ as $k \to \infty$ in $L^2(0, T; H^1(\Omega)^n)$ and assume that $u_k(t) \in L_\sigma^2(\Omega_k)$ for a.a. $t \in]0, T[$ and all $k \in \mathbb{N}$. Then $u(t) \in H_0^1(\Omega)^n$ for a.a. $t \in]0, T[$.*

Proof. Define $\tilde{u}(t) := 1_{[0,T[}(t)u(t), t \in \mathbb{R}$, for every $k \in \mathbb{N}$ define $\widetilde{u_k}(t) := 1_{[0,T[}(t)u_k(t)$ for $t \in \mathbb{R}$. For $\delta > 0$ let $u^\delta \in C^\infty(\mathbb{R}; H^1(\Omega)^n)$ and $u_k^\delta \in C^\infty(\mathbb{R}; H^1(\Omega_k)^n)$ be defined as in (7.81).

Fix $\delta > 0$ with $\delta < T - \delta$. From Lemma 7.17 it follows $u_k^\delta(t) \in L_\sigma^2(\Omega_k)$ for all $t \in [\delta, T-\delta[$ and $k \in \mathbb{N}$. Furthermore, by Lemma 7.16, there holds for all $t \in [\delta, T-\delta]$

that the sequence $(u_k^\delta(t))_{k \in \mathbb{N}}$ is bounded in $H^1(\Omega_k)^n$ and $u_k^\delta(t) \rightharpoonup u^\delta(t)$ as $k \to \infty$ in $H^1(\Omega)^n$. Therefore Theorem 7.25 (ii) implies $u^\delta(t) \in H_0^1(\Omega)^n$ for a.a. $t \in [\delta, T - \delta]$. By (7.82) and the Fischer-Riesz Theorem there is a sequence $(\epsilon_k)_{k \in \mathbb{N}}$ with $\epsilon_k \to 0$ as $k \to \infty$ such that $u^{\epsilon_k}(t) \to u(t)$ strongly in $H^1(\Omega)^n$ for a.a. $t \in [\delta, T - \delta]$. Thus $u(t) \in H_0^1(\Omega)$ for a.a. $t \in]0, T[$. $\qquad\square$

7.3 Main results

The goal of this section is to combine the results of Section 7.1 and Section 7.2 to formulate and prove the main results of this chapter.

Definition 7.27. Assume that the following assumptions on the data are fulfilled:

(i) Let $n \in \{2,3\}$, let $h_k : \mathbb{R}^{n-1} \to \mathbb{R}_0^+$, $k \in \mathbb{N}$, be non-negative functions such that the sequence $(h_k)_{k \in \mathbb{N}}$ is equi-Lipschitz continuous and $h_k \to 0$ as $k \to \infty$ uniformly on all compact subsets of $\mathbb{R}^{n-1}$. Further let $h : \mathbb{R}^{n-1} \to \mathbb{R}$ be a Lipschitz continuous function. Define the domains

$$\Omega := \{ (x', x_n) \in \mathbb{R}^n; x_n > h(x'), x' \in \mathbb{R}^{n-1} \}, \tag{7.122}$$
$$\Omega_k := \{ (x', x_n) \in \mathbb{R}^n; x_n > h(x') - h_k(x'), x' \in \mathbb{R}^{n-1} \}, \quad k \in \mathbb{N}. \tag{7.123}$$

(ii) Let $0 < T < \infty$, let $u_0 \in L^2(\mathbb{R}^n)^n$, $\theta_0 \in L^2(\mathbb{R}^n)$, and let $f_1 \in L^2(0, T; L^2(\mathbb{R}^n)^n)$, $f_2 \in L^2(0, T; L^2(\mathbb{R}^n))$. Further let $g \in L^\infty(]0, T[\times \mathbb{R}^n)^n$, $h_0 \in L^2(0, T; H^1(\mathbb{R}^n))$, let $\nu, \alpha, \beta, \kappa$ be positive, real constants, and let $\delta \in [0, \infty[$.

(iii) Moreover, assume that the sequence $(\nabla h_k)_{k \in \mathbb{N}}$ generates the Young measure $\nu = (\nu_{x'})_{x' \in \mathbb{R}^{n-1}} \in \mathcal{Y}(\mathbb{R}^{n-1}; \mathcal{P}(\mathbb{R}^{n-1}))$. Define $\Lambda : \partial\Omega \to \mathbb{R}$ by

$$\Lambda(x', h(x')) := \frac{1}{\sqrt{1 + |\nabla h(x')|^2}} \int_{\mathbb{R}^{n-1}} \sqrt{1 + |\nabla h(x') - \lambda|^2} \, d\nu_{x'}(\lambda) \tag{7.124}$$

for a.a. $x = (x', h(x')) \in \partial\Omega$ with $x' \in \mathbb{R}^{n-1}$.

The following theorem combines Theorem 7.9 and Theorem 7.19. This result will play an essential role in Chapter 8 and in the progress of this section.

Theorem 7.28. *Assume that the assumptions formulated in Definition 7.27 are fulfilled. Let $(u_k)_{k \in \mathbb{N}}$ be a bounded sequence in $L^\infty(0, T; L^2_\sigma(\Omega_k)) \cap L^2(0, T; H^1(\Omega_k)^n)$ and let $(\theta_k)_{k \in \mathbb{N}}$ be a bounded sequence in $L^\infty(0, T; L^2(\Omega_k)) \cap L^2(0, T; H^1(\Omega_k))$. Assume that*

$$-\int_0^T \langle u_k, w_t \rangle_\Omega \, dt + \nu \int_0^T \langle \nabla u_k, \nabla w \rangle_\Omega \, dt + \int_0^T \langle u_k \cdot \nabla u_k, w \rangle_\Omega \, dt$$
$$= \beta \int_0^T \langle \theta_k g, w \rangle_\Omega \, dt + \int_0^T \langle f_1, w \rangle_\Omega \, dt + \langle u_0, w(0) \rangle_\Omega \tag{7.125}$$

for all $w \in C_0^\infty([0, T[; C_{0,\sigma}^\infty(\Omega))$ and

$$-\int_0^T \langle \theta_k, \phi_t \rangle_{\Omega_k} \, dt + \kappa \int_0^T \langle \nabla \theta_k, \nabla \phi \rangle_{\Omega_k} \, dt + \int_0^T \langle u_k \cdot \nabla \theta_k, \phi \rangle_{\Omega_k} \, dt$$
$$= -\alpha\kappa \int_0^T \langle \theta_k - h_0, \phi \rangle_{\partial\Omega_k} \, dt + \int_0^T \langle f_2, \phi \rangle_{\Omega_k} \, dt + \langle \theta_0, \phi(0) \rangle_{\Omega_k} \tag{7.126}$$

for all $\phi \in C_0^\infty([0,T[\times\overline{\Omega_k})$. Further, consider

$$u \in L^\infty(0,T;L^2(\Omega)^n) \cap L^2(0,T;H^1(\Omega)^n),$$
$$\theta \in L^2(0,T;L^2(\Omega)) \cap L^2(0,T;H^1(\Omega))$$

such that

$$u_k \xrightarrow[k\to\infty]{*} u \quad in \ L^\infty(0,T;L^2(\Omega)^n), \quad u_k \xrightarrow[k\to\infty]{} u \quad in \ L^2(0,T;H^1(\Omega)^n),$$
$$\theta_k \xrightarrow[k\to\infty]{*} \theta \quad in \ L^\infty(0,T;L^2(\Omega)), \quad \theta_k \xrightarrow[k\to\infty]{} \theta \quad in \ L^2(0,T;H^1(\Omega)). \tag{7.127}$$

Then there holds $u(t) \in L_\sigma^2(\Omega)$ for a.a. $t \in]0,T[$, and

$$- \int_0^T \langle u, w_t \rangle_\Omega \, dt + \nu \int_0^T \langle \nabla u, \nabla w \rangle_\Omega \, dt + \int_0^T \langle u \cdot \nabla u, w \rangle_\Omega \, dt$$
$$= \beta \int_0^T \langle \theta g, w \rangle_\Omega \, dt + \int_0^T \langle f_1, w \rangle_\Omega \, dt + \langle u_0, w(0) \rangle_\Omega \tag{7.128}$$

for all $w \in C_0^\infty([0,T[;C_{0,\sigma}^\infty(\Omega))$. Moreover we have

$$- \int_0^T \langle \theta, \phi_t \rangle_\Omega \, dt + \kappa \int_0^T \langle \nabla\theta, \nabla\phi \rangle_\Omega \, dt + \int_0^T \langle u \cdot \nabla\theta, \phi \rangle_\Omega \, dt$$
$$= -\alpha\kappa \int_0^T \langle \Lambda(\theta - h_0), \phi \rangle_{\partial\Omega} \, dt + \int_0^T \langle f_2, \phi \rangle_\Omega \, dt + \langle \theta_0, \phi(0) \rangle_\Omega \tag{7.129}$$

for all $\phi \in C_0^\infty([0,T[\times\overline{\Omega})$ where the function $\Lambda : \partial\Omega \to \mathbb{R}$ is defined as in (7.124).

Proof. By Lemma 7.15 (ii) there holds $u(t) \in L_\sigma^2(\Omega)$ for a.a. $t \in]0,T[$. We see that the restrictions $u_k|_{]0,T[\times\Omega}$ and $\theta_k|_{]0,T[\times\Omega}$ satisfy all requirements of Theorem 7.9. Therefore, by this theorem, we get that (7.128) holds for all $w \in C_0^\infty([0,T[;C_{0,\sigma}^\infty(\Omega))$. Moreover

$$\theta_k \xrightarrow[k\to\infty]{} \theta \quad \text{strongly in } L^2(0,T;L^2(G)), \tag{7.130}$$
$$\theta_k \xrightarrow[k\to\infty]{} \theta \quad \text{strongly in } L^2(0,T;L^2(\partial G)) \tag{7.131}$$

for all bounded Lipschitz domains G with $G \subseteq \Omega$. Fix $\phi \in C_0^\infty([0,T[\times\overline{\Omega})$. Due to the definition of the space $C_0^\infty([0,T[\times\overline{\Omega})$ we can assume $\phi \in C_0^\infty([0,T[\times\mathbb{R}^n)$. Consequently, by (7.96), (7.97) we get

$$\lim_{k\to\infty} \int_0^T \langle \theta_k - h_0, \phi \rangle_{\partial\Omega_k} \, dt = \int_0^T \langle \Lambda(\theta - h_0), \phi \rangle_{\partial\Omega} \, dt \tag{7.132}$$

with the function $\Lambda : \partial\Omega \to \mathbb{R}$ defined in (7.124).

We use Hölder's inequality, the equi-Lipschitz continuity of $(h_k)_{k\in\mathbb{N}}$ and Lemma 4.12 to obtain that

$$
\begin{aligned}
\|\theta_k(t)u_k(t)\|_{2,\Omega_k} &\leq \|\theta_k(t)\|_{4,\Omega_k}\|u_k(t)\|_{4,\Omega_k} \\
&\leq c\|\theta_k(t)\|_{2,\Omega_k}^{1/4}\|\theta_k(t)\|_{H^1(\Omega_k)}^{3/4}\|u_k(t)\|_{2,\Omega_k}^{1/4}\|u_k(t)\|_{H^1(\Omega_k)^n}^{3/4} \\
&\leq c\|\theta_k\|_{2,\infty;\Omega_k;T}^{1/4}\|u_k\|_{2,\infty;\Omega_k;T}^{1/4}\big(\|\theta_k(t)\|_{H^1(\Omega_k)}^{3/2}+\|u_k(t)\|_{H^1(\Omega_k)^n}^{3/2}\big)
\end{aligned}
\tag{7.133}
$$

holds for a.a. $t\in]0,T[$ and all $k\in\mathbb{N}$ with a constant $c>0$ independent of k. Integrating (7.133) with respect to $t\in]0,T[$ and making use of (7.127) we get that the sequence

$$(\theta_k u_k)_{k\in\mathbb{N}} \text{ is bounded in } L^{4/3}(0,T;L^2(\Omega)^3). \tag{7.134}$$

Fix $v\in C_0^\infty(]0,T[\times\Omega)^3$. Choose a bounded Lipschitz domain G with $G\subseteq\Omega$ such that $\mathrm{supp}(v)\subseteq]0,T[\times G$. By (7.127), (7.130)

$$\lim_{k\to\infty}\int_0^T\langle\theta_k u_k,v\rangle_\Omega\,d\tau = \int_0^T\langle\theta u,v\rangle_\Omega\,d\tau. \tag{7.135}$$

Combining (7.134), (7.135) yields

$$\theta_k u_k \underset{k\to\infty}{\rightharpoonup} \theta u \quad \text{in } L^{4/3}(0,T;L^2(\Omega)^3). \tag{7.136}$$

Integration by parts in space and employing (7.75), (7.136) implies

$$
\begin{aligned}
\lim_{k\to\infty}\int_0^T\langle u_k\cdot\nabla\theta_k,\phi\rangle_{\Omega_k}\,dt &= -\lim_{k\to\infty}\int_0^T\langle\theta_k u_k,\nabla\phi\rangle_{\Omega_k}\,dt \\
&= -\int_0^T\langle\theta u,\nabla\phi\rangle_\Omega\,dt \\
&= \int_0^T\langle u\cdot\nabla\theta,\phi\rangle_\Omega\,dt.
\end{aligned}
\tag{7.137}
$$

Passing to the limit in (7.126) and making use of (7.132), (7.137) and of (7.127) in combination with (7.75) yield that (7.129) is satisfied for all $\phi\in C_0^\infty([0,T[\times\overline{\Omega})$. $\square$

For $u\in H^1(\Omega)$ let us introduce by $D(u):=\frac{1}{2}\big(\nabla u+(\nabla u)^T\big)$ the *symmetric part of the gradient*. Let $\delta\in[0,\infty[$ denote a friction coefficient. The *Navier slip* boundary condition reads as follows:

$$u\cdot N=0\,, \quad [\delta u+2\nu D(u)\cdot N]_{\mathrm{tan}}=0 \quad \text{on }]0,T[\times\partial\Omega. \tag{7.138}$$

In the special case $\delta=0$ the condition above is the *complete slip* boundary condition

$$u\cdot N=0\,, \quad [D(u)\cdot N]_{\mathrm{tan}}=0 \quad \text{on }]0,T[\times\partial\Omega.$$

We give the following

Definition 7.29. Let $\Omega \subseteq \mathbb{R}^n$, $n \in \{2,3\}$, be a domain satisfying the uniform Lipschitz condition, let $0 < T < \infty$, $g \in L^\infty(]0,T[\times\Omega)^n$, let $h_0 \in L^2(0,T;L^2(\partial\Omega))$. Further let $f_1 \in L^2(0,T;L^2(\Omega)^n)$, $f_2 \in L^2(0,T;L^2(\Omega))$, let $u_0 \in L^2(\Omega)^n$, $\theta_0 \in L^2(\Omega)$, and let $\alpha, \nu, \beta, \kappa$ be positive, real constants, and let $\delta \in [0,\infty[$.

A pair

$$
\begin{aligned}
&u \in L^\infty(0,T;L^2_\sigma(\Omega)) \cap L^2(0,T;H^1(\Omega)^n), \\
&\theta \in L^\infty(0,T;L^2(\Omega)) \cap L^2(0,T;H^1(\Omega))
\end{aligned}
\tag{7.139}
$$

is called a weak solution of the Boussinesq equations (7.1) with Navier slip boundary condition

$$
u \cdot N = 0, \quad [\delta u + 2\nu D(u) \cdot N]_{\text{tan}} = 0 \quad \text{on }]0,T[\times\partial\Omega \tag{7.140}
$$

and Robin boundary condition

$$
\frac{\partial \theta}{\partial N} + \alpha(\theta - h_0) = 0 \quad \text{on }]0,T[\times\partial\Omega \tag{7.141}
$$

if the following variational identities are satisfied:

(i) There holds

$$
\begin{aligned}
&-\int_0^T \langle u, w_t \rangle_\Omega \, dt + \nu \int_0^T \langle \nabla u, \nabla w \rangle_\Omega \, dt + \int_0^T \langle u \cdot \nabla u, w \rangle_\Omega \, dt + \delta \int_0^T \langle u, w \rangle_{\partial\Omega} \, dt \\
&= \beta \int_0^T \langle \theta g, w \rangle_\Omega \, dt + \int_0^T \langle f_1, w \rangle_\Omega \, dt + \langle u_0, w(0) \rangle_\Omega
\end{aligned}
\tag{7.142}
$$

for all $w \in C_0^\infty([0,T[\times\overline{\Omega})^n$ with $\operatorname{div} w = 0$ in $]0,T[\times\Omega$ and $w \cdot N = 0$ on $]0,T[\times\partial\Omega$.

(ii) We have

$$
\begin{aligned}
&-\int_0^T \langle \theta, \phi_t \rangle_\Omega \, dt + \kappa \int_0^T \langle \nabla\theta, \nabla\phi \rangle_\Omega \, dt + \int_0^T \langle u \cdot \nabla\theta, \phi \rangle_\Omega \, dt \\
&= -\alpha\kappa \int_0^T \langle \theta - h_0, \phi \rangle_{\partial\Omega} \, dt + \int_0^T \langle f_2, \phi \rangle_\Omega \, dt + \langle \theta_0, \phi(0) \rangle_\Omega
\end{aligned}
\tag{7.143}
$$

for all $\phi \in C_0^\infty([0,T[\times\overline{\Omega})$.

Theorem 7.30. *Assume that the assumptions on the data formulated in Definition 7.27 are fulfilled. For every $k \in \mathbb{N}$ let (u_k, θ_k) be a weak solution of the Boussinesq equations (7.1) in $[0,T[\times\Omega_k$ (with data $f_1, f_2, g, h_0|_{]0,T[\times\Omega_k}$ and $P_{\Omega_k} u_0, \theta_0|_{\Omega_k}$) satisfying the following boundary conditions: Let u_k satisfy the Navier slip boundary condition*

$$
u_k \cdot N = 0, \quad [\delta u + 2\nu D(u_k) \cdot N]_{tan} = 0 \quad \text{on }]0,T[\times\partial\Omega_k \tag{7.144}
$$

and let θ_k satisfy the Robin boundary condition

$$\frac{\partial \theta_k}{\partial N} + \alpha(\theta_k - h_0) = 0 \quad on \ ]0, T[\times \partial\Omega_k. \tag{7.145}$$

Further let (u_k, θ_k) satisfy the energy inequalities

$$\frac{1}{2}\|u_k(t)\|_{2,\Omega_k}^2 + \nu \int_0^t \|\nabla u_k\|_{2,\Omega_k}^2 \, d\tau$$

$$\leq \frac{1}{2}\|u_0\|_2^2 + \beta \int_0^t \langle \theta_k g, u_k \rangle_{\Omega_k} \, d\tau + \int_0^t \langle f_1, u_k \rangle_{\Omega_k} \, d\tau \,,$$

$$\frac{1}{2}\|\theta_k(t)\|_{2,\Omega_k}^2 + \kappa \int_0^t \|\nabla \theta_k\|_{2,\Omega_k}^2 \, d\tau + \alpha\kappa \int_0^t \|\theta_k\|_{\partial\Omega_k}^2 \, d\tau \tag{7.146}$$

$$\leq \frac{1}{2}\|\theta_0\|_2^2 + \alpha\kappa \int_0^t \langle h_0, \theta_k \rangle_{\partial\Omega_k} \, d\tau + \int_0^t \langle f_2, \theta_k \rangle_{\Omega_k} \, d\tau$$

for almost all $t \in [0, T[$ and all $k \in \mathbb{N}$. Assume that the Young measure ν in Definition 7.27 is non-degenerate. Then the following statements are satisfied:

(i) There holds

$$(u_k)_{k \in \mathbb{N}} \text{ is bounded in } L^\infty(0, T; L_\sigma^2(\Omega_k)) \cap L^2(0, T; H^1(\Omega_k)^n) \,,$$
$$(\theta_k)_{k \in \mathbb{N}} \text{ is bounded in } L^\infty(0, T; L^2(\Omega_k)) \cap L^2(0, T; H^1(\Omega_k)). \tag{7.147}$$

(ii) Assume that there is

$$u \in L^\infty(0, T; L^2(\Omega)^n) \cap L^2(0, T; H^1(\Omega)^n) \,,$$
$$\theta \in L^2(0, T; L^2(\Omega)) \cap L^2(0, T; H^1(\Omega))$$

and a subsequence $(u_{m_k}, \theta_{m_k})_{k \in \mathbb{N}}$ such that

$$u_{m_k} \underset{k\to\infty}{\overset{*}{\rightharpoonup}} u \quad in \ L^\infty(0, T; L^2(\Omega)^n), \quad u_{m_k} \underset{k\to\infty}{\rightharpoonup} u \quad in \ L^2(0, T; H^1(\Omega)^n),$$

$$\tag{7.148}$$

$$\theta_{m_k} \underset{k\to\infty}{\overset{*}{\rightharpoonup}} \theta \quad in \ L^\infty(0, T; L^2(\Omega)), \quad \theta_{m_k} \underset{k\to\infty}{\rightharpoonup} \theta \quad in \ L^2(0, T; H^1(\Omega)). \tag{7.149}$$

Then (u, θ) is a weak solution of the Boussinesq equations (7.1) in $[0, T[\times\Omega$ (with data $f_1, f_2, g, h_0|_{]0,T[\times\Omega}$ and $P_\Omega u_0, \theta_0|_\Omega$) satisfying the following boundary conditions: There holds that u satisfies the no slip boundary condition

$$u = 0 \quad on \ ]0, T[\times \partial\Omega \tag{7.150}$$

and θ satisfies the Robin boundary condition

$$\frac{\partial \theta}{\partial N} + \alpha\Lambda(\theta - h_0) = 0 \quad on \ ]0, T[\times \partial\Omega \tag{7.151}$$

where the function $\Lambda : \partial\Omega \to \mathbb{R}$ is defined as in (7.124).

Remark. We see that the boundary behaviour of u on $]0,T[\times\partial\Omega$ in (ii) of the theorem does not depend on the friction coefficient $\delta \in [0,\infty[$ fulfilled by the tangential component $[u_k]_{\tan}$ in the Navier slip condition (7.144) on $]0,T[\times\partial\Omega_k$.

Proof. Step 1. Due to the assumptions on h, $(h_k)_{k\in\mathbb{N}}$ there exists a constant $L > 0$ such that $\mathrm{Lip}(h) \leq L$ and $\mathrm{Lip}(h_k) \leq L$ for all $k \in \mathbb{N}$. By (4.17) there holds

$$\|h_0\|_{2,2;\partial\Omega_k;T} \leq c\|h_0\|_{L^2(0,T;H^1(\Omega_k))} \leq c\|h_0\|_{L^2(0,T;H^1(\mathbb{R}^n))} \tag{7.152}$$

with a constant $c = c(L) > 0$ independent of $k \in \mathbb{N}$. Looking at (4.126), (4.128), (7.152) and making use of $0 < T < \infty$ we obtain

$$
\begin{aligned}
\frac{1}{4}&\|\theta_k\|_{2,\infty;\Omega_k;T}^2 + \frac{\kappa}{2}\|\nabla\theta_k\|_{2,2;\Omega_k;T}^2 + \alpha\kappa\|\theta_k\|_{2,2;\partial\Omega_k;T}^2 \\
&\leq \|\theta_0\|_{2,\Omega_k}^2 + c\|h_0\|_{2,1;\partial\Omega_k;T}^2 + c\|h_0\|_{2,2;\partial\Omega_k;T}^2 + 4\|f_2\|_{2,1;\Omega_k;T}^2 \\
&\leq c\big(\|\theta_0\|_{2,\mathbb{R}^n}^2 + \|h_0\|_{L^2(0,T;H^1(\mathbb{R}^n))}^2 + \|f_2\|_{2,1;\mathbb{R}^n;T}^2\big)
\end{aligned}
\tag{7.153}
$$

for all $k \in \mathbb{N}$ with a constant $c = c(L,\alpha,\beta,\kappa,T) > 0$ independent of k. Thus

$$(\theta_k)_{k\in\mathbb{N}} \text{ is bounded in } L^\infty(0,T;L^2(\Omega_k)) \cap L^2(0,T;H^1(\Omega_k)). \tag{7.154}$$

Furthermore, by (4.127) there holds

$$
\begin{aligned}
\frac{1}{4}&\|u_k\|_{2,\infty;\Omega_k;T}^2 + \nu\|\nabla u_k\|_{2,2;\Omega_k;T}^2 \\
&\leq \|u_0\|_{2,\Omega_k}^2 + 4\beta^2\|g\|_{L^\infty(]0,T[\times\Omega_k)^n}^2\|\theta_k\|_{2,1;\Omega_k;T}^2 + 4\|f_1\|_{2,1;\Omega_k;T}^2 \\
&\leq c\big(\|u_0\|_{2,\mathbb{R}^n}^2 + \|g\|_{L^\infty(]0,T[\times\mathbb{R}^n)^n}^2\|\theta_k\|_{2,2;\Omega_k;T}^2 + \|f_1\|_{2,1;\mathbb{R}^n;T}^2\big)
\end{aligned}
\tag{7.155}
$$

for all $k \in \mathbb{N}$ with a constant $c = c(\beta,T) > 0$. Consequently, with (7.154) it follows

$$(u_k)_{k\in\mathbb{N}} \text{ is bounded in } L^\infty(0,T;L^2_\sigma(\Omega_k)) \cap L^2(0,T;H^1(\Omega_k)^n). \tag{7.156}$$

Step 2. Let the function $\Lambda : \partial\Omega \to \mathbb{R}$ be defined as in (7.124). We see that conditions (7.125), (7.126), (7.127) formulated in Theorem 7.28 are fulfilled. Therefore $u(t) \in L^2_\sigma(\Omega)$ for a.a. $t \in]0,T[$, and

$$
\begin{aligned}
-\int_0^T \langle u, w_t\rangle_\Omega \, dt &+ \nu\int_0^T \langle\nabla u, \nabla w\rangle_\Omega \, dt + \int_0^T \langle u\cdot\nabla u, w\rangle_\Omega \, dt \\
&= \beta\int_0^T \langle\theta g, w\rangle_\Omega \, dt + \int_0^T \langle f_1, w\rangle_\Omega \, dt + \langle u_0, w(0)\rangle_\Omega
\end{aligned}
\tag{7.157}
$$

for all $w \in C_0^\infty([0,T[;C_{0,\sigma}^\infty(\Omega))$. Furthermore

$$
\begin{aligned}
-\int_0^T \langle\theta, \phi_t\rangle_\Omega \, dt &+ \kappa\int_0^T \langle\nabla\theta, \nabla\phi\rangle_\Omega \, dt + \int_0^T \langle u\cdot\nabla\theta, \phi\rangle_\Omega \, dt \\
&= -\alpha\kappa\int_0^T \langle\Lambda(\theta - h_0), \phi\rangle_{\partial\Omega} \, dt + \int_0^T \langle f_2, \phi\rangle_\Omega \, dt + \langle\theta_0, \phi(0)\rangle_\Omega
\end{aligned}
\tag{7.158}
$$

for all $\phi \in C_0^\infty([0, T[\times\overline{\Omega})$. By Theorem 7.26 we get $u(t) \in H_0^1(\Omega)$ for almost all $t \in]0, T[$. New we obtain from $u(t) \in L_\sigma^2(\Omega) \cap H_0^1(\Omega)$, a.a. $t \in]0, T[$, and (2.15) that $u(t) \in W_{0,\sigma}^{1,2}(\Omega)$ holds for a.a. $t \in]0, T[$. Therefore

$$
\begin{aligned}
&u \in L^\infty(0, T; L_\sigma^2(\Omega)) \cap L^2(0, T; W_{0,\sigma}^{1,2}(\Omega)), \\
&\theta \in L^\infty(0, T; L^2(\Omega)) \cap L^2(0, T; H^1(\Omega)).
\end{aligned}
\tag{7.159}
$$

Thus, looking at (7.157), (7.158), (7.159) it follows that (u, θ) is a weak solution of (7.1) in $[0, T[\times\Omega$ with boundary conditions (7.150), (7.151). $\qquad\square$

In the case $n = 2$ we can use the uniqueness statement formulated in Theorem 4.23 to obtain the following version of Theorem 7.30.

Theorem 7.31. *Consider data as in Definition 7.27 and assume $n = 2$. As in Theorem 7.30 let $(u_k, \theta_k)_{k\in\mathbb{N}}$ be a sequence of weak solution of the Boussinesq equations (7.1) in $[0, T[\times\Omega_k$ satisfying the boundary conditions (7.144), (7.145) and the energy inequalities (7.146). Assume that the Young measure $\nu = (\nu_{x'})_{x'\in\mathbb{R}}$ in Definition 7.27 is non-degenerate. We denote by (u, θ) the unique weak solution of the Boussinesq equations (7.1) in $[0, T[\times\partial\Omega$ (with data $f_1, f_2, g, h_0|_{]0,T[\times\Omega}$ and $P_\Omega u_0, \theta_0|_\Omega$) satisfying the boundary conditions*

$$
u = 0 \quad and \quad \frac{\partial\theta}{\partial N} + \alpha\Lambda(\theta - h_0) = 0 \quad on \ ]0, T[\times\partial\Omega
$$

where the function $\Lambda : \partial\Omega \to \mathbb{R}$ is defined as in (7.124). Then we have

$$
\begin{aligned}
&(u_k)_{k\in\mathbb{N}} \text{ is bounded in } L^\infty(0, T; L_\sigma^2(\Omega_k)) \cap L^2(0, T; H^1(\Omega_k)^2), \\
&(\theta_k)_{k\in\mathbb{N}} \text{ is bounded in } L^\infty(0, T; L^2(\Omega_k)) \cap L^2(0, T; H^1(\Omega_k))
\end{aligned}
\tag{7.160}
$$

and there holds

$$
\begin{aligned}
u_k &\underset{k\to\infty}{\overset{*}{\rightharpoonup}} u \quad in \ L^\infty(0, T; L^2(\Omega)^2), \qquad u_k \underset{k\to\infty}{\rightharpoonup} u \quad in \ L^2(0, T; H^1(\Omega)^2), \\
\theta_k &\underset{k\to\infty}{\overset{*}{\rightharpoonup}} \theta \quad in \ L^\infty(0, T; L^2(\Omega)), \qquad \theta_k \underset{k\to\infty}{\rightharpoonup} \theta \quad in \ L^2(0, T; H^1(\Omega)).
\end{aligned}
\tag{7.161}
$$

Remark. For necessary and sufficient criteria for the non-degeneracy condition to be fulfilled in the two-dimensional case we refer to Lemma 7.22 and Lemma 7.23.

Proof. Due to Theorem 4.23 there exists a unique weak solution (u, θ) of the Boussinesq equations (7.1) in $[0, T[\times\partial\Omega$ (with data $f_1, f_2, g, h_0|_{]0,T[\times\Omega}$ and $P_\Omega u_0, \theta_0|_\Omega$) satisfying the boundary conditions

$$
u = 0 \quad and \quad \frac{\partial\theta}{\partial N} + \alpha\Lambda(\theta - h_0) = 0 \quad on \ ]0, T[\times\partial\Omega
$$

From Theorem 7.30 we immediately obtain that (7.160) is fulfilled. To show (7.161) we consider an arbitrary subsequence $(u_{m_k}, \theta_{m_k})_{k\in\mathbb{N}}$. Combining Theorem 7.30 and

Theorem 4.23 (i) we see that there exists a subsequence $(u_{m_l}, \theta_{m_l})_{l \in \mathbb{N}}$ of $(u_{m_k}, \theta_{m_k})_{k \in \mathbb{N}}$ such that

$$
\begin{aligned}
&u_{m_l} \underset{l \to \infty}{\overset{*}{\rightharpoonup}} u \quad \text{in } L^\infty(0, T; L^2(\Omega)^2), \quad u_{m_l} \underset{l \to \infty}{\rightharpoonup} u \quad \text{in } L^2(0, T; H^1(\Omega)^2), \\
&\theta_{m_l} \underset{l \to \infty}{\overset{*}{\rightharpoonup}} \theta \quad \text{in } L^\infty(0, T; L^2(\Omega)), \quad \theta_{m_l} \underset{l \to \infty}{\rightharpoonup} \theta \quad \text{in } L^2(0, T; H^1(\Omega)).
\end{aligned}
\tag{7.162}
$$

Due to the arbitrary choice of the subsequence $(u_{m_k}, \theta_{m_k})_{k \in \mathbb{N}}$ we obtain from (7.162) that (7.161) is fulfilled. $\qquad \square$

8 Construction of some special domains with rough boundaries

Looking at Theorem 7.30 we have observed the transition from the boundary condition (7.145) to the boundary condition (7.151) where $\Lambda : \partial\Omega \to [1, \infty[$ is a 'weight function' which reflects the rugosity of the boundaries. To describe the guiding ideas of the present chapter we assume in the following that $n = 2$. Therefore (see Theorem 4.23) there holds that weak solution of the Boussinesq equations (7.1) satisfying the boundary conditions (8.1) are unique. Let $\Omega \subseteq \mathbb{R}^2$ be the half space. We fix a function $\Lambda : \partial\Omega \to [1, \infty[$. Let (u, θ) denote the unique weak solution of the Boussinesq equations (7.1) in $]0, T[\times\Omega$ satisfying the boundary conditions

$$u = 0 \quad \text{and} \quad \frac{\partial\theta}{\partial N} + \alpha\Lambda(\theta - h_0) = 0 \quad \text{on }]0, T[\times\partial\Omega. \tag{8.1}$$

In Theorem 8.4 we will construct a sequence of domains $(\Omega_k)_{k\in\mathbb{N}}$ with $\Omega \subseteq \Omega_k$, $k \in \mathbb{N}$, and $\Omega_k \to \Omega$ such that the following properties are satisfied: For fixed $k \in \mathbb{N}$ let (u_k, θ_k) denote the unique weak solution of the Boussinesq equations (7.1) in $]0, T[\times\Omega_k$ satisfying the boundary conditions

$$u_k = 0 \quad \text{and} \quad \frac{\partial\theta_k}{\partial N} + \alpha(\theta_k - h_0) = 0 \quad \text{on }]0, T[\times\partial\Omega_k. \tag{8.2}$$

Then there holds that $(u_k, \theta_k)_{k\in\mathbb{N}}$ converges in a weak sense to (u, θ) in $]0, T[\times\Omega$. For details of the ideas sketched above we refer to Theorem 8.4.

In Theorem 8.3 we will formulate a 'weaker' version of the result sketched above which does not need the uniqueness of weak solutions of the Boussineq equations. Consequently, we can allow $n = 3$ in this theorem.

We begin with a theorem which will be crucial for the construction of the domains $(\Omega_k)_{k\in\mathbb{N}}$. It is essentially based on the existence of the gradient Young measures constructed in Theorem 6.20.

Theorem 8.1. *Let $n \in \mathbb{N}$, $n \geq 2$, and let $\tilde{\Lambda} : \mathbb{R}^{n-1} \to \mathbb{R}$ be a bounded, measurable function satisfying $\tilde{\Lambda}(x') \geq 1$ for a.a. $x' \in \mathbb{R}^{n-1}$. Then there exists a sequence $(h_k)_{k\in\mathbb{N}}$ of equi-Lipschitz continuous functions $h_k : \mathbb{R}^{n-1} \to \mathbb{R}$, $k \in \mathbb{N}$, and a Young measure $\nu = (\nu_{x'})_{x'\in\mathbb{R}^{n-1}} \in \mathcal{Y}(\mathbb{R}^{n-1}; \mathcal{P}(\mathbb{R}^{n-1}))$ satisfying the following properties:*

    (i) There holds $h_k(x') \geq 0$ for all $x' \in \mathbb{R}^{n-1}$, $k \in \mathbb{N}$, and $h_k \to 0$ uniformly on $\mathbb{R}^{n-1}$ as $k \to \infty$.

(ii) The sequence $(\nabla h_k)_{k\in\mathbb{N}}$ generates ν and

$$\tilde{\Lambda}(x') = \int_{\mathbb{R}^{n-1}} \sqrt{1+|\lambda|^2}\, d\nu_{x'}(\lambda) \quad \text{for a.a. } x' \in \mathbb{R}^{n-1}. \tag{8.3}$$

Proof. Choose an arbitrary vector $w \in \mathbb{R}^{n-1}$ such that $|w| = 1$. Introduce the function $z \in L^\infty(\mathbb{R}^{n-1})^{n-1}$ by

$$z(x') := \begin{cases} \sqrt{\left(\tilde{\Lambda}(x')\right)^2 - 1}\, w\,, & \text{if } \tilde{\Lambda}(x') \geq 1\,, \\ 0\,, & \text{otherwise.} \end{cases}$$

Consequently,

$$\tilde{\Lambda}(x') = \sqrt{1+|z(x')|^2} \tag{8.4}$$

for a.a. $x' \in \mathbb{R}^{n-1}$. Define the Young measure $\nu = (\nu_{x'})_{x'\in\mathbb{R}^{n-1}} \in \mathcal{Y}(\mathbb{R}^{n-1};\mathcal{P}(\mathbb{R}^{n-1}))$ by

$$\nu_x := \frac{1}{2}\big(\delta_{z(x)} + \delta_{-z(x)}\big)\,, \quad x' \in \mathbb{R}^{n-1}.$$

By Theorem 6.20 there exist non-negative functions $h_k : \mathbb{R}^{n-1} \to \mathbb{R}_0^+$, $k \in \mathbb{N}$, such that the following properties are fulfilled:

(1) The sequence $(h_k)_{k\in\mathbb{N}}$ is bounded in $W^{1,\infty}(\mathbb{R}^{n-1})$ and $(\nabla h_k)_{k\in\mathbb{N}}$ generates ν.

(2) There holds $h_k \to 0$ strongly in $L^\infty(\mathbb{R}^{n-1})$ as $k \to \infty$.

Looking at Lemma 2.24 we see that for every $k \in \mathbb{N}$ the function h_k can be redefined on a null set of $\mathbb{R}^{n-1}$ such that $h_k : \mathbb{R}^{n-1} \to \mathbb{R}$ is Lipschitz continuous. Consequently, from (1), (2) we get that the sequence $(h_k)_{k\in\mathbb{N}}$ is equi-Lipschitz continuous and that $h_k \to 0$ uniformly on $\mathbb{R}^{n-1}$ as $k \to \infty$. The construction of ν and (8.4) yield

$$\int_{\mathbb{R}^{n-1}} \sqrt{1+|\lambda|^2}\, d\nu_{x'}(\lambda) = \frac{1}{2}\left(\sqrt{1+|z(x')|^2} + \sqrt{1+|-(z(x'))|^2} \right) = \tilde{\Lambda}(x')$$

for a.a. $x' \in \mathbb{R}^{n-1}$. Altogether (i), (ii) are fulfilled. $\qquad\square$

The following Lemma will be needed to show the conservation of the no slip boundary condition in Theorem 8.3.

Lemma 8.2. *Let $n \in \mathbb{N}, n \geq 2$, let $0 < T < \infty$, let $h_k : \mathbb{R}^{n-1} \to \mathbb{R}_0^+$, $k \in \mathbb{N}$, be non-negative functions such that the sequence $(h_k)_{k\in\mathbb{N}}$ is equi-Lipschitz continuous and $h_k \to 0$ as $k \to \infty$ uniformly on all compact subsets of $\mathbb{R}^{n-1}$. Consider the domains*

$$\Omega := \{ (x',x_n) \in \mathbb{R}^n; x_n > 0\,, x' \in \mathbb{R}^{n-1} \}\,,$$
$$\Omega_k := \{ (x',x_n) \in \mathbb{R}^n; x_n > -h_k(x')\,, x' \in \mathbb{R}^{n-1} \}\,, \quad k \in \mathbb{N}.$$

Let $(u_k)_{k\in\mathbb{N}}$ be a bounded sequence in $L^2(0,T;H_0^1(\Omega_k)^n)$, let $u \in L^2(0,T;H^1(\Omega)^n)$ such that

$$u_k \underset{k\to\infty}{\rightharpoonup} u \quad \text{in } L^2(0,T;H^1(\Omega)^n).$$

Then $u(t) \in H_0^1(\Omega)^n$ for a.a. $t \in]0,T[$.

Proof. The proof is divided into two steps. In the first step the result will be proved for the stationary case and in the second step the instationary case will be reduced to the stationary case.

Step 1. The goal of this step is the prove of the following assertion.

Assertion. *Let* $(v_k)_{k \in \mathbb{N}}$ *be a bounded sequence in* $H_0^1(\Omega_k)^n$ *and let* $v \in H^1(\Omega)^n$ *such that* $v_k \rightharpoonup v$ *as* $k \to \infty$ *in* $H^1(\Omega)^n$. *Then* $v \in H_0^1(\Omega)^n$.

To prove the assertion consider an open cube $W \subseteq \mathbb{R}^{n-1}$ and define the bounded Lipschitz domain

$$G := \{ (x', x_n) \in \mathbb{R}^n; \, 0 < x_n < 1 \,, x' \in W \,\}.$$

From $v_k \rightharpoonup v$ as $k \to \infty$ in $H^1(G)^n$ it follows with Theorem 2.6 that $v_k \to v$ as $k \to \infty$ strongly in $L^2(\partial G)^n$ holds. Therefore

$$\lim_{k \to \infty} \int_W |v(x', 0) - v_k(x', 0)| \, dx' = 0. \tag{8.5}$$

Moreover, by (7.70)

$$\lim_{k \to \infty} \int_W |v_k(x', -h_k(x')) - v_k(x', 0)| \, dx' = 0. \tag{8.6}$$

We use (8.5), (8.6) and $v_k = 0$ on $\partial\Omega_k$ to obtain

$$\int_W |v(x', 0)| \, dx'$$
$$\leq \underbrace{\int_W |v(x', 0) - v_k(x', 0)| \, dx' + \int_W |v_k(x', 0) - v_k(x', -h_k(x))| \, dx'}_{\to 0 \text{ as } k \to \infty}. \tag{8.7}$$

Thus, from (8.7) it follows

$$\int_{\mathbb{R}^{n-1}} |v(x', 0)| \, dx' = 0$$

and consequently $v|_{\partial\Omega} = 0$. By (2.16) we see that the assertion is fulfilled. $\qquad \square$

Step 2. In this step we consider the general case: Let u, $(u_k)_{k \in \mathbb{N}}$ be as in Lemma 8.2. Define $\tilde{u}(t) := 1_{[0,T[}(t) u(t) \,, t \in \mathbb{R}$, for every $k \in \mathbb{N}$ define $\widetilde{u}_k(t) := 1_{[0,T[}(t) u_k(t) \,, t \in \mathbb{R}$. For $\delta > 0$ let $u^\delta \in C^\infty(\mathbb{R}; H^1(\Omega)^n)$ and $u_k^\delta \in C^\infty(\mathbb{R}; H^1(\Omega_k)^n)$ be defined as in (7.81).

Fix $\delta > 0$ with $\delta < T - \delta$. From Lemma 7.16 and Lemma 7.17 (ii) it follows that for all $t \in [\delta, T - \delta[$ the sequence $(u_k^\delta(t))_{k \in \mathbb{N}}$ is bounded in $H_0^1(\Omega_k)^n$ and there holds $u_k^\delta(t) \rightharpoonup u^\delta(t)$ as $k \to \infty$ in $H^1(\Omega)^n$. Therefore step 1 implies $u^\delta(t) \in H_0^1(\Omega)^n$ for a.a. $t \in [\delta, T - \delta]$. By (7.82) and the Fischer-Riesz Theorem there exists a sequence $(\epsilon_k)_{k \in \mathbb{N}}$ with $\epsilon_k \to 0$ as $k \to \infty$ such that $u^{\epsilon_k}(t) \to u(t)$ as $k \to \infty$ strongly in $H^1(\Omega)^n$ for a.a. $t \in [\delta, T - \delta]$. Thus $u(t) \in H_0^1(\Omega)^n$ for a.a. $t \in]0, T[$. $\qquad \square$

The first main result of this chapter reads as follows.

Theorem 8.3. *Consider the following data:*

- *Let $n \in \{2,3\}$, let $\Omega := \{ (x', x_n) \in \mathbb{R}^n;\, x_n > 0\,, x' \in \mathbb{R}^{n-1} \}$, let $0 < T < \infty$*

- *Let $\Lambda \in L^\infty(\partial\Omega)$ with $\Lambda(x) \geq 1$ for a.a. $x \in \partial\Omega$*

- *Let $g \in L^\infty(]0,T[\times\mathbb{R}^n)^n$, $h_0 \in L^2(0,T; H^1(\mathbb{R}^n))$*

- *Let $f_1 \in L^2(0,T; L^2(\mathbb{R}^n)^n)$, $f_2 \in L^2(0,T; L^2(\mathbb{R}^n))$*

- *Let $u_0 \in L^2(\mathbb{R}^n)^n$, let $\theta_0 \in L^2(\mathbb{R}^n)$*

- *Let $\alpha, \beta, \nu, \kappa$ be positive, real constants*

- *In the case $n = 3$ we need the additional assumption $g \in L^4(0,T; L^2(\Omega)^3)$*

Then there exists a sequence $(h_k)_{k\in\mathbb{N}}$ of non-negative, equi-Lipschitz continuous functions $h_k : \mathbb{R}^{n-1} \to \mathbb{R}_0^+$, $k \in \mathbb{N}$, with $h_k \to 0$ as $k \to \infty$ uniformly on $\mathbb{R}^{n-1}$ and a sequence of domains $(\Omega_k)_{k\in\mathbb{N}}$ defined by

$$\Omega_k := \{ (x', x_n) \in \mathbb{R}^n; x_n > -h_k(x')\,, x' \in \mathbb{R}^{n-1} \}, \quad k \in \mathbb{N}, \tag{8.8}$$

such that the following two statements are fulfilled:

(i) *For every $k \in \mathbb{N}$ there exists a weak solution (u_k, θ_k) of the Boussinesq equations (7.1) in $[0,T[\times\Omega_k$ (with data $f_1, f_2, g, h_0|_{]0,T[\times\Omega_k}$ and $P_{\Omega_k}u_0, \theta_0|_{\Omega_k}$) satisfying the energy inequalities (7.146) and the boundary conditions*

$$u_k = 0 \quad and \quad \frac{\partial\theta_k}{\partial N} + \alpha(\theta_k - h_0) = 0 \quad on\ ]0,T[\times\partial\Omega_k. \tag{8.9}$$

There holds

$$\begin{aligned} &(u_k)_{k\in\mathbb{N}}\ is\ bounded\ in\ L^\infty(0,T; L^2_\sigma(\Omega_k)) \cap L^2(0,T; W^{1,2}_{0,\sigma}(\Omega_k)),\\ &(\theta_k)_{k\in\mathbb{N}}\ is\ bounded\ in\ L^\infty(0,T; L^2(\Omega_k)) \cap L^2(0,T; H^1(\Omega_k)). \end{aligned} \tag{8.10}$$

(ii) *There exists a weak solution (u,θ) of the Boussinesq equations (7.1) in $[0,T[\times\Omega$ (with data $f_1, f_2, g, h_0|_{]0,T[\times\Omega}$ and $P_\Omega u_0, \theta_0|_\Omega$) satisfying the boundary conditions*

$$u = 0 \quad and \quad \frac{\partial\theta}{\partial N} + \alpha\Lambda(\theta - h_0) = 0 \quad on\ ]0,T[\times\partial\Omega \tag{8.11}$$

such that

$$\begin{aligned} u_k &\overset{*}{\underset{k\to\infty}{\rightharpoonup}} u \quad in\ L^\infty(0,T; L^2(\Omega)^n), \quad u_k \underset{k\to\infty}{\rightharpoonup} u \quad in\ L^2(0,T; H^1(\Omega)^n),\\ \theta_k &\overset{*}{\underset{k\to\infty}{\rightharpoonup}} \theta \quad in\ L^\infty(0,T; L^2(\Omega)), \quad \theta_k \underset{k\to\infty}{\rightharpoonup} \theta \quad in\ L^2(0,T; H^1(\Omega)). \end{aligned} \tag{8.12}$$

Proof. Define

$$\tilde{\Lambda} : \mathbb{R}^{n-1} \to \mathbb{R}, \quad \tilde{\Lambda}(x') := \Lambda(x', 0), \quad \text{for a.a. } x' \in \mathbb{R}^{n-1}. \tag{8.13}$$

By Theorem 8.1 there exists a sequence $(h_k)_{k \in \mathbb{N}}$ of non-negative, equi-Lipschitz continuous functions $h_k : \mathbb{R}^{n-1} \to \mathbb{R}_0^+$, $k \in \mathbb{N}$, with $h_k \to 0$ as $k \to \infty$ uniformly on $\mathbb{R}^{n-1}$, and there exists a Young measure $\nu = (\nu_{x'})_{x' \in \mathbb{R}^{n-1}} \in \mathcal{Y}(\mathbb{R}^{n-1}; \mathcal{P}(\mathbb{R}))$ such that the following properties are satisfied:

(1) $(\nabla h_k)_{k \in \mathbb{N}}$ generates ν.

(2) We have

$$\tilde{\Lambda}(x') = \int_{\mathbb{R}^{n-1}} \sqrt{1 + \lambda^2} \, d\nu_{x'}(\lambda) \tag{8.14}$$

for a.a. $x' \in \mathbb{R}^{n-1}$.

Define

$$\Omega_k := \{ (x', x_n) \in \mathbb{R}^n; x_n > -h_k(x'); x' \in \mathbb{R}^{n-1} \}, \quad k \in \mathbb{N}. \tag{8.15}$$

Making use of Theorem 4.18 if $n = 3$ and of Theorem 4.23 in the case $n = 2$ we obtain for every $k \in \mathbb{N}$ a weak solution (u_k, θ_k) of the Boussinesq equations (7.1) in $[0, T[\times \Omega_k$ (with data $f_1, f_2, g, h_0|_{]0,T[\times \Omega_k}$ and $P_{\Omega_k} u_0, \theta_0|_{\Omega_k}$) satisfying the energy inequalities (7.146) and the following boundary conditions: There holds $u_k = 0$ on $]0, T[\times \partial \Omega_k$ and

$$\frac{\partial \theta_k}{\partial N} + \alpha(\theta_k - h_0) = 0 \quad \text{on }]0, T[\times \partial \Omega_k.$$

Exactly as in the proof of Theorem 7.30 (i) we can show that

$$\begin{aligned}
&(u_k)_{k \in \mathbb{N}} \text{ is bounded in } L^\infty(0, T; L_\sigma^2(\Omega_k)) \cap L^2(0, T; H^1(\Omega_k)^n), \\
&(\theta_k)_{k \in \mathbb{N}} \text{ is bounded in } L^\infty(0, T; L^2(\Omega_k)) \cap L^2(0, T; H^1(\Omega_k))
\end{aligned} \tag{8.16}$$

holds.

From (8.16) we get the existence of

$$\begin{aligned}
&u \in L^\infty(0, T; L^2(\Omega)^n) \cap L^2(0, T; H^1(\Omega)^n), \\
&\theta \in L^\infty(0, T; L^2(\Omega)) \cap L^2(0, T; H^1(\Omega))
\end{aligned}$$

and of a subsequence $(u_{m_k}, \theta_{m_k})_{k \in \mathbb{N}}$ such that

$$\begin{aligned}
&u_{m_k} \underset{k \to \infty}{\overset{*}{\rightharpoonup}} u \quad \text{in } L^\infty(0, T; L^2(\Omega)^n), \quad u_{m_k} \underset{k \to \infty}{\rightharpoonup} u \quad \text{in } L^2(0, T; H^1(\Omega)^n), \\
&\theta_{m_k} \underset{k \to \infty}{\overset{*}{\rightharpoonup}} \theta \quad \text{in } L^\infty(0, T; L^2(\Omega)), \quad \theta_{m_k} \underset{k \to \infty}{\rightharpoonup} \theta \quad \text{in } L^2(0, T; H^1(\Omega))
\end{aligned} \tag{8.17}$$

are satisfied. With no loss of generality we may assume $m_k = k$, $k \in \mathbb{N}$. To complete the proof we have to show that (u, θ) is a weak solution of the Boussinesq equations (7.1) in $[0, T[\times \Omega$ satisfying the boundary conditions (8.11). By construction it follows that $(u_k, \theta_k)_{k \in \mathbb{N}}$ and (u, θ) satisfy (7.125), (7.126), (7.127). Consequently, Theorem 7.28 yields $u(t) \in L^2_\sigma(\Omega)$ for a.a. $t \in]0, T[$ and

$$
-\int_0^T \langle u, w_t \rangle_\Omega \, dt + \nu \int_0^T \langle \nabla u, \nabla w \rangle_\Omega \, dt + \int_0^T \langle u \cdot \nabla u, w \rangle_\Omega \, dt
$$
$$
= \beta \int_0^T \langle \theta g, w \rangle_\Omega \, dt + \int_0^T \langle f_1, w \rangle_\Omega \, dt + \langle u_0, w(0) \rangle_\Omega
$$

(8.18)

for all $w \in C_0^\infty([0, T[; C_{0,\sigma}^\infty(\Omega))$. Furthermore we have

$$
-\int_0^T \langle \theta, \phi_t \rangle_\Omega \, dt + \kappa \int_0^T \langle \nabla \theta, \nabla \phi \rangle_\Omega \, dt + \int_0^T \langle u \cdot \nabla \theta, \phi \rangle_\Omega \, dt
$$
$$
= -\alpha \kappa \int_0^T \langle \Pi(\theta - h_0), \phi \rangle_{\partial\Omega} \, dt + \int_0^T \langle f_2, \phi \rangle_\Omega \, dt + \langle \theta_0, \phi(0) \rangle_\Omega
$$

(8.19)

for all $\phi \in C_0^\infty([0, T[\times \overline{\Omega})$ where the function $\Pi : \partial\Omega \to \mathbb{R}$ is defined by

$$
\Pi(x', 0) := \int_{\mathbb{R}^{n-1}} \sqrt{1 + \lambda^2} \, d\nu_{x'}(\lambda) \quad \text{for a.a. } x' \in \mathbb{R}^{n-1}.
$$

(8.20)

Comparing (8.13), (8.14), (8.20) we see that $\Lambda(x) = \Pi(x)$ for a.a. $x \in \partial\Omega$. Finally, Lemma 8.2 yields $u(t) \in H_0^1(\Omega)^n$ for a.a. $t \in]0, T[$. Looking at (2.15) we get that $u(t) \in W_{0,\sigma}^{1,2}(\Omega)$ for a.a. $t \in]0, T[$.

Altogether, (u, θ) is a weak solution of the Boussinesq equations (7.1) in $[0, T[\times \Omega$ (with data $f_1, f_2, g, h_0|_{]0,T[\times\Omega}$ and $P_\Omega u_0, \theta_0|_\Omega$) such that u satisfies the no slip condition $u = 0$ on $]0, T[\times \partial\Omega$ and θ satisfies the Robin boundary condition

$$
\frac{\partial \theta}{\partial N} + \alpha \Lambda(\theta - h_0) = 0 \quad \text{on }]0, T[\times \partial\Omega.
$$

$\square$

In the two-dimensional case, due to the uniqueness of weak solutions of the Boussinesq equations (7.1) with boundary conditions (8.1), we obtain the following 'stronger version' of Theorem 8.3.

Theorem 8.4. *Consider the following data:*

- *Let $\Omega := \{ (x', x_2) \in \mathbb{R}^2; \, x_2 > 0, x' \in \mathbb{R} \}$, let $0 < T < \infty$*

- *Let $\Lambda \in L^\infty(\partial\Omega)$ with $\Lambda(x) \geq 1$ for a.a. $x \in \partial\Omega$*

- *Let $g \in L^\infty(]0, T[\times \mathbb{R}^2)^2$, $h_0 \in L^2(0, T; H^1(\mathbb{R}^2))$*

- *Let $f_1 \in L^2(0, T; L^2(\mathbb{R}^2)^2)$, $f_2 \in L^2(0, T; L^2(\mathbb{R}^2))$*

- *Let $u_0 \in L^2(\mathbb{R}^2)^2$, let $\theta_0 \in L^2(\mathbb{R}^2)$*

- *Let $\alpha, \beta, \nu, \kappa$ be positive, real constants*

Let (u, θ) be the unique weak solution of the Boussinesq equations (7.1) in $[0, T[\times \Omega$ (with data $f_1, f_2, g, h_0|_{]0,T[\times\Omega}$ and $P_\Omega u_0, \theta_0|_\Omega$) satisfying the boundary conditions

$$u = 0 \quad \text{and} \quad \frac{\partial \theta}{\partial N} + \alpha \Lambda(\theta - h_0) = 0 \quad \text{on }]0, T[\times \partial\Omega. \qquad (8.21)$$

Then there exists a sequence $(h_k)_{k\in\mathbb{N}}$ of non-negative, equi-Lipschitz continuous functions $h_k : \mathbb{R} \to \mathbb{R}_0^+$, $k \in \mathbb{N}$, with $h_k \to 0$ uniformly on $\mathbb{R}$ as $k \to \infty$ and a sequence of domains $(\Omega_k)_{k\in\mathbb{N}}$ defined by

$$\Omega_k := \left\{ (x', x_2) \in \mathbb{R}^2; x_2 > -h_k(x'), x' \in \mathbb{R} \right\}, \quad k \in \mathbb{N}, \qquad (8.22)$$

such that the properties stated below are fulfilled:

For every $k \in \mathbb{N}$ we denote by (u_k, θ_k) the unique weak solution of the Boussinesq equations (7.1) in $[0, T[\times\Omega_k$ (with data $f_1, f_2, g, h_0|_{]0,T[\times\Omega_k}$ and $P_{\Omega_k} u_0, \theta_0|_{\Omega_k}$) satisfying the boundary conditions

$$u_k = 0 \quad \text{and} \quad \frac{\partial \theta_k}{\partial N} + \alpha(\theta_k - h_0) = 0 \quad \text{on }]0, T[\times \partial\Omega_k. \qquad (8.23)$$

Then there holds

$$\begin{aligned}
&(u_k)_{k\in\mathbb{N}} \text{ is bounded in } L^\infty(0, T; L^2_\sigma(\Omega_k)) \cap L^2(0, T; W^{1,2}_{0,\sigma}(\Omega_k)), \\
&(\theta_k)_{k\in\mathbb{N}} \text{ is bounded in } L^\infty(0, T; L^2(\Omega_k)) \cap L^2(0, T; H^1(\Omega_k)),
\end{aligned} \qquad (8.24)$$

and

$$\begin{aligned}
u_k \underset{k\to\infty}{\overset{*}{\rightharpoonup}} u \quad &\text{in } L^\infty(0, T; L^2(\Omega)^2), \quad u_k \underset{k\to\infty}{\rightharpoonup} u \quad \text{in } L^2(0, T; H^1(\Omega)^2), \\
\theta_k \underset{k\to\infty}{\overset{*}{\rightharpoonup}} \theta \quad &\text{in } L^\infty(0, T; L^2(\Omega)), \quad \theta_k \underset{k\to\infty}{\rightharpoonup} \theta \quad \text{in } L^2(0, T; H^1(\Omega)).
\end{aligned} \qquad (8.25)$$

Proof. By Theorem 4.23 we have that (u, θ) considered above is indeed the unique weak solution (u, θ) of the Boussinesq equations (7.1) in $[0, T[\times\Omega$ satisfying the boundary conditions (8.21). Choose $(h_k)_{k\in\mathbb{N}}$ and let $(\Omega_k)_{k\in\mathbb{N}}$ be as in Theorem 8.3. Let $(u_k, \theta_k)_{k\in\mathbb{N}}$ be the unique weak solutions of the Boussinesq equations (7.1) in $[0, T[\times\Omega_k$ (with data $f_1, f_2, g, h_0|_{]0,T[\times\Omega_k}$ and $P_{\Omega_k} u_0, \theta_0|_{\Omega_k}$) satisfying the boundary conditions

$$u_k = 0 \quad \text{and} \quad \frac{\partial \theta_k}{\partial N} + \alpha(\theta_k - h_0) = 0 \quad \text{on }]0, T[\times \partial\Omega_k. \qquad (8.26)$$

By (8.10) we see that (8.24) is fulfilled. By Theorem 8.3 (ii) there exists a weak solution $(\bar{u}, \bar{\theta})$ of (7.1) in $[0, T[\times\Omega$ (with data $f_1, f_2, g, h_0|_{]0,T[\times\Omega}$ and $P_\Omega u_0, \theta_0|_\Omega$) satisfying the boundary conditions

$$\bar{u} = 0 \quad \text{and} \quad \frac{\partial \bar{\theta}}{\partial N} + \alpha \Lambda(\bar{\theta} - h_0) = 0 \quad \text{on }]0, T[\times \partial\Omega \qquad (8.27)$$

such that

$$
\begin{aligned}
u_k \underset{k\to\infty}{\overset{*}{\rightharpoonup}} \overline{u} \quad & \text{in } L^\infty(0,T;L^2(\Omega)^n), \quad u_k \underset{k\to\infty}{\rightharpoonup} \overline{u} \quad \text{in } L^2(0,T;H^1(\Omega)^n), \\
\theta_k \underset{k\to\infty}{\overset{*}{\rightharpoonup}} \overline{\theta} \quad & \text{in } L^\infty(0,T;L^2(\Omega)), \quad \theta_k \underset{k\to\infty}{\rightharpoonup} \overline{\theta} \quad \text{in } L^2(0,T;H^1(\Omega)).
\end{aligned}
\tag{8.28}
$$

Since (u,θ) is a weak solution of the Boussinesq equations (7.1) satisfying the same boundary conditions as $(\overline{u}, \overline{\theta})$ on $]0,T[\times\partial\Omega$ we obtain from Theorem 4.23 (i) that $u(t) = \overline{u}(t)$ and $\theta(t) = \overline{\theta}(t)$ are satisfied for almost all $t \in]0,T[$. Therefore we get from (8.28) that (8.25) is satisfied. $\qquad\square$

9 Optimization of the heat energy transferred to the exterior

As already outlined in the introduction, this chapter deals with optimizing the heat transfer. Throughout this chapter we supplement the Boussinesq equations

$$
\begin{aligned}
u_t - \nu \Delta u + (u \cdot \nabla)u + \frac{1}{\rho_0}\nabla p &= \beta \theta g + f_1 \quad &\text{in }]0, T[\times\Omega\,, \\
\operatorname{div} u &= 0 \quad &\text{in }]0, T[\times\Omega\,, \\
\theta_t - \kappa \Delta \theta + (u \cdot \nabla)\theta &= f_2 \quad &\text{in }]0, T[\times\Omega\,, \\
u(0) &= u_0 \quad &\text{in } \Omega\,, \\
\theta(0) &= \theta_0 \quad &\text{in } \Omega
\end{aligned}
\tag{9.1}
$$

with the no slip boundary condition

$$
u = 0 \quad \text{on }]0, T[\times\partial\Omega
\tag{9.2}
$$

and the Robin boundary condition

$$
\frac{\partial\theta}{\partial N} + \Lambda(\theta - h_0) = 0 \quad \text{on }]0, T[\times\partial\Omega.
\tag{9.3}
$$

In the following section we formulate the optimization problems which will be our topic in the sequel.

9.1 Mathematical formulation of the optimization problems

We begin with the following

Definition 9.1. Given the following data:

- Let $n \in \{2, 3\}$, let $\Omega \subseteq \mathbb{R}^n$ be a domain satisfying the uniform Lipschitz condition, let $0 < T < \infty$, and let $0 < L < \infty$

- Let $\Gamma \in \mathcal{L}(\partial\Omega)$ be such such that there exists a bounded Lipschitz domain $D \subseteq \mathbb{R}^n$ with $D \subseteq \Omega$ and $\Gamma \subseteq \partial D$

- Let $g \in L^\infty(]0, T[\times\Omega)^n$, $h_0 \in L^2(0, T; L^2(\partial\Omega))$
- Let $f_1 \in L^2(0, T; L^2(\Omega)^n)$, $f_2 \in L^2(0, T; L^2(\Omega))$
- Let $u_0 \in L^2_\sigma(\Omega)$, $\theta_0 \in L^2(\Omega)$
- Let β, ν, κ be positive, real constants

Consider data as in Definition 9.1. Define the sets

$$U_{\mathrm{ad}} := \{ \Lambda \in L^\infty(\partial\Omega); 0 \leq \Lambda(x) \leq L \text{ for a.a. } x \in \partial\Omega \}, \tag{9.4}$$

$$F_{\mathrm{ad}} := \{ (u, \theta, \Lambda); \Lambda \in U_{\mathrm{ad}}; (u, \theta, \Lambda) \text{ is a weak solution of (9.1) with (9.2), (9.3)}$$
$$\text{and } (u, \theta, \Lambda) \text{ satisfies the energy inequalities (4.23), (4.24)} \}. \tag{9.5}$$

Motivated by (1.16) we define for $(u, \theta, \Lambda) \in F_{\mathrm{ad}}$ the total heat energy transferred through Γ to the exterior by

$$E(u, \theta, \Lambda) := \int_0^T \int_\Gamma \Lambda(\theta - h_0)\, dS\, dt. \tag{9.6}$$

The topic of this chapter is to prove existence of solutions of the optimization problems

$$\min E(u, \theta, \Lambda) \quad \text{subject to } (u, \theta, \Lambda) \in F_{\mathrm{ad}}, \tag{9.7}$$

$$\max E(u, \theta, \Lambda) \quad \text{subject to } (u, \theta, \Lambda) \in F_{\mathrm{ad}}. \tag{9.8}$$

To clarify these notations, we call $(\overline{u}, \overline{\theta}, \overline{\Lambda}) \in F_{\mathrm{ad}}$ a solution of the minimization problem (9.7), if $E(\overline{u}, \overline{\theta}, \overline{\Lambda}) \leq E(u, \theta, \Lambda)$ for all $(u, \theta, \Lambda) \in F_{\mathrm{ad}}$. Analogously, we deal with (9.8).

Let us sketch our approach to prove existence of solutions of the minimization problem (9.7). The first step (see Lemma 9.5) is to prove $\inf_{(u,\theta,\Lambda)\in F_{\mathrm{ad}}} E(u, \theta, \Lambda) \in \mathbb{R}$. Consequently, we are able to choose a *minimizing sequence* $(u_k, \theta_k, \Lambda_k)_{k\in\mathbb{N}}$, i.e. let $(u_k, \theta_k, \Lambda_k) \in F_{\mathrm{ad}}$, $k \in \mathbb{N}$, with

$$\lim_{k\to\infty} E(u_k, \theta_k, \Lambda_k) = \inf_{(u,\theta,\Lambda)\in F_{\mathrm{ad}}} E(u, \theta, \Lambda).$$

The crucial point is to prove the existence of a subsequence $(u_{m_k}, \theta_{m_k}, \Lambda_{m_k})_{k\in\mathbb{N}}$ and of $(\overline{u}, \overline{\theta}, \overline{\Lambda}) \in F_{\mathrm{ad}}$ such that

$$\lim_{k\to\infty} E(u_k, \theta_k, \Lambda_k) = E(\overline{u}, \overline{\theta}, \overline{\Lambda}). \tag{9.9}$$

Making use of Lemma 9.4 we obtain the existence of a weak solution $(\overline{u}, \overline{\theta}, \overline{\Lambda})$ of (9.1) satisfying the boundary conditions (9.2), (9.3) such that (9.13)-(9.17) are fulfilled. In the case $n = 2$ (with no additional assumptions on Ω) or in the case that $\Omega \subseteq \mathbb{R}^3$ is a bounded Lipschitz domain it can be proven that $(\overline{u}, \overline{\theta}, \overline{\Lambda}) \in F_{\mathrm{ad}}$ is satisfied. Making use of the strong convergence (9.17) we finally obtain that (9.9) is fulfilled.

A reformulation in the two-dimensional case

To the end of this section we assume $n = 2$. We know from Theorem 4.23 that weak solutions of the Boussinesq equations (9.1) with boundary conditions (9.2), (9.3) are uniquely determined. Furthermore, every weak solution (u, θ) with these boundary conditions fulfils the energy equalities (4.131), (4.132). Consequently

$$F_{\text{ad}} = \big\{ (u, \theta, \Lambda); \Lambda \in U_{\text{ad}}; (u, \theta, \Lambda) \text{ is a weak solution of (9.1) with (9.2), (9.3)} \big\}. \tag{9.10}$$

Making use of the uniqueness of weak solutions we can reformulate the optimization problems (9.7), (9.8) in the two-dimensional case. Consider $\Lambda \in U_{\text{ad}}$. By Theorem 4.23 there exists exactly one weak solution (u, θ) of (9.1) satisfying the boundary conditions (9.2), (9.3). There holds $(u, \theta) \in W^2(0, T; W^{1,2}_{0,\sigma}(\Omega), L^2_\sigma(\Omega)) \times W^2(0, T; H^1(\Omega), L^2(\Omega))$ (see (2.36) for a definition of these spaces). Therefore, we can define the solution map

$$R : U_{\text{ad}} \to W^2(0, T; W^{1,2}_{0,\sigma}(\Omega), L^2_\sigma(\Omega)) \times W^2(0, T; H^1(\Omega), L^2(\Omega)), \quad \Lambda \mapsto (u, \theta).$$

We can use the solution map R to rewrite the minimization problem (9.7) in the equivalent form as the *reduced minimization problem*

$$\min \hat{E}(\lambda) := E(R(\Lambda), \Lambda) \quad \text{subject to } \Lambda \in U_{\text{ad}}$$

where E is defined by (9.6). Analogously, the maximization problem (9.8) can be reformulated in the equivalent form

$$\max \hat{E}(\lambda) := E(R(\Lambda), \Lambda) \quad \text{subject to } \Lambda \in U_{\text{ad}}.$$

9.2 Preliminary lemmata

In this section we formulate the results needed for the proof of Theorem 9.6 which is the main result of the present chapter. We start with the lemma below which will be used to obtain the boundedness of a minimizing sequence $(u_k, \theta_k, \Lambda_k)_{k \in \mathbb{N}}$ in suitable function spaces.

Lemma 9.2. *Let $n \in \{2, 3\}$, let $\Omega \subseteq \mathbb{R}^n$ be a domain satisfying the uniform Lipschitz condition, let $0 < T < \infty$, $g \in L^\infty(]0, T[\times \Omega)^n$, let $\Lambda \in L^\infty(\partial\Omega)$ with $\Lambda(x) \geq 0$ for a.a. $x \in \partial\Omega$, and let $h_0 \in L^2(0, T; L^2(\partial\Omega))$. Further assume $f_1 \in L^2(0, T; L^2(\Omega)^n)$, $f_2 \in L^2(0, T; L^2(\Omega))$, assume $u_0 \in L^2_\sigma(\Omega)$, $\theta_0 \in L^2(\Omega)$, and assume that β, ν, κ are positive, real constants. Let (u, θ) be a weak solution of (9.1) with (9.2), (9.3). Then the estimate*

$$\|u_t\|_{L^{4/3}(0,T;W^{1,2}_{0,\sigma}(\Omega)')} \leq \|f_1\|_{2,\frac{4}{3};\Omega;T} + \nu \|\nabla u\|_{2,\frac{4}{3};\Omega;T}$$
$$+ c\|u\|^{1/2}_{2,\infty;\Omega;T} \|u\|^{3/2}_{L^2(0,T;H^1(\Omega)^n)} + \beta\|g\|_{L^\infty(]0,T[\times\Omega)^n} \|\theta\|_{2,\frac{4}{3};\Omega;T} \tag{9.11}$$

is satisfied with a constant $c = c(n) > 0$.

Proof. By Theorem 4.16 there holds

$$u_t = f_1 + \nu \Delta u - B(u,u) + \beta \theta g \quad \text{in } L^{4/3}(0,T;W_{0,\sigma}^{1,2}(\Omega)').$$

Consequently, by Lemma 4.15 we get

$$\|u_t\|_{L^{4/3}(0,T;W_{0,\sigma}^{1,2}(\Omega)')} \leq \|f_1\|_{2,\frac{4}{3};\Omega;T} + \nu \|\nabla u\|_{2,\frac{4}{3};\Omega;T}$$
$$+ c\|B(u,u)\|_{L^{4/3}(0,T;W_{0,\sigma}^{1,2}(\Omega)')} + \beta \|g\|_{L^{\infty}(]0,T[\times\Omega)^n} \|\theta\|_{2,\frac{4}{3};\Omega;T}$$

Using Lemma 4.14 we obtain

$$\|B(u,u)\|_{L^{4/3}(0,T;W_{0,\sigma}^{1,2}(\Omega)')} \leq c(n)\|u\|_{2,\infty;\Omega;T}^{1/2} \|u\|_{L^{2}(0,T;H^{1}(\Omega)^n)}^{3/2}.$$

Thus (9.11) holds. $\qquad \square$

The next Lemma states that the set U_{ad} defined in (9.4) is *weak* sequentially closed* in $L^{\infty}(\partial\Omega)$. By definition, a set $V \subseteq L^{\infty}(\partial\Omega)$ is called weak* sequentially closed, if for all sequences $(\Lambda_k)_{k\in\mathbb{N}}$ and $\Lambda \in L^{\infty}(\partial\Omega)$ such that $\Lambda_k \in V, k \in \mathbb{N}$, and $\Lambda_k \rightharpoonup^* \Lambda$ as $k \to \infty$ in $L^{\infty}(\partial\Omega)$ we have that $\Lambda \in V$ is satisfied.

Lemma 9.3. *Let $n \in \mathbb{N}, n \geq 2$, let $\Omega \subseteq \mathbb{R}^n$ be a domain satisfying the uniform Lipschitz condition, let $M_1, M_2 \in \mathbb{R}$ with $M_1 \leq M_2$. Define the set*

$$V := \{\Lambda \in L^{\infty}(\partial\Omega); M_1 \leq \Lambda(x) \leq M_2 \text{ for a.a. } x \in \partial\Omega\}.$$

Then V is weak* sequentially closed in $L^{\infty}(\partial\Omega)$.

Proof. Let $(U_j)_{j\in M}$ be a locally finite open cover of $\partial\Omega$ as in Definition 4.2. Fix $j \in M$, consider

$$\widetilde{\Gamma} := U_j \cap \partial\Omega = \{(x', x_n) \in \mathbb{R}^n; x_n = h(x'); \|x'\|_{\infty} < \alpha\} \tag{9.12}$$

where $h : \mathbb{R}^{n-1} \to \mathbb{R}$ is a Lipschitz continuous function and $\alpha \in]0,\infty[$. (To simplify the notation we assume $\phi_j = \mathrm{id}$.) We denote by $L^2(\widetilde{\Gamma})$ the L^2-space with respect to the measure space $(\widetilde{\Gamma}, \mathcal{L}^{n-1}(\widetilde{\Gamma}), \mathcal{H}^{n-1})$. We refer to Section 4.1 for a definition of these spaces. Define

$$\widetilde{V} := \{f \in L^2(\widetilde{\Gamma}); M_1 \leq f(x) \leq M_2 \quad \text{for a.a. } x \in \widetilde{\Gamma}\}.$$

It is easily seen that $\widetilde{V}$ is convex and that $\widetilde{V}$ is (strongly) closed in $L^2(\widetilde{\Gamma})$. Therefore, by a well known functional analytic result it follows that $\widetilde{V}$ is weak sequentially closed in $L^2(\widetilde{\Gamma})$.

To prove that V is weak* sequentially closed, let $\Lambda_k \in V$, $k \in \mathbb{N}$, and $\Lambda \in L^{\infty}(\partial\Omega)$ such that $\Lambda_k \rightharpoonup^* \Lambda$ in $L^{\infty}(\partial\Omega)$. Since $\mathcal{H}^{n-1}(\widetilde{\Gamma}) < \infty$ it follows $\Lambda_k \in \widetilde{V}, k \in \mathbb{N}$ and $\Lambda_k \rightharpoonup \Lambda$ as $k \to \infty$ in $L^2(\widetilde{\Gamma})$. Thus $\Lambda \in \widetilde{V}$ and consequently $M_1 \leq \Lambda(x) \leq M_2$ for a.a. $x \in \widetilde{\Gamma}$. Altogether $M_1 \leq \Lambda(x) \leq M_2$ for a.a. $x \in \partial\Omega$. Hence $\Lambda \in V$. $\qquad \square$

The following lemma will be the main tool to prove existence of solutions to the optimization problems (9.7), (9.8).

Lemma 9.4. *Consider data as in Definition 9.1. Let $(u_k, \theta_k, \Lambda_k)_{k \in \mathbb{N}} \subseteq F_{ad}$ be any sequence.*

(i) *(**General case.**) Then there exists a subsequence $(u_{m_k}, \theta_{m_k}, \Lambda_{m_k})_{k \in \mathbb{N}}$ and a weak solution (u, θ, Λ) of (9.1) satisfying the boundary conditions (9.2), (9.3) such that*

$$\Lambda_{m_k} \underset{k \to \infty}{\overset{*}{\rightharpoonup}} \Lambda \quad in \ L^\infty(\partial\Omega), \tag{9.13}$$

$$u_{m_k} \underset{k \to \infty}{\overset{*}{\rightharpoonup}} u \quad in \ L^\infty(0, T; L^2_\sigma(\Omega)), \quad u_{m_k} \underset{k \to \infty}{\rightharpoonup} u \quad in \ L^2(0, T; W^{1,2}_{0,\sigma}(\Omega)), \tag{9.14}$$

$$\theta_{m_k} \underset{k \to \infty}{\overset{*}{\rightharpoonup}} \theta \quad in \ L^\infty(0, T; L^2(\Omega)), \quad \theta_{m_k} \underset{k \to \infty}{\rightharpoonup} \theta \quad in \ L^2(0, T; H^1(\Omega)). \tag{9.15}$$

Moreover

$$\theta_{m_k} \underset{k \to \infty}{\rightarrow} \theta \quad strongly \ in \ L^2(0, T; L^2(G)), \tag{9.16}$$

$$\theta_{m_k} \underset{k \to \infty}{\rightarrow} \theta \quad strongly \ in \ L^2(0, T; L^2(\partial G)) \tag{9.17}$$

for all bounded Lipschitz domains G with $\overline{G} \subseteq \Omega$.

(ii) *(**Case Ω bounded.**) Assume that Ω is bounded and $n \in \{2, 3\}$. Then (u, θ, Λ) satisfies the energy inequalities (4.23), (4.24) and consequently $(u, \theta, \Lambda) \in F_{ad}$. Moreover*

$$u_{m_k} \underset{k \to \infty}{\rightarrow} u \quad strongly \ in \ L^2(0, T; L^2(\Omega)). \tag{9.18}$$

(iii) *(**Case $n=2$.**) Assume $n = 2$ and $\Omega \subseteq \mathbb{R}^2$ as in Definition 9.1. Then (u, θ, Λ) satisfies the energy inequalities (4.23), (4.24) (in fact they are satisfied as equalities) and consequently $(u, \theta, \Lambda) \in F_{ad}$. Moreover (u, θ) is the unique weak solution of (9.1) with the boundary conditions*

$$u = 0 \quad and \quad \frac{\partial\theta}{\partial N} + \Lambda(\theta - h_0) = 0 \quad on \ ]0, T[\times \partial\Omega. \tag{9.19}$$

Proof. **Proof of (i)** By definition of U_{ad} it follows that

$$(\Lambda_k)_{k \in \mathbb{N}} \ is \ bounded \ in \ L^\infty(\partial\Omega). \tag{9.20}$$

By (9.20), (4.126) we see

$$(\theta_k)_{k \in \mathbb{N}} \ is \ bounded \ in \ L^\infty(0, T; L^2(\Omega)) \cap L^2(0, T; H^1(\Omega)). \tag{9.21}$$

By (9.20), (9.21), (4.127) it follows that

$$(u_k)_{k \in \mathbb{N}} \ is \ bounded \ in \ L^\infty(0, T; L^2_\sigma(\Omega)) \cap L^2(0, T; W^{1,2}_{0,\sigma}(\Omega)). \tag{9.22}$$

Altogether we find a subsequence $(u_{m_k}, \theta_{m_k}, \Lambda_{m_k})_{k \in \mathbb{N}}$ and (u, θ, Λ) such that (9.13), (9.14), (9.15) are satisfied. To simplify notations assume $m_k = k$, $k \in \mathbb{N}$.

By Lemma 9.3 we get $\Lambda \in U_{\mathrm{ad}}$. We see that $(u_k, \theta_k)_{k \in \mathbb{N}}$ satisfy all requirements of Theorem 7.9. Therefore, by this theorem, we get

$$
\begin{aligned}
&-\int_0^T \langle u, w_t \rangle_\Omega \, dt + \nu \int_0^T \langle \nabla u, \nabla w \rangle_\Omega \, dt + \int_0^T \langle u \cdot \nabla u, w \rangle_\Omega \, dt \\
&= \beta \int_0^T \langle \theta g, w \rangle_\Omega \, dt + \int_0^T \langle f_1, w \rangle_\Omega \, dt + \langle u_0, w(0) \rangle_\Omega
\end{aligned}
\tag{9.23}
$$

for all $w \in C_0^\infty([0, T[; C_{0,\sigma}^\infty(\Omega))$. Furthermore

$$
\theta_k \underset{k \to \infty}{\to} \theta \quad \text{strongly in } L^2(0, T; L^2(G)), \tag{9.24}
$$

$$
\theta_k \underset{k \to \infty}{\to} \theta \quad \text{strongly in } L^2(0, T; L^2(\partial G)) \tag{9.25}
$$

for all bounded Lipschitz domains G with $G \subseteq \Omega$.

Since (u_k, θ_k) is a weak solution of (9.1) with (9.2), (9.3) there holds

$$
\begin{aligned}
&-\int_0^T \langle \theta_k, \phi_t \rangle_\Omega \, dt + \kappa \int_0^T \langle \nabla \theta_k, \nabla \phi \rangle_\Omega \, dt + \int_0^T \langle u_k \cdot \nabla \theta_k, \phi \rangle_\Omega \, dt \\
&= -\kappa \int_0^T \langle \Lambda_k(\theta_k - h_0), \phi \rangle_{\partial\Omega} \, dt + \int_0^T \langle f_2, \phi \rangle_\Omega \, dt + \langle \theta_0, \phi(0) \rangle_\Omega
\end{aligned}
\tag{9.26}
$$

for all $k \in \mathbb{N}$ and all $\phi \in C_0^\infty([0, T[\times \overline{\Omega})$. We use (9.13), (9.25) to obtain

$$
\lim_{k \to \infty} \int_0^T \langle \Lambda_k(\theta_k - h_0), \phi \rangle_{\partial\Omega} \, dt = \int_0^T \langle \Lambda(\theta - h_0), \phi \rangle_{\partial\Omega} \, dt \tag{9.27}
$$

for all $\phi \in C_0^\infty([0, T[\times \overline{\Omega})$. We integrate by parts and use (9.14), (9.24) to obtain

$$
\begin{aligned}
\lim_{k \to \infty} \int_0^T \langle u_k \cdot \nabla \theta_k, \phi \rangle_\Omega \, dt &= -\lim_{k \to \infty} \int_0^T \langle \theta_k u_k, \nabla \phi \rangle_\Omega \, dt \\
&= -\int_0^T \langle \theta u, \nabla \phi \rangle_\Omega \, dt \\
&= \int_0^T \langle u \cdot \nabla \theta, \phi \rangle_\Omega \, dt
\end{aligned}
\tag{9.28}
$$

for all $\phi \in C_0^\infty([0, T[\times \overline{\Omega})$. Passing to the limit in (9.26) and employing (9.15), (9.27), (9.28) we get

$$
\begin{aligned}
&-\int_0^T \langle \theta, \phi_t \rangle_\Omega \, dt + \kappa \int_0^T \langle \nabla \theta, \nabla \phi \rangle_\Omega \, dt + \int_0^T \langle u \cdot \nabla \theta, \phi \rangle_\Omega \, dt \\
&= -\kappa \int_0^T \langle \Lambda(\theta - h_0), \phi \rangle_{\partial\Omega} \, dt + \int_0^T \langle f_2, \phi \rangle_\Omega \, dt + \langle \theta_0, \phi(0) \rangle_\Omega
\end{aligned}
\tag{9.29}
$$

for all $\phi \in C_0^\infty([0, T[\times \overline{\Omega})$. Looking at (9.23), (9.29) it follows that (u, θ) is a weak solution of (9.1) with (9.2), (9.3).

Proof of (ii). Assume that Ω is bounded. Combining (9.11), (9.21), (9.22) and making use of $0 < T < \infty$ imply

$$\left(\frac{d}{dt} u_k\right)_{k \in \mathbb{N}} \text{ is bounded in } L^{4/3}(0, T; W_{0,\sigma}^{1,2}(\Omega)'). \tag{9.30}$$

Consider the imbedding scheme

$$W_{0,\sigma}^{1,2}(\Omega) \underset{\text{compact}}{\hookrightarrow} L_\sigma^2(\Omega) \underset{\text{continuous}}{\hookrightarrow} W_{0,\sigma}^{1,2}(\Omega)'. \tag{9.31}$$

Theorem 2.25 applied to (9.30), (9.31) yields (9.18). To show $(\overline{u}, \overline{\theta}, \overline{\Lambda}) \in F_{\text{ad}}$, it remains to prove the validity of the energy inequalities (4.23), (4.24).

From (9.18) we get

$$u_k(t) \underset{k \to \infty}{\to} u(t) \text{ strongly in } L^2(\Omega) \tag{9.32}$$

for almost all $t \in]0, T[$. We get with (9.14), (9.18), (9.32) and the fact that (u_k, θ_k) satisfy the energy inequalities that

$$\begin{aligned}
\frac{1}{2}\|u(t)\|_2^2 + \nu \int_0^t \|\nabla u\|_2^2 \, d\tau &\leq \frac{1}{2} \lim_{k \to \infty} \|u_k(t)\|_2^2 + \nu \liminf_{k \to \infty} \int_0^t \|\nabla u_k\|_2^2 \, d\tau \\
&= \liminf_{k \to \infty} \left(\frac{1}{2}\|u_k(t)\|_2^2 + \nu \int_0^t \|\nabla u_k\|_2^2 \, d\tau \right) \\
&\leq \liminf_{k \to \infty} \left(\frac{1}{2}\|u_0\|_2^2 + \int_0^t \langle f_1, u_k \rangle_\Omega + \beta \int_0^t \langle \theta_k g, u_k \rangle_\Omega \, d\tau \right) \\
&= \frac{1}{2}\|u_0\|_2^2 + \int_0^t \langle f_1, u \rangle_\Omega + \beta \int_0^t \langle \theta g, u \rangle_\Omega \, d\tau
\end{aligned}$$

holds for almost all $t \in [0, T[$. Analogously, we prove that θ satisfies (4.24) for almost all $t \in [0, T[$.

Proof of (iii). Assume $n = 2$ and that Ω is a (bounded or unbounded) domain satisfying the uniform Lipschitz condition. Then we obtain from $\Lambda \in U_{\text{ad}}$ and (9.10) that $(u, \theta, \Lambda) \in F_{\text{ad}}$ is satisfied. The uniqueness of a weak solution of (9.1) with (9.2), (9.3) is also a consequence of this theorem. $\square$

Furthermore, the lemma below is needed.

Lemma 9.5. *Consider data as in Definition 9.1. If $n = 3$ assume additionally $g \in L^4(0, T; L^2(\Omega)^3)$. Let $E : F_{ad} \to \mathbb{R}$ be defined as in (9.6). Then*

$$\inf_{(u,\theta,\Lambda) \in F_{ad}} E(u, \theta, \Lambda) \in \mathbb{R}, \tag{9.33}$$

$$\sup_{(u,\theta,\Lambda) \in F_{ad}} E(u, \theta, \Lambda) \in \mathbb{R}. \tag{9.34}$$

Proof. If $n = 2$ we get from Theorem 4.23 that $F_{\text{ad}} \neq \varnothing$. In the case $n = 3$ we use the additional assumption on g and Theorem 4.18 to conclude that $F_{\text{ad}} \neq \varnothing$.

By (4.126) there is a constant $c > 0$ such that

$$\|\theta\|_{L^2(0,T;H^1(\Omega))} \leq c$$

for all $(u, \theta, \Lambda) \in F_{\text{ad}}$. Due to the assumptions on Γ it follows $|\Gamma| := \mathcal{H}^{n-1}(\Gamma) < \infty$. We obtain with Hölder's inequality and the continuity of the trace operator that

$$
\begin{aligned}
&\left| \int_0^T \int_\Gamma \Lambda(\theta - h_0) \, dS \, dt \right| \\
&\leq \|\Lambda\|_{\infty,\partial\Omega} \sqrt{T|\Gamma|} \, \|\theta - h_0\|_{2,2;\partial\Omega;T} \\
&\leq c\sqrt{T|\Gamma|} \left(\|\theta\|_{L^2(0,T;H^1(\Omega))} + \|h_0\|_{2,2;\partial\Omega;T} \right) \\
&\leq c
\end{aligned}
\tag{9.35}
$$

holds for all $(u, \theta, \Lambda) \in F_{\text{ad}}$ with a constant $c > 0$ independent of (u, θ, Λ). Since $F_{\text{ad}} \neq \varnothing$ we get that (9.33) and (9.34) are fulfilled. □

9.3 Existence of solutions of the optimization problems

Having all ingredients at hand we can prove existence of solutions to the optimization problems (9.7), (9.8) in the two-dimensional case or in the case that $\Omega \subseteq \mathbb{R}^3$ is a bounded Lipschitz domain.

Theorem 9.6. *Consider data as in Definition 9.1 and assume that one of the following additional conditions is satisfied:*

1. *There holds $n = 2$.*

2. *There holds $n = 3$ and $\Omega \subseteq \mathbb{R}^3$ is a bounded Lipschitz domain.*

Then the following statements are satisfied:

(i) *The optimization problem*

$$\min E(u, \theta, \Lambda) \quad \text{subject to } (u, \theta, \Lambda) \in F_{ad}, \tag{9.36}$$

where E is defined by

$$E(u, \theta, \Lambda) := \int_0^T \int_\Gamma \Lambda(\theta - h_0) \, dS \, dt, \quad (u, \theta, \Lambda) \in F_{ad}, \tag{9.37}$$

has a solution $(\overline{u}, \overline{\theta}, \overline{\Lambda}) \in F_{ad}$. If $n = 2$, we have that $(\overline{u}, \overline{\theta})$ is the unique weak solution of (9.1) satisfying the boundary conditions

$$\overline{u} = 0 \quad \text{and} \quad \frac{\partial \overline{\theta}}{\partial N} + \overline{\Lambda}(\overline{\theta} - h_0) = 0 \quad \text{on }]0, T[\times \partial\Omega. \tag{9.38}$$

(ii) The same conclusions as in (i) remain true for the optimization problem

$$\max E(u, \theta, \Lambda) \quad \text{subject to } (u, \theta, \Lambda) \in F_{ad}, \tag{9.39}$$

where E is defined as in (9.37).

Proof. **Proof of (i).** Let us prove that the optimization problem (9.36) has a solution. By (9.33) there holds $\inf_{(u, \theta, \Lambda) \in F_{ad}} E(u, \theta, \Lambda) \in \mathbb{R}$. Consequently, there exists a minimizing sequence $(u_k, \theta_k, \Lambda_k)_{k \in \mathbb{N}}$, i.e. choose $(u_k, \theta_k, \Lambda_k) \in F_{ad}, k \in \mathbb{N}$, such that

$$\lim_{k \to \infty} E(u_k, \theta_k, \Lambda_k) = \inf_{(u, \theta, \Lambda) \in F_{ad}} E(u, \theta, \Lambda). \tag{9.40}$$

By Lemma 9.4 there exists $(\overline{u}, \overline{\theta}, \overline{\Lambda}) \in F_{ad}$ such that (9.13)-(9.17) are satisfied (where (u, θ, Λ) is replaced by $(\overline{u}, \overline{\theta}, \overline{\Lambda})$). Due to the assumptions on Γ we choose a bounded Lipschitz domain $D \subseteq \mathbb{R}^n$ with $D \subseteq \Omega$ and $\Gamma \subseteq \partial D$. It follows with (9.13), (9.17) that

$$\begin{aligned}
\lim_{k \to \infty} E(u_k, \theta_k, \Lambda_k) &= \lim_{k \to \infty} \int_0^T \int_\Gamma \Lambda_k (\theta_k - h_0) \, dS \, dt \\
&= \int_0^T \int_\Gamma \overline{\Lambda} (\overline{\theta} - h_0) \, dS \, dt \\
&= E(\overline{u}, \overline{\theta}, \overline{\Lambda})
\end{aligned} \tag{9.41}$$

is satisfied. Altogether $(\overline{u}, \overline{\theta}, \overline{\Lambda}) \in F_{ad}$ is a solution of (9.36).

Proof of (ii). For the proof of statement (ii) we use (9.34) and choose a sequence $(u_k, \theta_k, \Lambda_k)_{k \in \mathbb{N}}$ with $(u_k, \theta_k, \Lambda_k) \in F_{ad}, k \in \mathbb{N}$, satisfying

$$\lim_{k \to \infty} E(u_k, \theta_k, \Lambda_k) = \sup_{(u, \theta, \Lambda) \in F_{ad}} E(u, \theta, \Lambda). \tag{9.42}$$

Looking at the proof of statement (i) above, it follows that there exists $(\overline{u}, \overline{\theta}, \overline{\Lambda}) \in F_{ad}$ such that (9.13)-(9.17) are fulfilled. Identities (9.41) and (9.42) yield

$$E(\overline{u}, \overline{\theta}, \overline{\Lambda}) = \max_{(u, \theta, \Lambda) \in F_{ad}} E(u, \theta, \Lambda).$$

$\square$

10 Appendix

The goal of this appendix is to prove existence and continuity of the trace operator in domains satisfying the uniform Lipschitz condition. Moreover it will be proved that the norm of this operator depends only on the constants (α, β, N, L) appearing in the definition of the uniform Lipschitz condition. To the best of our knowledge this result is not available in the classical literature. Therefore we present a full and self contained treatment of this topic.

Before we prove this result we need some preparation. Let $n, m \in \mathbb{N}$ be given with $n \leq m$, let $\Omega \subseteq \mathbb{R}^n$ be an open set, let $f : \Omega \to \mathbb{R}^m$ be a Lipschitz continuous function. From Rademacher's Theorem it follows that for almost all $x \in \Omega$ the derivative $Df(x)$ exists as a linear map $Df(x) : \mathbb{R}^n \to \mathbb{R}^m$. We will identify $Df(x)$ in the usual way with the matrix $Df(x) \in \mathbb{R}^{m \times n}$. We define the *Jacobian* of f as

$$J_f(x) := \sqrt{\det\left(Df(x)^T Df(x)\right)} \tag{10.1}$$

for a.a. $x \in \Omega$. We remark that if $n = m$ then $J_f(x) = |\det\left(Df(x)\right)|$ for a.a. $x \in \Omega$. The next theorem is the change of variable formula for Lipschitz continuous functions.

Theorem 10.1. *Let $n \in \mathbb{N}$, let $\Omega \subseteq \mathbb{R}^n$ be an open set, let $f : \Omega \to \mathbb{R}^n$ be an injective Lipschitz continuous function, let $u : f(\Omega) \to \mathbb{R}$ be a measurable function, and let $E \subseteq \Omega$ be a measurable subset. Then the following statements are satisfied:*

(i) The function $u \circ f \cdot J_f : E \to \mathbb{R}$ is (after a redefinition on a null set) well defined and measurable.

(ii) If $u(x) \geq 0$ for a.a. $x \in \Omega$ then

$$\int_E u(f(x)) \, J_f(x) \, dx = \int_{f(E)} u(x) \, dx. \tag{10.2}$$

(iii) The function $(u \circ f) \cdot J_f$ is integrable over E if and only if u is integrable over $f(E)$. In this case (10.2) holds.

Proof. This follows from [Ha93, Theorem 3]. $\square$

The proof of the trace theorem 4.6 is based on the following lemma.

Lemma 10.2. *Let $\alpha, \beta \in]0, \infty[$, and let $h : \mathbb{R}^{n-1} \to \mathbb{R}$ be a Lipschitz continuous function. Define the bounded domains*

$$\Omega_{\alpha,\beta}^+ := \{ (y', y_n) \in \mathbb{R}^n; h(y') < y_n < h(y') + \beta, |y'| < \alpha \},$$
$$W_{\alpha,\beta}^+ := \{ (y', y_n) \in \mathbb{R}^n; 0 < y_n < \beta, |y'| < \alpha \}. \tag{10.3}$$

Let $\psi : W_{\alpha,\beta}^+ \to \Omega_{\alpha,\beta}^+$ be the bijective transformation defined by

$$\psi(y', y_n) := \big(y', y_n + h(y') \big). \tag{10.4}$$

Furthermore define

$$\Gamma := \{ (y', y_n) \in \mathbb{R}^n; y_n = 0, |y'| < \alpha \}. \tag{10.5}$$

(i) For $u \in C^1(\mathbb{R}^n)$ consider the function

$$u \circ \psi : W_{\alpha,\beta}^+ \to \mathbb{R}, \quad (u \circ \psi)(y', y_n) = u(y', y_n + h(y')).$$

Then $u \circ \psi$ is continuous and bounded on the set $W_{\alpha,\beta}^+$ and $u \circ \psi \in H^1(W_{\alpha,\beta}^+)$.

(ii) The estimate

$$\|u \circ \psi\|_{L^2(\Gamma)} \le c \|u \circ \psi\|_{H^1(W_{\alpha,\beta}^+)} \tag{10.6}$$

is satisfied for all $u \in C^1(\mathbb{R}^n)$ with a constant $c = c(\beta) > 0$.

(iii) We have

$$\|u \circ \psi\|_{H^1(W_{\alpha,\beta}^+)} \le c \|u\|_{H^1(\Omega_{\alpha,\beta}^+)} \tag{10.7}$$

for all $u \in C^1(\mathbb{R}^n)$ with a constant $c = c(\mathrm{Lip}(h)) > 0$.

Proof. **Proof of (i).** Applying [Al02, Bemerkung A 6.11] we get $u \circ \psi \in H^1(W_{\alpha,\beta}^+)$. Moreover the following equations are fulfilled for almost all $y = (y', h(y')) \in W_{\alpha,\beta}^+$:

$$\partial_i(u \circ \psi) = (\partial_i u) \circ \psi + (\partial_n u) \circ \psi \cdot \partial_i h, \quad 1 \le i \le n - 1, \tag{10.8}$$
$$\partial_n(u \circ \psi) = (\partial_n u) \circ \psi. \tag{10.9}$$

It is easily seen that the function $u \circ \psi$ is continuous and bounded on $W_{\alpha,\beta}^+$.

Proof of (ii). Define

$$v : W_{\alpha,\beta}^+ \to \mathbb{R}, \quad v(y', y_n) := (u \circ \psi)(y', y_n),$$

and $D := \{y' \in \mathbb{R}^{n-1}; |y'| < \alpha\}$. We already know $v \in H^1(W_{\alpha,\beta}^+)$. Applying Theorem 2.19, we can redefine v on a null set of $W_{\alpha,\beta}^+$, such that the function $v(y', \cdot) :]0, \beta[\to \mathbb{R}$ is locally absolutely continuous for all $y' \in D \setminus N$ where $N \subseteq D$ is a null set. Consequently, by Theorem 2.17, there holds

$$v(y', 0) = v(y', \tau) - \int_0^\tau \partial_n v(y', s) \, ds.$$

for all $0 < \tau < \beta$. Therefore, an elementary estimate and Hölder's inequality imply

$$v^2(y',0) \le 2v^2(y',\tau) + 2\left(\int_0^\beta |\partial_n v(y',s)|\,ds\right)^2 \le 2v^2(y',\tau) + 2\beta \int_0^\beta \left(\partial_n v(y',s)\right)^2 ds \tag{10.10}$$

for almost all $y' \in D$ and all $0 < \tau < \beta$. Integration of (10.10) over $\tau \in [0,\beta]$ yields

$$\beta v^2(y',0) \le 2\int_0^\beta v^2(y',s)\,ds + 2\beta^2 \int_0^\beta \left(\partial_n v(y',s)\right)^2 ds \tag{10.11}$$

for almost all $y' \in D$. We integrate (10.11) over D to obtain

$$\int_D v^2(y',0)\,dy' \le \frac{2}{\beta}\int_D \int_0^\beta v^2(y',s)\,ds\,dy' + 2\beta \int_D \int_0^\beta \left(\partial_n v(y',s)\right)^2 ds\,dy'$$
$$\le c\big(\|v\|^2_{L^2(W^+_{\alpha,\beta})} + \|\partial_n v\|^2_{L^2(W^+_{\alpha,\beta})}\big) \tag{10.12}$$

with a constant $c = c(\beta) > 0$. From (10.12) we obtain (10.6).

Proof of (iii). By Theorem 2.22 the Lipschitz continuous function ψ is differentiable for a.a. $y = (y', y_n) \in W^+_{\alpha,\beta}$ with derivative

$$(D\psi)(y',y_n) = \begin{pmatrix} 1 & 0 & \ldots & 0 & 0 \\ 0 & 1 & \ldots & 0 & 0 \\ \ldots & \ldots & \ldots & \ldots & \ldots \\ 0 & 0 & \ldots & 1 & 0 \\ \frac{\partial h}{\partial y_1}(y') & \frac{\partial h}{\partial y_2}(y') & \ldots & \frac{\partial h}{\partial y_{n-1}}(y') & 1 \end{pmatrix}. \tag{10.13}$$

Especially $|\det D\psi(y)| = 1$ for a.a. $y \in W^+_{\alpha,\beta}$. We obtain with (10.2)

$$\|u \circ \psi\|^2_{L^2(W^+_{\alpha,\beta})} = \int_{W^+_{\alpha,\beta}} |u(\psi(y))|^2 |\det D\psi(y)|\,dy = \int_{\Omega^+_{\alpha,\beta}} |u(y)|^2\,dy = \|u\|^2_{L^2(\Omega^+_{\alpha,\beta})}. \tag{10.14}$$

For every $i \in \{1,\ldots,n\}$ we obtain with (10.8) and Theorem 10.1 that

$$\|\partial_i(u \circ \psi)\|^2_{L^2(W^+_{\alpha,\beta})} = \int_D \int_0^\beta \big|(\partial_i u)(\psi(y',y_n)) + (\partial_n u)(\psi(y',y_n)) \cdot (\partial_i h)(y')\big|^2 dy_n\,dy'$$
$$\le c \int_D \int_0^\beta |(\partial_i u)(\psi(y',y_n))|^2 + |(\partial_n u)(\psi(y',y_n))|^2\,dy_n\,dy'$$
$$= c\Big(\|(\partial_i u) \circ \psi\|^2_{L^2(W^+_{\alpha,\beta})} + \|(\partial_n u) \circ \psi\|^2_{L^2(W^+_{\alpha,\beta})}\Big)$$
$$= c\Big(\|\partial_i u\|^2_{L^2(\Omega^+_{\alpha,\beta})} + \|\partial_n u\|^2_{L^2(\Omega^+_{\alpha,\beta})}\Big) \tag{10.15}$$

with a constant $c = c(\mathrm{Lip}(h)) > 0$. A similar calculation using (10.9) instead of (10.8) yields

$$\|\partial_n(u \circ \psi)\|_{L^2(W^+_{\alpha,\beta})} = \|\partial_n u\|_{L^2(\Omega^+_{\alpha,\beta})}. \tag{10.16}$$

We combine (10.14), (10.15), (10.16) to get (10.7). $\qquad\square$

Now we can prove the main theorem of this chapter.

Theorem 10.3. *Let $\Omega \subseteq \mathbb{R}^n$ be a domain satisfying the uniform Lipschitz condition with parameters (α, β, N, L). Then there exists a continuous, linear trace operator*

$$T : H^1(\Omega) \to L^2(\partial\Omega) \tag{10.17}$$

with the property

$$Tu = u|_{\partial\Omega} \quad \text{for all } u \in C_0^1(\overline{\Omega}). \tag{10.18}$$

There is a constant $c = c(\beta, L, N) > 0$ such that

$$\|Tu\|_{L^2(\partial\Omega)} \le c\|u\|_{H^1(\Omega)} \tag{10.19}$$

for all $u \in H^1(\Omega)$.

Remark. In the special case that

$$\Omega = \left\{ (x', x_n) \in \mathbb{R}^n; x_n > h(x'), x' \in \mathbb{R}^{n-1} \right\} \tag{10.20}$$

where $h : \mathbb{R}^n \to \mathbb{R}$ is a Lipschitz continuous function, it follows that there is a constant $c = c(\text{Lip}(h)) > 0$ such that

$$\|Tu\|_{L^2(\partial\Omega)} \le c\|u\|_{H^1(\Omega)} \tag{10.21}$$

for all $u \in H^1(\Omega)$.

Proof. Let $U_j = U_{x_j, \alpha, \beta, h_j, \phi_j}$, $j \in M$, be a locally finite open cover as in Definition 4.2.

Step 1. The local situation. Fix $j \in M$ and consider $U_j = U_{x_j, \alpha, \beta, h_j, \phi_j}$. After a rotation and translation of the coordinate system we may assume $\phi_j = \text{id}$. We want to apply Lemma 10.2. Therefore let $W_{\alpha,\beta}^{j,+}$, $\Omega_{\alpha,\beta}^{j,+}$ and ψ_j be defined as in (10.3), (10.4) with respect to h_j and let Γ be defined as in (10.5). We get with Lemma 10.2 (ii), (iii) that

$$
\begin{aligned}
\|u\|_{L^2(\partial\Omega \cap U_j)}^2 &= \int_{\{y' \in \mathbb{R}^{n-1}; |y'| < \alpha\}} |u(y', h_j(y'))|^2 \sqrt{1 + |\nabla h_j(y')|^2} \, dy' \\
&\le c(L) \int_{\{y' \in \mathbb{R}^{n-1}; |y'| < \alpha\}} |u(y', h_j(y'))|^2 \, dy' \\
&= c(L)\|u \circ \psi_j\|_{L^2(\Gamma)}^2 \\
&\le c(\beta, L)\|u \circ \psi_j\|_{H^1(W_{\alpha,\beta}^{j,+})}^2 \\
&\le c(\beta, L)\|u\|_{H^1(\Omega_{\alpha,\beta}^{j,+})}^2 \\
&= c(\beta, L)\|u\|_{H^1(\Omega \cap U_j)}^2
\end{aligned}
\tag{10.22}
$$

is satisfied for all $u \in C_0^1(\mathbb{R}^n)$.

Step 2. The global situation. Using step 1 and the fact that at most N different sets of the U_j can have a nonempty intersection, we obtain

$$\int_{\partial\Omega} |u(x)|^2 \, dS(x) \leq \sum_{j\in M} \int_{\partial\Omega\cap U_j} |u(x)|^2 \, dS(x)$$

$$= \sum_{j\in M} \|u\|^2_{L^2(\partial\Omega\cap U_j)}$$

$$\leq c(\beta, L) \sum_{j\in M} \|u\|^2_{H^1(\Omega\cap U_j)} \tag{10.23}$$

$$\leq c(\beta, L) N \|u\|^2_{H^1(\Omega)}$$

for all $u \in C_0^1(\mathbb{R}^n)$.

Step 3. By definition, for $u \in C_0^1(\overline{\Omega})$ there exists $v \in C_0^1(\mathbb{R}^n)$ with $v|_{\overline{\Omega}} = u$. Define $Tu := v|_{\partial\Omega}$. By definition of $C_0^1(\overline{\Omega})$ and $L^2(\partial\Omega)$, this definition is independent of the choice of v. By (10.23) there holds

$$\|Tu\|_{L^2(\partial\Omega)} = \|v\|_{L^2(\partial\Omega)} \leq c\|v\|_{H^1(\Omega)} = c\|u\|_{H^1(\Omega)} \tag{10.24}$$

with a constant $c = c(\beta, N, L) > 0$. By (10.24) and the density of $C_0^1(\overline{\Omega})$ in $H^1(\Omega)$ (see Lemma 2.7) we see that T can be uniquely extended to a linear, continuous operator $T : H^1(\Omega) \to L^2(\partial\Omega)$ satisfying the estimate (4.17)

Assume that Ω is given as in (10.20). Define $\alpha := 1, \beta := 1, N := 3$. It is not difficult to show that Ω fulfils the uniform Lipschitz condition with parameters $(\alpha, \beta, N, \mathrm{Lip}(h))$. Consequently, since α, β, N are independent of $\mathrm{Lip}(h)$, we obtain that (10.21) is satisfied with a constant $c = c(\mathrm{Lip}(h)) > 0$. $\qquad\square$

Bibliography

[AdFo03] R. Adams and J. Fournier: *Sobolev Spaces*. Academic Press, New York, 2003

[Al02] H. W. Alt: *Lineare Funktionalanlysis*. Springer-Verlag, 2002

[Am95] H. Amann: *Linear and Quasilinear Parabolic Problems*. Birkhäuser Verlag, 1995

[Ba89] J. M. Ball: *A version of the fundamental theorem for Young measures*. Proceedings of conference on "Partial differential equations and continuum models of phase transitions", Springer Lecture Notes in Physics **359** (1989), 3-16

[BeMa02] A. Bertozzi and A. Majda: *Vorticity and Incompressible Flow*. Cambridge Texts in Applied Mathematics, Cambridge University Press, 2002

[Br10] J. Březina: *Asymptotic Properties of Solutions to the Equations of Incompressible Fluid Mechanics*. J. Math. Fluid Mech. **12** (2010), 536-553

[BFNW08] D. Bucur, E. Feireisl, Š. Nečasová and J. Wolf: *On the asymptotic limit of the Navier-Stokes system on domains with rough boundaries*. J. Differential Equations **244** (2008), 2890-2908

[BFN08] D. Bucur, E. Feireisl and Š. Nečasová: *On the Asymptotic Limit of Flows Past a Ribbed Boundary*. J. Math. Fluid Mech. **10** (2008), 554-568

[BFN10] D. Bucur, E. Feireisl and Š. Nečasová: *Boundary Behavior of Viscous Fluids: Influence of Wall Roughness and Friction-driven Boundary Conditions*. Arch. Rational Mech. Anal. **197** (2010), 117-138

[BoBu12] M. Bonnivard and D. Bucur: *The Uniform Rugosity Effect*. J. Math. Fluid Mech. **14** (2012), 201-215

[BrSc12] L. Brandolese and M. Schonbeck: *Large time decay and growth for solutions of a viscous Boussinesq system*. Trans. Amer. Math. Soc. **364** No.10 (2012), 5057-5090

[CFS03] J. Casado-Díaz, E. Fernández-Cara and J. Simon: *Why viscous fluids adhere to rugose walls: A mathematical explanation*. J. Differential Equations **189** (2003), 526-537

[CLS10] J. Casado-Díaz, M. Luna-Laynez and F. Suárez-Grau : *Asymptotic Behavior of a viscous fluid with slip boundary conditions on a slightly rough wall*. Mathematical Models and Methods in Applied Sciences **20**, No. 1 (2010), 121-156

[CaDiB80] J. Canon and E. DiBenedetto: *The initial value problem for the Boussinesq equations with data in L^p.* Approximation Methods for Navier-Stokes Problems, ed. by R. Rautman, Lecture Notes in Mathematics **771**, Springer-Verlag, 1980

[CKM92] Z. Chen, Y. Kagei and T. Miyakawa: *Remarks on stability of purely conductive steady states to the exterior Boussinesq problem.* Adv. Math. Sci. Appl. **1**, No. 2 (1992), 411-430

[CiDo99] D. Cioranescu and P. Donato: *An Introduction to Homogenization.* Oxford Lecture Series in Mathematics and its Applications, Oxford University Press, 1999

[DHP03] R. Denk, M. Hieber and J. Prüss: *$\mathcal{R}$-Boundedness, Fourier Multipliers and Problems of Elliptic and Parabolic Type.* Memoirs Amer. Math. Soc. **788**, 2003

[deS64] L. deSimon: *Un'applicazione della teoria degli integrali singolari allo studio delle equazioni differenziali lineari astratte del primo ordine.* Rend. Sem. Mat. Univ. Padova **34** (1964), 205-223

[EnNa99] K. Engel and R. Nagel: *One Parameter Semigroups for Linear Evolution Equations.* Graduate Texts in Mathematics, Springer-Verlag, 1999

[El05] J. Elstrodt: *Maß- und Integrationstheorie.* Springer-Verlag, 2005

[Ev98] L. Evans: *Partial Differential Equations.* American Mathematical Society, Providence, 1998

[EvGa92] L. Evans and R. Gariepy: *Measure Theory and Fine Properties of Functions.* CRC Press, 1992

[Fa93] R. Farwig: *The weak Neumann problem and the Helmholtz decomposition in general aperture domains.* Progress in Partial Differential Equations: the Metz Surveys 2, ed. by M. Chipot. Pitman Research Notes in Mathematics **296** (1993), 86-96

[FSV09] R. Farwig, H.Sohr and W. Varnhorn: *On optimal initial value condiditions for the Existence of Local Strong Solutions of the Navier-Stokes equations.* Ann. Univ. Ferrara Sez. VII Sci. Mat. **55** (2009), 89-110

[FSV11] R. Farwig, H. Sohr and W. Varnhorn: *Necessary and sufficient conditions on local strong solvability of the Navier-Stokes system.* Appl. Anal. **90** (2011), 47-58

[FS10] R. Farwig and H. Sohr: *On the existence of local strong solutions for the Navier-Stokes equations in completely general domains.* Nonlinear Analysis: Theory, Methods and Applications **73**, Issue 6 (2010), 1459-1465

[Fe69] H. Federer: *Geometric Measure Theory.* Springer-Verlag, 1969

[FeNe08] E. Feireisl and Š. Nečasová:*The effective boundary conditions for vector fields on domains with rough boundaries: Applications to fluid mechanics.* Applications of Mathematics **56** , No. 1 (2011), 39-49

[FoLe07] I. Fonseca and G. Leoni: *Modern Methods in the Calculus of Variations: L^p Spaces.* Springer New York, 2007

[GGZ74] H. Gajewski, K. Gröger and K. Zacharias. *Nichtlineare Operatorgleichungen und Operatordifferentialgleichungen.* Akademie-Verlag, 1974

[Ga98] G. P. Galdi: *An Introduction to the Mathematical Theory of the Navier Stokes Equations, Volume I.* Springer Verlag, New York, 1998

[Ga00] G. P. Galdi: *An Introduction to the Navier-Stokes Initial-Boundary Value Problem.* Fundamental Directions in Mathematical Fluid Mechanics, editors G. P. Galdi, J. Heywood, R. Rannacher, series "Advances in Mathematical Fluid Mechanics", Vol. **1**, Birkhäuser-Verlag (2000), 1-98

[Ha93] P. Hajłasz: *Change of variables formula under minimal assumptions.* Colloq. Math. **64** (1993), 93-101

[Hi91] T. Hishida: *Existence and Regularizing Properties of Solutions for the Nonstationary Convection Problem.* Funkc. Ekvac. **34** (1991), 449-474

[HiYa92] T. Hishida and Y. Yamada: *Global solutions for the Heat Convection equations in an Exterior Domain.* Tokyo J. Math. **15**, No. 1 (1992), 135-151

[HiPh57] E. Hille and R. Phillips: *Functional Analysis and Semigroups.* American Mathematical Society, Providence, 1957

[HPUU09] M. Hinze, R. Pinnau, M. Ulbrich and S. Ulbrich: *Optimization with PDE Constraints.* Mathematical Modelling: Theory and Applications 23, Springer-Verlag, 2009

[Hu97] H. Hungerbühler: *A refinement of Ball's Theorem on Young Measures.* New York J. Math. **3** (1997), 48-53

[Jo98] J. Jost: *Partielle Differentialgleichungen.* Springer, Heidelberg, 1998

[Ka93] Y. Kagei: *On weak solutions of nonstationary Boussinesq equations.* Differential Integral Equations **6**, oo. 3 (1993), 587-611

[KaPr08] G. Karch and N. Prioux: *Self-similarity in viscous Boussinesq equation.* Proc. Amer. Math. Soc. **136** No. 3 (2008), 879-888

[Ko67] Y. Komura: *Nonlinear semi-groups in Hilbert spaces.* J. Math. Soc. Japan, **19** (1967), 493-507

[Le34] J. Leray: *Sur le movement d' un liquide visqueux emplissant l' espace.* Acta Math. 63, 193. 248(1934)

[Li96] P.-L. Lions: *Mathematical Topics in Fluid Mechanics, Volume 1: Incompressible Models.* Oxford Lecture Series in Mathematics and its Applications, Oxford University Press, 1996

[Lu09] A. Lunardi: *Interpolation Theory.* Edizoni della Normale, Pisa, 2009

[Mo92] H. Morimoto: *Non-stationary Boussinesq equations.* J. Fac. Sci. Univ. Tokyo Sect. IA, Math. **39** (1992), 61-75

[Pa11] M. Padula: *Asymptotic Stability of Steady Compressible Fluids*. Lecture Notes in Mathematics 2024, Springer, 2011

[Pe97] P. Pedregal: *Parametrized Measures and Variational Principles*. Progress in Nonlinear Differential Equations and Their Applications, Birkhäuser, 1997

[Ru04] M Růžička: *Nichtlineare Funktionalanalysis*. Springer-Verlag, 2004

[SaTa04] O. Sawada and Y. Taniuchi: *On the Boussinesq Flow with Nondecaying Initial Data*. Funkc. Ekvac. **47** (2004), 225-250

[SoWa84] H. Sohr and W. van Wahl: *On the singular set and the uniqueness of weak solutions of the Navier-Stokes equations*. Manuscripta Math. **49** (1984), 27-59

[So01] H. Sohr: *The Navier-Stokes-Equations: An elementary functional analytic approach*. Birkhäuser Verlag, Basel, 2001

[St70] E. Stein: *Singular Integrals and Differentiability Properties of Functions*. Princeton University Press, Princeton, 1970

[Te77] R. Temam: *Navier-Stokes Equations*. North-Holland, 1977

[Tri78] H. Triebel: *Interpolation Theory, Function Spaces, Differential Operators*. North-Holland, 1978

[Wa00] W. Walter: *Gewöhnliche Differentialgleichungen*. Springer Verlag, 2000

[Wl82] J. Wloka: *Partielle Differentialgleichungen*. B.G. Teubner, Stuttgart, 1982

[Wo07] J. Wolf: *Existence of Weak Solutions to the Equations of Non-Stationary Motion of Non-Newtonian Fluids with Shear Rate Dependent Viscosity*. J. Math. Fluid Mech. **9** (2007), 104-138

[Yos80] K. Yosida. *Functional analysis*. Springer, 1980

[Ze03] R. Zeytounian: *Joseph Boussinesq and his approximation: a contemporary view*. C.R. Mécanique **331** (2003), 575-586

[Zi89] W. P. Ziemer: *Weakly differentiable functions*. Graduate Texts in Mathematics, Springer-Verlag, 1989

Akademischer Werdegang

7. April 1984   Geburt in Offenbach am Main

2003 - 2009   Studium der Mathematik mit Nebenfach Informatik an der
Technischen Universität Darmstadt

2009   Abschluss als Diplom-Mathematiker. Titel der Diplomarbeit:
Regularität von schwachen Lösungen der Navier-Stokes-Gleichungen im Außengebiet

2009 - 2013   Promotion und Tätigkeit als wissenschaftlicher Mitarbeiter in der
Arbeitsgruppe Analysis an dem Fachbereich Mathematik
der Technischen Universität Darmstadt

11. Juli 2013   Abschluss der Promotion. Titel:
Weak solutions of the Boussinesq equations in domains with rough boundaries